Introduction
to
Global Analysis

Pure and Applied Mathematics

A Series of Monographs and Textbooks

Editors **Samuel Eilenberg and Hyman Bass**

Columbia University, New York

Introduction to Global Analysis

DONALD W. KAHN

School of Mathematics
University of Minnesota
Minneapolis, Minnesota

1980

ACADEMIC PRESS
A Subsidiary of Harcourt Brace Jovanovich, Publishers
New York   London   Toronto   Sydney   San Francisco

ACADEMIC PRESS, INC.
111 Fifth Avenue, New York, New York 10003

United Kingdom Edition published by
ACADEMIC PRESS, INC. (LONDON) LTD.
24/28 Oval Road, London NW1 7DX

Library of Congress Cataloging in Publication Data

Kahn, Donald W   Date
    Introduction to global analysis.

    (Pure and applied mathematics)
    Bibliography: p.
    1. Global analysis (Mathematics) I. Title.
II. Series: Pure and applied mathematics, a series of
monographs and textbooks ;

QA3.P8   [QA614]   510s [514'.74]   79–8858
ISBN 0–12–394050–8

PRINTED IN THE UNITED STATES OF AMERICA

80 81 82 83   9 8 7 6 5 4 3 2 1

Contents

v

Preface

The classical tradition of mathematical analysis studies functions, vector fields, differential equations, etc., in Euclidean space $\mathbb{R}^n$. It has become apparent in recent years that there is great interest in the methods of mathematical analysis carried out on manifolds, that is, spaces that look like $\mathbb{R}^n$ in the small. They have found concrete applications to the fields of physics, engineering, and economics; possible applications are under investigation in many other areas.

The present book aims at making much of this material available to anyone who is at the level of a beginning graduate student in mathematics. The first half of the book requires little more than a good foundation in undergraduate mathematics. In later chapters we use some integration, Fourier analysis, and homology theory. At appropriate places, we outline all the theory that we need, and we list many references. In the Introduction we sketch the history of the subject and give a detailed description of the individual chapters. A full list of references is given at the end of the book.

The material presented in this book was first exposed by the author in a graduate course at the University of Minnesota in 1976–1977. The course was very successful, attracting both students and faculty from mathematics and the sciences. I want to thank my listeners and friends for helping me to fix my ideas on the subjects presented here. I am also grateful for the remarkable typing talents of Peggy Gendron. As usual, responsibility for accuracy, interest, etc., lies entirely with me.

Introduction

The term *global analysis* has come into general mathematical usage during the past decade. It covers a variety of topics, some of which are not new but have achieved a certain prominence, perhaps owing to a broadening of the general mathematical viewpoint. In addition, certain topics naturally have attracted attention once an important theorem was proved. These processes have worked in various ways to create this new field.

For example, the general study of differentiable manifolds moved into the spotlight largely because of an increased awareness among the general mathematical public of such topics as dynamical systems, Lie groups, and transformation groups. The more specific topic of differential operators defined in a global setting attracted wide attention after the famous theorem of Atiyah and Singer (see [5, 6]). By July 1968, when the American Mathematical Society held a symposium on global analysis at Berkeley (see [19]), the existence of this field was no longer in question. A quick glance at the three volumes of papers generated by that symposium [19] shows an enormous breadth of topics, from classical partial differential equations to ergodic theory. The amount of algebraic topology that is required to understand these papers varies considerably, but buried in the midst of these varying topics a solid core field had begun to emerge.

Unknown to many mathematicians, a strong impetus for research in this field has come from applications, potential or real, to a multitude of subjects.

1

In the *Batelle Rencontres* [8], a collection of papers dating from 1967, based on lectures in mathematics and physics, there is a substantial amount of what we might now call global analysis. More recently, the field has been shown to be of interest in engineering; for example, the survey article of Brockett [14] for the Institute of Electrical and Electronics Engineers is deeply involved with manifolds, vector fields, etc. I think, however, that almost everyone would agree that it is the catastrophe theory of Thom (e.g., [13, 108]) that has most caught the imagination of the general public. Surely some of this theory is controversial, and some of its more flamboyant applications will have to be reformulated or restrained to conform with the accepted rigor of science, but I doubt if anyone will deny·that there is a certain amount of new mathematics, with a significant potential for application.

Where, then, is the middle ground of global analysis, and why would I want to risk writing a book in this young field? The foundation of any rational view of this field must be the study of smooth manifolds and the maps between them. Because I want to make this subject accessible to many people and am not afraid to do something once and then generalize it later (when there is a clear advantage in good exposition), I begin this text with a basic discussion of (finite-dimensional) differentiable manifolds. The first three chapters all are concerned with that area. I begin with those aspects of this theory that flow out of the usual advanced calculus, and I carry it through to prove versions of the Whitney embedding theorem, the theorem of Sard on the measure of the set of critical values, and finally, the transversality lemma of Thom.

With the foundations set, we proceed to look at the tangent bundle to a manifold, and more generally the theory of vector bundles (which I treat fully in the real case over a compact space). I feel that a certain amount of bundle theory is essential here, but that it would be a perversion of the general purpose of this text to include a full study of fiber bundles. For this reason I limit myself as indicated, rather in the spirit of the beautiful book of Atiyah [4].

This leads up to what may be considered the first topic belonging properly to global analysis, the general study of differential operators on manifolds. This text being an introduction, my goal here is to bring the student to where he or she may begin to learn the Atiyah–Singer index theorem. I make no attempt to offer any competition to the existing excellent, but rather more advanced, expositions in that area. It goes without saying that by that point, I must cover the basic topics of differentiation and integration on manifolds, including of course the algebra of differential forms, Stokes' theorem, the Poincaré lemma, and the basic definition of deRham cohomology. But no material from algebraic topology is presumed to be known by the reader up to this place in the book.

It is my belief that this is the proper place to extend some of our earlier material on differentiable manifolds to the infinite-dimensional case. To say something meaningful here would require a certain knowledge of functional analysis. We include some short references to the relevant parts of that field, hoping to guide the reader to appropriate places in the literature. Much of the earlier material about manifolds, smooth maps, tangent bundles, etc. carries over without serious difficulty to the more general case of infinite-dimensional manifolds. We do prove one basic result, which exhibits the function space of maps between manifolds as an infinite-dimensional manifold. We also discuss, and prove part of, Kuiper's well-known result concerning the contractability of the infinite-dimensional linear group. But to keep within the general scope I have in mind here, I must stop short of more recent developments, such as the work of Kuiper, Eells, Elworthy, and others on Hilbert manifolds.

The remainder of this text consists of a series of four topics, all closely related to analysis on manifolds:

   (i)   Morse theory, the study of smooth functions at critical points;
   (ii)   Lie groups and their actions on manifolds;
  (iii)   dynamical systems and structural stability;
   (iv)   a descriptive introduction to singularities and catastrophies.

We aim here for a solid description of the basic results in these areas. For example, one needs some basic facts from topology (rational homology groups, Euler characteristic, etc.) to discuss the Morse inequalities. We include an outline of this material, along with many references, at the appropriate point. In the case of topic (iv) we aim only to describe the general ideas; we mention the seven elementary catastrophics, but do not prove that they are exhaustive in the dimensions in question.

No author would be entirely honest if he did not clearly indicate what relevant material has been deliberately omitted from the book. First of all, we have stressed the compact case. Much of what we have said about manifolds, vector bundles, etc. is not, in fact, limited by that restriction. Unfortunately, the methods of proof in the general cases are frequently much more involved. Second, we never enter properly into the domain of K-theory, a topic that is both beautiful and naturally related to the material of this text. It is unfortunate that K-theory presumes much more algebraic topology than I felt could be safely assumed as a prerequisite for this text. Third, for many of the later topics in the book, I strive to give a clear idea of what is involved in the area rather than give maximal known generality. Examples of this are the sections on infinite-dimensional manifolds, the discussion of Lie groups (minimal reference to Lie algebras), and the frequent restriction to the real case (with at most a remark about the equally important complex case). Of course I have tried to give full lists of references, so that a serious reader will not have too much trouble tracking things down.

The original text was intended to be available to a well-prepared first-year graduate student. I had to assume a full foundation in basic analysis (still often taught under the misleading name of "advanced calculus"), linear algebra, and point-set topology. As I indicated above, I have had to stray from this at a few points, but I think a serious and patient student can survive these hurdles. I do not think that the basic material from algebraic topology (as in the chapter on Morse theory) or the ocassional reference to Lebesgue measure, initially restricted to a discussion of sets of Lebesgue measure zero, need be regarded as a major obstacle. Most students who are at the level of this book are simultaneously learning some of these important topics. On occassion, some of these additional topics may be taken on faith for a short period of time to accommodate the contingencies of one's program. In a similar way, I have felt free to use the basic existence and uniqueness theorems from differential equations.

I have decided against a formal list of problems at the end of every section. I have specifically indicated where good sources of relevant problems may be found, and have otherwise indicated some projects relevant to the material at hand. At some places, however, I do state specific problems. Naturally, working some problems is essential in getting an understanding of the subject.

On the other side of the coin, mature mathematicians may well find some interest in portions of this book. For them the basic material will probably be boring, and it will be necessary to feel their way around to find where new and interesting material first occurs. I hope that nobody will be angry; it is a lot easier to skip pages than to fill in material that one does not know. I will personally feel that I have succeeded if I can bring students, as well as faculty whose interests lie elsewhere, to the point of understanding collo-quium lectures in this beautiful field.

1

Manifolds and Their Maps

The purpose of this chapter is to lay the groundwork for the study of smooth manifolds and their maps. We begin with their general properties and then review quickly the important special case of Euclidean spaces. No political treaty has ever been drawn-up to delineate clearly between advanced calculus and the elementary theory of manifolds. I shall attempt, toward the end of this chapter, to give a concise presentation of this borderline material, hoping to be helpful to some without offending others.

DIFFERENTIABLE MANIFOLDS AND THEIR MAPS

A manifold is a nice topological space that in the small is just like Euclidean space (a rigorous definition will follow). To do mathematical analysis on manifolds, one imposes some condition of smoothness or differentiability concerning the relation of any two nearby pieces, each of which is like Euclidean space. To assure that the thing does not get too large, one usually imposes some mild set-theoretic conditions. A typical example would be the ordinary sphere (the surface of the earth). A nice small region—such as a polar ice cap—is virtually indistinguishable from a small region in the plane, while the entire manifold, the sphere, is fundamentally different from the plane. We shall define a manifold first, and after some examples turn to differentiable manifolds.

Definition 1.1   An n-manifold (or manifold of dimension n) is a topological space M^n that satisfies the following:

(a) M^n is a Hausdorff space (satisfying the separation axiom T_2 of Hausdorff).

(b)   If x is a point of M^n, then there is an open set $O \subseteq M^n$, $x \in O$, with O homeomorphic to the Euclidean space $\mathbb{R}^n$ ($\mathbb{R}^n = \mathbb{R} \times \cdots \times \mathbb{R}$, n factors).

(c) M^n has a countable basis for its topology; i.e., there is a countable family of open sets $\{O_m\}$ such that every open set is a union of some of the O_m's.

Examples   *1.* The Euclidean spaces $\mathbb{R}^n$ are trivial examples (take a single open set $O = \mathbb{R}^n$).

2. Any open set $U \subseteq \mathbb{R}^n$ is clearly an n-dimensional manifold because, as is easily seen, an open ball of fixed radius about a point in $\mathbb{R}^n$ is homeomorphic to all of $\mathbb{R}^n$. (Using polar coordinates, the required homeomorphism is the identity on all angular coordinates and on the radial coordinate can be chosen to be any homeomorphism from the interval in question to the entire real numbers, and is easily constructed from functions such as the tangent.)

More generally, if M^n is an n-manifold and $U \subseteq M^n$ is an open subset, then U is an n-manifold.

3. Let $S^n = \{X \in \mathbb{R}^{n+1} \mid d(\mathbb{O}, X) = 1\}$, where $d(\mathbb{O}, X)$ means the distance from X to the origin $\mathbb{O}$. This is the standard n-dimensional sphere. If $X = (x_1, \ldots, x_{n+1})$, then we set

$$U = \{X \in S^n \mid x_{n+1} > -1\}, \qquad V = \{X \in S^n \mid x_{n+1} < +1\}.$$

Clearly U and V are a covering of S^n by two open sets.

Now, if $X \in U$, let l_X denote the straight line in $\mathbb{R}^{n+1}$ through the two points $(0, 0, \ldots, -1)$ and X. Then clearly l_X meets the hyperplane

$$H_1 = \{X \in \mathbb{R}^{n+1} \mid x_{n+1} = 1\}$$

at a single point, called $\phi(X)$. Then ϕ is trivially checked to be a homeomorphism from U to H_1. But the projection onto the first n coordinates yields a homeomorphism between H_1 and $\mathbb{R}^n$. Thus U is homeomorphic to $\mathbb{R}^n$.

Similarly, V is homeomorphic to $\mathbb{R}^n$. S^n is obviously a compact Hausdorff space with a countable basis; thus it is an n-manifold

4. If M^n and N^k are manifolds of dimension n and k, respectively, then $M^n \times N^k$ is easily seen to be a manifold of dimension $n + k$. The torus $S^1 \times S^1$ is an easy example.

5. Let $Gl(n; \mathbb{R})$ be the group of $n \times n$ matrices over the real numbers with nonvanishing determinant. It is a topological space because it is a subspace of all $n \times n$ matrices that is topologized exactly as Euclidean space of dimension n^2. Alternatively, we may define the norm of a matrix (a_{ij}),

$$\|(a_{ij})\| = \sqrt{\sum_{i,j} (a_{ij})^2},$$

and then a metric by

$$d((a_{ij}),(b_{ij})) = \|(a_{ij} - b_{ij})\|.$$

If we set

$$D = \{(a_{ij}) \mid \det(a_{ij}) = 0\},$$

then $\mathrm{Gl}(n;\mathbb{R})$ is clearly the complement of D in the set of all $n \times n$ matrices. But D is trivially checked to be closed, so $\mathrm{Gl}(n,\mathbb{R})$ is open and hence a manifold of dimension n^2.

This should suffice to show that the collection of manifolds is very substantial; we shall meet many other examples as we proceed through this book. Before turning to differentiable manifolds, it is important that we look for a moment at a few technical points connected with the definition of a manifold.

(1) All of the hypotheses made in the definition of an n-manifold are independent of one another. For example, condition (b) in Definition 1.1 does not imply that M^n is Hausdorff. Consider the following space in the form of a letter Y

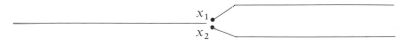

where the left branch is an open interval and the two right branches are half open and contain their left-hand endpoints. The topology around every point other than x_1 and x_2 is the usual topology of the real line. An open neighborhood of x_1 consists of the union of a half-open interval in the upper-right segment with x_1 as leftmost point and an open subset of the left branch consisting of all points in that branch that are to the right of a given point. Similarly, an open neighborhood of x_2 is defined using points of the lower-right segment. In particular, any open neighborhood of either x_1 or x_2 must contain all points to the right of some given point in the left-hand branch.

This topological space has the property that every point has an open neighborhood homeomorphic to $\mathbb{R}^1$, but the two points x_1 and x_2 do not have disjoint open neighborhoods, so that it is not a Hausdorff space.

There are classic examples of "manifolds" for which condition (c) fails, the so-called long line, for example (see [5]).

(2)   An n-manifold in the sense of Definition 1.1 is ocasionally called a topological manifold, to distinguish it, for example, from a differentiable manifold (Definition 1.2).

(3)   An n-manifold M^n enjoys various point-set theoretic properties in a somewhat automatic fashion. For example,

   (i)   A manifold M^n is locally compact. If $x \in M^n$, choose an open neighborhood O of x that is homeomorphic to $\mathbb{R}^n$. Around the point x in O we may then select an open ball of finite radius (regarding O as $\mathbb{R}^n$). The closure of such an open ball is, of course, compact.

   (ii) M^n is separable (that is, it has a countable dense subset). This follows at once from Definition 1.1c.

   (iii) M^n is regular; that is, if x and C are a point and a closed subset of M^n with $x \notin C$, then x and C are contained in disjoint open sets. For this it suffices to handle the case where both x and C are contained in an open neighborhood of x homeomorphic to $\mathbb{R}^n$. This is trivial, for $\mathbb{R}^n$ is a metric space.

   (iv)   With slightly more work, M^n may be shown to be a normal space.

   (v)   Combining Definition 1.1c with (iv) or (iii) above and using the Urysohn metrization theorem, we see that an n-manifold in the sense of Definition 1.1 is always a metric space. This will be greatly strengthened in the case of a smooth manifold when we construct a Riemann metric. Furthermore, we shall prove, in the case of a compact smooth manifold, that M^n is always homeomorphic to a subspace of $\mathbb{R}^{2n+1}$. This embedding theorem, which may also be established without compactness, trivially implies metrization.

(4)   The number n, the dimension of an n-manifold, is in fact an invariant. That is, if two manifolds are homeomorphic, then they have the same dimension. This result is significantly deeper than one might guess on first looking at it. Even when the manifolds are restricted to be simple Euclidean spaces, it is nontrivial. The assertion that $\mathbb{R}^n$ and $\mathbb{R}^m$ can only be homeomorphic when $n = m$ is the famous theorem on invariance of dimension from algebraic topology. The general fact that the dimension of a manifold is an invariant follows by similar methods of algebraic topology (see, e.g., [25, 33]).

In this context, we might also mention another relevant theorem from algebraic topology, the invariance of domain. This asserts that if U_1 and U_2 are homeomorphic subsets of a given manifold M^n, and U_1 is open in M^n, then so is U_2. (Note that the homeomorphism between these subsets is not assumed to extend to a homeomorphism of the entire manifold to itself, in which case the theorem would, of course, be trivial.)

On the other hand, for differentiable manifolds (soon to be defined) the fact that dimension is an invariant is rather trivial, and we shall prove it later in this chapter (Corollary 1.5).

(5) Naturally, a manifold need not be connected. For example, $\mathrm{Gl}(n, \mathbb{R})$ has two connected components, corresponding to the positive and the negative determinants. But if a manifold is connected, it is automatically pathwise connected. For let x_0 be a given point in the manifold M^n and let M_0 be the set of points that may be connected to x_0 by a continuous path. Using the fact that the manifold is locally like Euclidean space, it is easy to see that M_0 is both open and closed. Since M^n is connected, we must have $M_0 = M^n$, proving that M^n is pathwise connected.

Our principal interest in this text will be manifolds with some differentiability or smoothness. This can be viewed in two ways. First, any point lies in an open set that is homeomorphic to $\mathbb{R}^n$. If more than one such open set overlap or mesh badly, the change of coordinates obtained by viewing a given point in two different open sets might be a bad sort of function. But if the open cover by sets homeomorphic to $\mathbb{R}^n$ can be chosen nicely, then the possible change of coordinates arising from viewing a given point as lying in two overlapping open sets might always be a differentiable function from a subset of $\mathbb{R}^n$ to another. This is the first (and most common) way of getting at the notion of a differentiable manifold.

A second possible approach is in terms of functions. Given a function $f: M^n \to \mathbb{R}$ and a point $x \in M^n$ contained in an open set O that is homeomorphic to $\mathbb{R}^n$, we might wish to call f differentiable at x, provided f restricted to O is differentiable at x, in terms of the usual notion of differentiability of functions defined on Euclidean space. Unfortunately, this will not be an intrinsic property of the function f and the point x, but will depend on the choice of open set O. If there is a cover by suitable open sets, each homeomorphic to $\mathbb{R}^n$, such that the question of whether a function is differentiable at any given point x has a consistent answer, not dependent on the choice of a particular open set, then the manifold clearly ought to be thought of as differentiable. This is an alternative and equivalent approach, first exposed in [20], that we shall also sketch.

Here then is the official definition.

Definition 1.2 A manifold is called smooth (or infinitely differentiable, or sometimes just differentiable) if there is a family of open sets $\{O_\alpha\}$ that cover the manifold, each of which is homeomorphic to $\mathbb{R}^n$ by a given homeomorphism

$$\phi_\alpha: O_\alpha \to \mathbb{R}^n$$

such that, whenever $O_\alpha \cap O_\beta$ is nonempty, the composite map

$$\phi_\beta(O_\alpha \cap O_\beta) \xrightarrow{\;(\phi_\beta | O_\alpha \cap O_\beta)^{-1}\;} O_\alpha \cap O_\beta \xrightarrow{\;\phi_\alpha | O_\alpha \cap O_\beta\;} \mathbb{R}^n,$$

whose domain is visibly an open subset of $\mathbb{R}^n$, has continuous partial derivatives of all orders.

The pair (O_α, ϕ_α) is often called a chart or a coordinate chart.

Remarks (1) Our previous examples furnish many examples of differentiable or smooth manifolds. $\mathbb{R}^n$ and all its open subsets are obviously smooth manifolds. One may easily check that S^n, with the cover by two open sets given above, is a smooth manifold.

(2) Obviously, a smooth or differentiable manifold is a manifold, but it is remarkable that the reverse assertion is false. Kervaire [63], and later Smale and others, have shown that a compact manifold need not be differentiable. Kervaire constructs a 10-dimensional manifold and shows that it is not possible to select an open cover by sets meeting the condition of Definition 1.2. (Actually, the proof is rather indirect and depends on considerable algebraic topology.)

(3) Given a topological manifold, Remark (2) shows that there is an existence problem, the question whether one can find a differentiable structure on that manifold. Precisely, one may ask whether there is a smooth manifold that is homeomorphic to the original manifold. Similarly, there is a uniqueness problem, which we shall formulate precisely below. In a very famous paper, Milnor [76] has shown that the 7-dimensional sphere has a plurality of differentiable structures.

(4) Occasionally one needs to study various degrees of differentiability. If we require that the maps in Definition 1.2 have continuous partial derivatives of all orders less than or equal to some nonnegative integer k, then we say that the manifold is C^k, or differentiable of class k. In this terminology a smooth manifold is a C^∞ manifold, while an ordinary or topological manifold (Definition 1.1) is a C^0 manifold. A key theorem of Whitney [116] asserts that if a manifold has a C^k differentiable structure—that is, a cover by coordinate charts such that the maps in Definition 1.2, $(\phi_\alpha | O_\alpha \cap O_\beta) \circ (\phi_\beta | O_\alpha \cap O_\beta)^{-1}$, have continuous partials of order less than or equal to k (with $k > 0$)—then the manifold has a compatible C^∞ differentiable structure, that is, new charts may be found to make it C^∞. It is for that reason that I deemphasize the various degrees of differentiability and focus primarily on the smooth or C^∞ case.

(5) There is an undue amount of emphasis in Definition 1.2 on the choice of the open cover of the manifold. After we consider mappings between smooth manifolds, we will be able to introduce a natural notion of equiva-

lence between manifolds, and then we shall know when manifolds are equivalent, regardless of these covers.

Definition 1.3   Let M^n and N^k be smooth manifolds and $f: M^n \to N^k$ be a continuous map. Then f is called smooth (or on occasion just differentiable) if for every $x \in M^n$ and every pair of coordinate charts $\phi_\alpha: O_\alpha \to \mathbb{R}^n$, with $x \in O_\alpha$, and $\psi_\beta: U_\beta \to \mathbb{R}^k$, with $f(x) \in U_\beta$, on these manifolds, the composite mapping $\psi_\beta \circ f \circ \phi_\alpha^{-1}$ has continuous partial derivatives of all orders at the point $\phi_\alpha(x)$.

Remarks   (1)   The mapping $\psi_\beta \circ f \circ \phi_\alpha^{-1}$ is easily seen to be defined on an open neighborhood of $\phi_\alpha(x)$, so it makes sense to talk about partial derivatives. In addition, one easily checks that this is independent of coordinate charts.

(2)   In Definition 1.3, if we require that there be continuous partial derivatives of orders less than or equal to some nonnegative integer k, then the map f would be called C^k. As before, smooth is then the same as C^∞.

(3)   The intuitive content of this definition is that we use the coordinate charts to transfer the notion from manifolds to the easily understood notion in Euclidean spaces.

Before looking into diffeomorphisms, submanifolds, etc., I would like to mention that there is one further type of manifold, which one might encounter in the literature. This is an analytic (or, rarely, C^ω) manifold. To define such manifolds, one requires that the maps in Definition 1.2

$$(\phi_\alpha | O_\alpha \cap O_\beta) \circ (\phi_\beta | O_\alpha \cap O_\beta)^{-1}$$

be given by convergent power series in n variables in a neighborhood of any point of their domains.

Similarly, we have an analytic map between analytic manifolds whenever the map $\psi_\beta \circ f \circ \phi_\alpha^{-1}$ is always represented by convergent power series (k series, each in n variables) in a neighborhood of any such point $\phi_\alpha(x)$.

It is a generally accepted notion that a map between two mathematical objects is an *isomorphism* or *equivalence* if there is a map going in the reverse direction that when composed with the original map—on either side—yields the identity map. This is also the idea behind a diffeomorphism.

Definition 1.4   Two smooth manifolds M_1^n and M_2^n are *diffeomorphic* (or occasionally *equivalent*) if there are smooth maps

$$f: M_1^n \to M_2^n \qquad \text{and} \qquad g: M_2^n \to M_1^n$$

such that $g \circ f = 1_{M_1^n}$ and $f \circ g = 1_{M_2^n}$, the identity maps on M_1^n and M_2^n, respectively.

The maps f and g are called *diffeomorphisms*.

Remarks (1) If the manifolds and maps are C^k, $k > 0$, one also calls them diffeomorphic (or diffeomorphisms). In the C^0 case, a diffeomorphism is just a homeomorphism.

We do not actually need to assume that the manifolds have the same dimension. This follows from the existence of f and g [see Remark (4) following Definition 1.1].

(2) We may now compare differentiable structures on a given manifold. Two differentiable structures on a given differentiable manifold, that is to say, two different open covers each meeting the condition of Definition 1.2 are called equivalent when there is a diffeomorphism from the manifold with the first differentiable structure to that with the second differentiable structure. It is known (see [64, 70]) that any 2- or 3-dimensional manifold has a unique differentiable structure, whereas the 7-dimensional sphere has 28 inequivalent differentiable structures [30].

One might suspect from analogous situations in real analysis that the notions of a smooth or a differentiable manifold could be made to rest on the properties of the set (or algebra) of functions on that manifold. This idea first arose in book form in [20], and we would like (for completeness) to offer a sketch of it here.

If $U \subseteq M^n$ is open and $f: U \rightarrow \mathbb{R}^n$ is continuous, we write (U, f) for short, and we call the set of all such pairs (U, f) by the symbol $\mathcal{D}$. Consider the following axioms on $\mathcal{D}$:

(i) If $(U, f) \in \mathcal{D}$ and $V \subseteq U$ is also open, then the restriction to V also belongs to $\mathcal{D}$, i.e., $(V, f|V) \in \mathcal{D}$.

(ii) Suppose that $U = \bigcup_i U_i$, with each U_i open, and $f: U \rightarrow \mathbb{R}^n$. If every $(U_i, f|U_i) \in \mathcal{D}$, then $(U, f) \in \mathcal{D}$.

(iii) For each $x \in M^n$, there is an open neighborhood U of x and a homeomorphism $\phi: U \rightarrow \mathbb{R}^n$ such that whenever $V \subseteq U$ is a smaller open set, the set of all (V, f), with V fixed, is precisely the set

$$\{g \circ (\phi|V): \mathbb{R}^n \rightarrow \mathbb{R}^n \,|\, g \text{ is a smooth map from the open subset } \phi(V) \text{ of } \mathbb{R}^n \text{ to } \mathbb{R}^n.\}$$

It is obvious that the sets of differentiable functions (in the sense of Definition 1.3) from the open U to $\mathbb{R}^n$ meet these conditions, and one may check without difficulty that if these axioms are satisfied, then there is an open cover that satisfies Definition 1.2.

One frequently encounters situations where one manifold lies inside another. This can happen in a pleasant way, as in the case of a line in a plane, for example, or in a complicated way, in which a (homeomorphic copy of a) line may have positive area in a plane. In bad cases a manifold can lie in another manifold of higher dimension with the first manifold being dense in the bigger one. The following notion of submanifold is carefully constructed to respect the differentiable structure.

Definition 1.5 Let M^n be a smooth manifold and $A \subseteq M^n$ a subset. We call A a k-dimensional submanifold, $k \le n$, if for each $a \in A \subseteq M^n$ there is an open neighborhood O_α of a and a homeomorphism $\phi_\alpha : O_\alpha \to \mathbb{R}^n$ that is either a coordinate chart from the differentiable structure on M^n or is consistent with it in the sense of Definition 1.2, meaning that for any chart O_β of the manifold intersecting O_α, $(\phi_\beta | O_\alpha \cap O_\beta) \circ (\phi_\alpha | O_\alpha \cap O_\beta)^{-1}$ and $(\phi_\alpha | O_\alpha \cap O_\beta) \circ (\phi_\beta | O_\alpha \cap O_\beta)^{-1}$ have continuous partials, so that

$$A \cap O_\alpha = \phi_\alpha^{-1}(\mathbb{R}^k),$$

where we regard $\mathbb{R}^k = \{(x_1, \ldots, x_n) | x_{k+1} = x_{k+2} = \cdots = x_n = 0\}$.

Remarks (1) In a word, a submanifold is a subset that, via the differentiable structure, looks locally like a linear subspace.

(2) A submanifold of a smooth manifold is clearly itself a smooth manifold. A similar definition may naturally be formulated for the C^r or the analytic cases.

(3) In order to appreciate the strength of this definition, it is important to realize that if $f : N^k \to M^n$ is a smooth map that is 1–1 (injective), then the image of f need not be a submanifold. As an example, let N^1 be the open unit interval $(0, 1)$ and M^2 be the plane. Consider the map $f : N^1 \to M^2$ defined by Fig. 1.1, where we have labeled some points in terms of the value in the unit interval to which f is applied. It is trivial to see that such a map can be

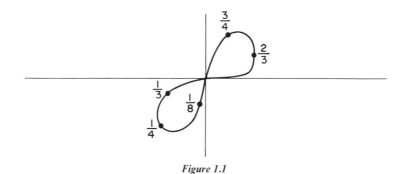

Figure 1.1

taken to be smooth. But for any chart at the origin, say O_α, $f(N^1) \cap O_\alpha$ is clearly not the set $\phi_\alpha^{-1}(\mathbb{R}^1)$. In fact, no local homeomorphism between open sets in the plane can take the crossed lines to a single line, for there is a point in the crossed lines which when removed leaves four connected components, while the single line has no such point.

However, we shall soon see that such problems cannot arise if the domain manifold is compact and the map f is nonsingular; we shall prove (Corollary 1.3) that in that case the image is always a submanifold.

THE CASE OF EUCLIDEAN SPACES

In order to get a deeper understanding of smooth manifolds, it is essential to know as much as possible about the local structure of such manifolds, which amounts to examining—in depth—the case where the manifold is Euclidean space. Much of this material overlaps with advanced calculus or elementary real analysis, and we shall attempt to give a concise modern treatment.

We shall attempt to adhere to what appears to be the standard notation from calculus of several variables, and I believe the reader will find most of the terminology familiar.

Definition 1.6 Let $U \subseteq \mathbb{R}^n$ be an open set, and let $f: U \to \mathbb{R}^m$ be a smooth (or at least $C^r, r \geq 1$) map. Let $x \in U$. The *differential* of f at x is the linear map

$$D_f(x): \mathbb{R}^n \to \mathbb{R}^m$$

defined by the rule

$$D_f(x)\begin{bmatrix} a_1 \\ \vdots \\ a_n \end{bmatrix} = \begin{bmatrix} \dfrac{\partial f_1}{\partial x_1} & \cdots & \dfrac{\partial f_1}{\partial x_n} \\ \vdots & & \\ \dfrac{\partial f_m}{\partial x_1} & \cdots & \dfrac{\partial f_m}{\partial x_n} \end{bmatrix}\begin{bmatrix} a_1 \\ \vdots \\ a_n \end{bmatrix},$$

where $f(x) = (f_1(x), \ldots, f_m(x))$, and the operation is standard matrix multiplication; the partial derivatives are to be taken at x.

The matrix is called the Jacobian $J_f(x)$.

The differential offers an optimal linear approximation to f at the point x in the sense of the following:

Proposition 1.1 With the above notation, if $y \in U$, then

$$\lim_{||y-x|| \to 0} \left. \left\| f(y) - f(x) - J_f(x)(y - x) \right\| \middle/ \|y - x\| \right. = 0.$$

Proof Set $y_i = x_i + h_i$ and calculate, by the mean-value theorem,

$$f_i(y_1, \ldots, y_n) - f_i(x_1, \ldots, x_n)$$
$$= f_i(x_1 + h_1, \ldots, x_n + h_n) - f_i(x_1, \ldots, x_n)$$
$$= f_i(x_1 + h_1, \ldots, x_n + h_n) - f_i(x_1, x_2 + h_2, \ldots, x_n + h_n)$$
$$+ f_i(x_1, x_2 + h_2, \ldots, x_n + h_n) - f_i(x_1, x_2, x_3 + h_3, \ldots) + \cdots$$
$$+ f_i(x_1, \ldots, x_{n-1}, x_n + h_n) - f_i(x_1, \ldots, x_n)$$
$$= \frac{\partial f_i}{\partial x_1}(x_1^*, \ldots)h_1 + \frac{\partial f_i}{\partial x_2}(\ldots, x_2^*, \ldots)h_2 + \cdots + \frac{\partial f_i}{\partial x_n}(\ldots, x_n^*)h_n$$

for suitable numbers x_j^* between x_j and $x_j + h_j$. In other words, the *i*th component of $f(y) - f(x)$ is

$$\sum_{j=1}^{n} \frac{\partial f_i}{\partial x_j}(x_1, \ldots, x_{j-1}, x_j^*, x_{j+1} + h_{j+1}, \ldots, x_n + h_n)h_j.$$

On the other hand,

$$J_f(x)(y - x)$$

has as *i*th component

$$\sum_{j=1}^{n} \frac{\partial f_i}{\partial x_j}(x_1, \ldots, x_n)h_j.$$

By the assumption that f is $C^r, r \geq 1$, or smooth, we see at once that each term

$$\left(\frac{\partial f_i}{\partial x_j}(x_1, \ldots, x_{j-1}, x_j^*, x_{j+1} + h_{j+1}, \ldots) - \frac{\partial f_i}{\partial x_j}(x_1, \ldots, x_n)\right)\frac{h_j}{\|y - x\|}$$

approaches zero as $\|y - x\|$ goes to zero (i.e., all $h_j \to 0$), from which the result follows.

On the other hand, one may use the formula in Proposition 1.1 to give a coordinate-free definition of D_f. The following proposition shows that the Jacobian is well behaved.

Proposition 1.2 Suppose that we are given $C^r (r \geq 1)$ maps $f_1, f_2 : \mathbb{R}^n \to \mathbb{R}^m$ and $g : \mathbb{R}^m \to \mathbb{R}^k$ and real numbers α and β; then

(a) $J_{\alpha f_1 + \beta f_2}(x) = \alpha J_{f_1}(x) + \beta J_{f_2}(x)$, and
(b) $J_{g \circ f_1}(x) = J_g(f_1(x))J_{f_1}(x)$, that is the composition of linear transformations (which, with respect to the given choice of bases, is matrix multiplication).

Proof This is standard direct calculation. The functions need not be defined on the entire spaces, but only on suitable open subsets.

We set $\|J_f(x)\| = \max_{\|v\|=1} \|J_f(x)v\|$, where v may be any n-dimensional vector of length 1. This clearly exists because it is the maximum of a continuous function on the compact space $\{v \in \mathbb{R}^n | \|v\| = 1\}$, which is a unit sphere.

Proposition 1.3 (The Finite Increments Lemma)   With the assumption that f is a C^r function, $r \geq 1$, defined on the open set $U \subseteq \mathbb{R}^n$, we suppose that U contains the entire line segment from x to y. We suppose that $\|J_f(a)\| < M$ for all a on the line segment from x to y.
Then

$$\|f(y) - f(x)\| < M\|y - x\|.$$

Proof   First we handle the local case. If $\varepsilon > 0$, we assert that for x and y close enough

$$\|f(y) - f(x) - J_f(x)(y - x)\| < \varepsilon\|y - x\|,$$

or

$$\|f(y) - f(x)\| < \|J_f(x)(y - x)\| + \varepsilon\|y - x\|.$$

Since

$$\|J_f(x)(y - x)\| < M\|y - x\|,$$

we may conclude that there is an open neighborhood of x such that if y is also in that neighborhood

$$\|f(y) - f(x)\| < M\|y - x\|.$$

It is clear that this remark is valid for any point a on the line from x to y, rather than just x. Now we choose (by compactness) a finite cover of open sets, in each of which the above inequality holds. We may easily select a succession of points on the line, say $x = x_0, \ldots, x_n = y$, such that x_i, for $1 \leq i \leq n - 1$, lies in the overlap of successive open sets. Then we may calculate

$$\|f(y) - f(x)\| = \|f(x_n) - f(x_{n-1}) + f(x_{n-1}) - \cdots + f(x_1) - f(x_0)\|$$
$$\leq \|f(x_n) - f(x_{n-1})\| + \cdots + |f(x_1) - f(x_0)\|$$
$$< M\|x_n - x_{n-1}\| + \cdots + M\|x_1 - x_0\|.$$

Since our points lie on a straight line, this is equal to

$$M\|x_n - x_0\| = M\|y - x\|.$$

Remarks   (1)   If f has continuous partials on the line, the Jacobian is continuous, and such a desired bound M may be found.

(2) If U is convex and bounded and $\|J_f(a)\| < M$ for any $a \in U$, then for all $x, y \in U$, we have

$$\|f(y) - f(x)\| < M\|y - x\|.$$

(3) If f is a map with continuous partial derivatives that vanish on a connected open set U, then f is a constant on U. For $\|J_f(a)\| = 0$, and any two points in U may be joined by a finite succession of lines, as in Proposition 1.3.

The following proposition, sometimes called the contraction lemma, is often a useful device.

Proposition 1.4 Let A be a closed subset of $\mathbb{R}^n$, and let $f : A \to A$ be a continuous map such that there is a real number M, $0 < M < 1$, with the property that if $a, b \in A$, then

$$\|f(b) - f(a)\| \leq M\|b - a\|.$$

Then there is a unique $x \in A$ such that $f(x) = x$.

Proof Take $a_0 \in A$, and put $a_1 = f(a_0)$, $a_n = f(a_{n-1})$. If $n < m$,

$$\begin{aligned}
\|a_m - a_n\| &= \|a_m - a_{m-1} + a_{m-1} - a_{m-2} + \cdots + a_{n+1} - a_n\| \\
&\leq \|a_m - a_{m-1}\| + \cdots + \|a_{n+1} - a_n\| \\
&= \|f^{m-1} \circ f(a_0) - f^{m-1}(a_0)\| + \cdots + |f^n \circ f(a_0) - f^n(a_0)\|,
\end{aligned}$$

where f^n means the n-fold composition of f with itself. If we call $\|f(a_0) - a_0\| = K$, then our hypothesis implies that this is less than $M^{m-1}K + \cdots + M^n K$, which clearly goes to zero as m and n increase. Thus, the sequence $\{a_n\}$ is a Cauchy sequence. Because A is closed, we know that it converges to an element of A, say $\lim_n a_n = x$. The reader may easily check that $f(x) = x$.

To prove uniqueness, suppose that, in addition, $f(y) = y$. Note that $\|x - y\| = \|f(x) - f(y)\| \leq M\|x - y\|$. Unless it were the case that $x = y$, we would have the inequality $M \geq 1$, contradicting our hypothesis. Hence $x = y$, and the proposition is proved.

The next result, the famous inverse function theorem, asserts that a differentiable map whose Jacobian does not vanish at a certain point behaves like a diffeomorphism sufficiently near that point.

Theorem 1.1 (Inverse Function Theorem) Let $f : \mathbb{R}^m \to \mathbb{R}^m$ be a C^r map ($r \geq 1$). For $a \in \mathbb{R}^m$ suppose that $J_f(a)$ is a nonsingular matrix, i.e., as a linear transformation $J_f(a)$ is onto. Then there are open subsets U and V of $\mathbb{R}^m$, and $a \in U$ and $b = f(a) \in V$, such that the restriction $f|U$ is a diffeomorphism from U onto V; i.e., there is a differentiable map $g: V \to U$, which is an inverse to $f \mid U$, and $J_g(b)$ is the inverse to the matrix $J_f(a)$.

Proof   It is easy to see that there is no loss of generality in assuming that $a = (0, \ldots, 0) = \mathbb{O} \in \mathbb{R}^m$ and $a = b$. Furthermore, it is clear that it would suffice to prove the theorem for f composed with a nonsingular linear transformation (i.e., a matrix with nonzero determinant), so that it is not a restriction of generality to assume that $J_f(a) = J_f(\mathbb{O})$ is the identity matrix.

Now, we set $h(x) = f(x) - x$, for any $x \in \mathbb{R}^m$. Clearly, $\|J_h(\mathbb{O})\| = 0$, so that in an open ball of some radius δ about $\mathbb{O}$, we may assume $\|J_h(x)\| < \frac{1}{2}$. By the increment lemma, since the ball is convex, we conclude that whenever x_1 and x_2 are closer to the origin than δ,

$$\|h(x_1) - h(x_2)\| \le \tfrac{1}{2}\|x_1 - x_2\|.$$

Or taking $x_2 = \mathbb{O}$, we see that if $\|x_1\| < \delta$, then

$$\|h(x_1)\| \le \tfrac{1}{2}\delta.$$

Now, if $\|y\| < \frac{1}{2}\delta$ and $\|x_1\| < \delta$, then

$$\|y - h(x_1)\| \le \|y\| + \|h(x_1)\| < \delta.$$

If we define $\phi(x) = y - h(x)$, then ϕ takes the ball of radius δ at the origin $\mathbb{O}$ into itself; in addition, when x_1 and x_2 are closer to $\mathbb{O}$ than δ,

$$\|y - h(x_1) - (y - h(x_2))\| = \|h(x_2) - h(x_1)\| \le \tfrac{1}{2}\|x_1 - x_2\|.$$

We conclude

(i) ϕ is a contraction mapping (as in Proposition 1.4), and thus ϕ has a unique fixed point x_0, $\|x_0\| \le \delta$.

(ii) $\phi(x_0) = x_0$ implies that $x_0 = y - (f(x_0) - x_0)$, or $y = f(x_0)$. In particular, f is onto the ball of radius $\frac{1}{2}\delta$ at $\mathbb{O}$.

(iii) $\|(x_0)\| < \delta$; for if $\|x_0\| = \delta$, then since $\|y\| < \frac{1}{2}\delta$ we easily have

$$\|x_0\| = \|f(x_0) - h(x_0)\| < \tfrac{1}{2}\delta + \tfrac{1}{2}\delta = \delta.$$

To construct the sets claimed in the theorem, we set $A = B_\delta(\mathbb{O}) \cap f^{-1}(B_{\delta/2}(\mathbb{O}))$. Conclusion (i)—that ϕ has a unique fixed point—means that f is 1–1 on the set A.

To verify that f is an open mapping when restricted to A, we note that if x_1 and x_2 belong to $B_\delta(\mathbb{O})$ then

$$\|f(x_1) - f(x_2)\| = \|x_1 + h(x_1) - x_2 - h(x_2)\| \ge \|x_1 - x_2\| - \|h(x_1) - h(x_2)\|$$
$$\ge \|x_1 - x_2\| - \tfrac{1}{2}\|x_1 - x_2\| = \tfrac{1}{2}\|x_1 - x_2\|.$$

But this says that points outside a small open ball are mapped to points outside a small open ball. It is then easy to see that the 1–1 map f will map sufficiently small open sets to open sets in the ball $B_{\delta/2}(\mathbb{O})$, and hence $f|A$ is an open mapping, or equivalently, $(f|A)^{-1}$ is continuous. (Of course, this can also be deduced from the standard facts that images of compact sets are compact and 1–1 mappings send complementary sets into complementary sets.)

We now need only check that the inverse function $(f|A)^{-1}$ is differentiable and has the correct Jacobian. For example, at the origin, we need check

$$\|[(f|A)^{-1}(z) - (f|A)^{-1}(\mathbb{O}) - J_f(\mathbb{O})^{-1}(z - f(\mathbb{O}))]\|/\|z - f(\mathbb{O})\| \to 0.$$

But since $J_f(\mathbb{O})$ is the identity and $f(\mathbb{O}) = \mathbb{O}$, this reduces to

$$\|[(f|A)^{-1}(z) - z]\|/\|z\| \to 0,$$

and if we write $z = f(w)$, this is

$$\|[w - f(w)]\|/\|f(w)\| \to 0.$$

But it follows easily from Proposition 1.1 that near enough to $\mathbb{O}$ we may assume that $\|f(w)\| > \frac{1}{2}\|w\|$, so that the desired limit follows at once.

The final assertion about the Jacobian of the inverse then follows from Proposition 1.2b.

A virtually equivalent theorem, frequently chosen in advanced calculus texts because of specific applications, is the famous implicit function theorem:

Theorem 1.2 (Implicit Function Theorem) Let $f: \mathbb{R}^n \times \mathbb{R}^m \to \mathbb{R}^m$ be a C^r $(r \geq 1)$ map. Suppose that for $(a, b) \in \mathbb{R}^n \times \mathbb{R}^m$ the mapping

$$i_a: \mathbb{R}^m \to \mathbb{R}^{n+m}$$

is defined by $i_a(b) = (a, b)$. (It is trivial to check that i_a is C^r.)

If $J_{f \circ i_a}(b)$ is a nonsingular $m \times m$ matrix, then there is a unique continuous map g from a sufficiently small open neighborhood of a in $\mathbb{R}^n$ to $\mathbb{R}^m$ such that

$$f(x, g(x)) = f(a, b)$$

with $g(a) = b$.

Proof With no loss of generality, we may assume that a, b, and $f(a, b)$ are all at the origins of their respective spaces. We define

$$\phi: \mathbb{R}^n \times \mathbb{R}^m \to \mathbb{R}^n \times \mathbb{R}^m$$

by $\phi(x, y) = (x, f(x, y))$. It is trivial to check that $J_\phi(\mathbb{O}, \mathbb{O})$ is nonsingular, so by the previous theorem the map ϕ locally has an inverse, which we call ψ.

Then in a neighborhood of the origin, we may clearly write

$$\psi(x, y) = (x, g(x, y))$$

and set $g(x) = g(x, \mathbb{O})$, defining a map $g: \mathbb{R}^n \to \mathbb{R}^m$. It is obvious that

$$f(x, g(x)) = f(x, g(x, \mathbb{O})) = f(\psi(x, \mathbb{O})).$$

Since $\phi(\psi(x, \mathbb{O})) = (x, \mathbb{O})$, we see at once, from the definition of the map ϕ, that $f(\psi(x, \mathbb{O}))$ must be $\mathbb{O}$.

Continuity is clear, and uniqueness follows, for example, from the facts that g is differentiable and that its value at $\mathbb{O}$ and the values of its derivatives near $\mathbb{O}$ are forced from our given data.

There are many traditional applications of these theorems, but we wish to apply them to get results on manifolds in general. To this end, we must first talk about linear approximations or tangential maps.

Definition 1.7   Let M^m and N^n be differentiable manifolds ($C^r, r \geq 1$) and let $f: M^m \to N^n$ be a differentiable (continuous partials of order through r) map. Let $a \in M^m$, $b = f(a) \in N^n$, with $a \in O_\alpha$ being the coordinate chart $\phi_\alpha: O_\alpha \to \mathbb{R}^m$ and $b \in U_\beta$ being the coordinate chart $\psi_\beta: U_\beta \to \mathbb{R}^n$. The linear map

$$J_{\alpha,\beta,f}(a): \mathbb{R}^m \to \mathbb{R}^n$$

described by the Jacobian matrix $J_{\psi_\beta \circ f \circ \phi_\alpha^{-1}}(\phi_\alpha(a))$ is called a *linear approximation of f at a*, or *a tangential map to f at a*.

If we take two different coordinate charts $O_{\alpha'}$ and $U_{\beta'}$ at a and b, then it is easily checked that there are nonsingular matrices P and Q such that

$$P \circ J_{\alpha',\beta',f}(a) \circ Q = J_{\alpha,\beta,f}(a).$$

In a certain sense (to be made precise later when we discuss tangent bundles) the linear approximation is well defined regardless of the choice of open sets. For the time being, however, we simply note that certain statements such as "$J_{\alpha,\beta,f}(a)$ is $1-1$ (or injective) as a linear transformation," or "$J_{\alpha,\beta,f}(a)$ is onto (or surjective)," are well defined, regardless of the choice of open sets, because these concepts are not affected by multiplying, either before or after, by nonsingular matrices.

We now apply the inverse function theorem to this situation.

Corollary 1.1   Let $f: M^n \to N^n$ be a differentiable map between differentiable manifolds, and let $a \in M^n$. Suppose that $J_{\alpha,\beta,f}(a)$ is nonsingular, for one, and hence every, choice of coordinate charts. Then there is an open set $U \subseteq M^n$, $a \in U$, that is mapped by f homeomorphically onto an open set containing $f(a)$.

Proof We apply Theorem 1.1 to the following composite map:

$$\phi_\alpha(O_\alpha \cap f^{-1}(U_\beta)) \xrightarrow{\phi_\alpha^{-1}} O_\alpha \cap f^{-1}(U_\beta) \xrightarrow{f} U_\beta \xrightarrow{\psi_\beta} \mathbb{R}^n.$$

$$\underset{\mathbb{R}^n}{\cap} \qquad\qquad \underset{M^n}{\cap} \qquad \underset{N^n}{\cap}$$

We can now study when the image of an injective map of manifolds is necessarily a submanifold.

Corollary 1.2 Let $m < n$ and let $g: M^m \to N^n$ be a differentiable map (C^r with r at least 1) that is 1–1. We suppose that for each $a \in M^m$ $J_{\alpha,\beta,g}(a)$ is 1–1, or equivalently has rank m, and we suppose that g is an open map, that is, g sends open sets in M into relatively open sets in the subset $g(M^m)$ of N^n. Then $g(M^m)$ is a submanifold (as in Definition 1.5) of N^n.

Proof We need to produce a suitable chart about each point of $g(M^m)$. Suppose that $g(a) = b$, that O_α is an open coordinate chart about a, and that U_β is an open chart about b. Because g is 1–1 and open, we may assume that O_α is $U \cap \text{Im } g$, for some open set $U \subseteq U_\beta$. Now in $\mathbb{R}^n$, we denote by $\mathbb{R}^{n-m}$ an $(n - m)$-dimensional subspace orthogonal to the image of $J_{\alpha,\beta,g}(a)$ at $\psi_\beta(b)$. We define a (locally) smooth map

$$g': O_\alpha \times \mathbb{R}^{n-m} \to U$$

by $g'(x, y) = \psi_\beta^{-1}(\psi_\beta(g(x)) + y)$, where of course, we regard $\mathbb{R}^n$ as $\mathbb{R}^m \times \mathbb{R}^{n-m}$.

It is trivial to verify that g' is nonsingular at $(a, 0)$ so that we may use Theorem 1.1 to get a smooth inverse locally around $g'(a, 0)$. This inverse is a diffeomorphism from an open neighborhood of b in U to an open neighborhood of $(a, 0)$ in $O_\alpha \times \mathbb{R}^{n-m}$. But then this diffeomorphism gives a new chart for the manifold N at b, for which the inverse image of the set

$$\{(x_1, \ldots, x_m, 0, \ldots, 0) \in \mathbb{R}^n\}$$

is locally the image of g.

Note that this new chart is compatible with the differentiable structure on N^n, since it is fabricated from a known chart with a diffeomorphism. Note also that the original structure on M^m is equivalent to the one that it inherits from N^n.

Corollary 1.3 If $m < n$, M^m is compact, $g: M^m \to N^n$ is a differentiable map of differentiable manifolds that is 1–1, and $J_{\alpha,\beta,g}$ has maximal rank, then $g(M^m)$ is a submanifold of N^n.

Proof We need only check that g is open in order to invoke the preceding corollary. But since g is 1–1, we need only check that g sends closed sets to closed sets. Since the domain is compact, this reduces to the well-known assertion that the continuous image of a compact space is compact.

The intuitive content of the above corollaries is that in a small neighborhood of any point, in the image of the 1–1 map g, the map g looks like the standard inclusion of a factor in Euclidean space, i.e.,

$$\mathbb{R}^m \subseteq \mathbb{R}^m \times \mathbb{R}^{n-m} \equiv \mathbb{R}^n.$$

More general conclusions are also possible; for example, the following corollary of Theorem 1.1, occasionally called the rank theorem.

Corollary 1.4 Let $f : M^m \to N^n$ be a smooth map with the property that in an open neighborhood of $a \in M^m$, all the maps $J_{\alpha,\beta,f}(b)$, at every point b of this neighborhood, have the same rank. Then we may find small open neighborhoods of a and $f(a)$, say U and V, with coordinate homeomorphisms $h : U \to \mathbb{R}^m$ and $k : V \to \mathbb{R}^n$ such that

$$k \circ f \circ h^{-1}(b) = J_{\alpha,\beta,f}(b)$$

when $b \in U$. In other words, the diagram

$$
\begin{array}{ccc}
U & \xrightarrow{\;f\,|\,U\;} & V \\
{\scriptstyle h}\downarrow & & \downarrow{\scriptstyle k} \\
\mathbb{R}^m & \xrightarrow{\;J_{\alpha,\beta,f}\;} & \mathbb{R}^n
\end{array}
$$

is commutative: that is the two compositions around the picture are the same map.

Proof This is a similar application of Theorem 1.1; details are left to the reader. For references, one may consult [27, 93].

We recall now that the dimension of a manifold is prescribed in the definition of the manifold as the dimension of the Euclidean space to which the manifold is locally homeomorphic. Our earlier remarks, which appealed to material from algebraic topology, indicated that this number, the dimension, was an invariant. However, to show that the dimension of a smooth manifold is invariant under diffeomorphisms is a rather simpler matter.

Corollary 1.5 Let $f : M^m \to N^n$ be a diffeomorphism. Then $m = n$.

Proof By Corollary 1.2 $f(M^m)$ must be a (smooth) submanifold of N^n. This shows at once that $m \leq n$. But a diffeomorphism has a two-sided inverse f^{-1}, and the same remarks about the inverse show that $n \leq m$.

In contrast to some of the above corollaries, there are some more practical applications of this machinery. For example,

Corollary 1.6 (The Fundamental Theorem of Algebra) Let $a_0, \ldots, a_{n-1}$ be fixed complex numbers, and we consider the polynomial

$$f(z) = z^n + a_{n-1}z^{n-1} + \cdots + a_1 z + a_0.$$

Then f, as a map of the complex numbers to itself, is onto. In particular, there is at least one z_0 such that $f(z_0) = 0$.

Proof f clearly takes an unbounded sequence of points into an unbounded sequence of points. It follows easily, from basic material about one-point compactifications, that f extends to a continuous map $\hat{f}: S^2 \to S^2$, which is easily checked to be smooth. Furthermore, since a polynomial of degree $n - 1$ has no more than $n - 1$ roots, the linear approximation to f, $J_{\alpha,\beta,f}(a)$, is nonsingular at all but a finite number of points. Then, by Theorem 1.1, $\hat{f}$ is locally a homeomorphism in a neighborhood of all but a finite number of points.

Several facts are now obvious:

  (i)   The image of $\hat{f}$ is closed (it is compact).
  (ii) The image of $\hat{f}$, $\hat{f}(S^2)$, is connected.
  (iii) Im $(\hat{f})$ contains infinitely many points.
  (iv) For each $b \in S$, $f^{-1}(b)$ is a finite set.

Now, staying away from the finite number of points where the image of the linear approximation to $\hat{f}$ is singular, it is clear that the number of points in $\hat{f}^{-1}(b)$, which we write as $|\hat{f}^{-1}(b)|$, is constant (for it is locally constant on a connected set). From (iii), there must be a point where the linear approximation is nonsingular for which $|\hat{f}^{-1}(b)|$ is nonzero. In other words, outside the finite singular set, $|\hat{f}^{-1}(b)|$ is a nonzero constant.

If $\hat{f}$ were not onto, then the complement of the image of f would be a nonempty open set (because $\hat{f}(S^2)$ is compact), thus containing an infinite number of points for which $|\hat{f}^{-1}(b)|$ is zero. Therefore f must be onto.

POWER SERIES

We now wish to look at some other basic points in real analysis that are relevant to the study of manifolds. To begin, we quote the standard Taylor's theorem.

Proposition 1.5 (Taylor's Theorem)   Let O be an open interval of real numbers, with $a \in O$. Suppose the real function $f : O \to \mathbb{R}$ has k continuous derivatives at each point of O. Then for any $x \in O$ we have

$$f(x) = \sum_{l=0}^{k-1} \frac{f^{(l)}(a)}{l!}(x - a)^l + \frac{f^{(k)}(\xi)}{k!}(x - a)^k,$$

for some ξ, with $|\xi - a| \le |x - a|$.

Proof   Standard.

Theorem 1.3 (Taylor's Theorem for $\mathbb{R}^m$)   We set the following notation: $\alpha = (\alpha_1, \ldots, \alpha_m)$ is a m-tuple of nonnegative integers. $|\alpha| = \alpha_1 + \cdots + \alpha_m$, $\alpha! = \alpha_1! \cdots \alpha_m!$, and $\binom{\alpha}{\beta} = \alpha!/\beta!(\alpha - \beta)!$ whenever β is another m-tuple with each $\beta_j \le \alpha_j$ (in which case we write $\beta \le \alpha$).

Let $O \subseteq \mathbb{R}^m$ be open and $f : O \to \mathbb{R}$ have continuous partials of order less than or equal to n throughout O. Suppose, finally, that for all t, $0 \le t \le 1$, $tx + (1 - t)y \in O$.

Then, writing

$$D^\alpha f(x) = \frac{\partial^{|\alpha|} f}{\partial x_1^{\alpha_1} \cdots \partial x_m^{\alpha_m}}(x) \qquad \text{and} \qquad x^\alpha = x_1^{\alpha_1} \cdots x_m^{\alpha_m},$$

we have

$$f(x) = \sum_{|\alpha| \le n-1} \frac{1}{\alpha!} D^\alpha f(y)(x - y)^\alpha + \sum_{|\alpha| = n} \frac{1}{\alpha!} D^\alpha f(\xi)(x - y)^\alpha$$

for some $\xi = \tau x + (1 - \tau)y$, $0 \le \tau \le 1$.

Proof   Set $g(t) = f(tx + (1 - t)y) = f(y + t(x - y))$. A trivial check with the chain rule will show that g meets the conditions of the previous theorem, Hence,

$$g(1) = \sum_{l=0}^{n-1} \frac{g^{(l)}(0)}{l!}(1 - 0)^l + \frac{g^{(n)}(\xi)}{n!}(1 - 0)^n = \sum_{l=0}^{n-1} \frac{g^{(l)}(0)}{l!} + \frac{g^{(n)}(\xi)}{n!}.$$

One must then calculate $g^{(l)}(0)$, for a given l. For example,

$$g'(0) = \sum_i \left(\frac{\partial f}{\partial x_i}(y) \right)(x_i - y_i).$$

The general result follows easily by a standard multinomial formula argument. (The concise notation utilized here was first proposed by Laurent Schwartz.) If, in case we take n to be ∞, the above series converges in a neighborhood of x, we speak of a Taylor series.

We now formalize an earlier remark.

Definition 1.8. A differentiable manifold M^n with coordinate charts $\phi_\alpha: O_\alpha \to \mathbb{R}^n$ is called analytic if the coordinates of the maps

$$\phi_\beta(O_\alpha \cap O_\beta) \xrightarrow{(\phi_\beta \,|\, O_\alpha \cap O_\beta)^{-1}} O_\alpha \cap O_\beta \xrightarrow{\phi_\alpha \,|\, O_\alpha \cap O_\beta} \mathbb{R}^n$$

are given by a convergent Taylor series in a neighborhood of every point. Similarly, one defines an analytic map between analytic manifolds.

Remark   All reasonable differentiable manifolds have analytic structures. However, in contrast to the smooth case, there have long been known analytic manifolds (without countable basis) with a multitude of distinct analytic structures. See [51] for references.

As a first application, of Taylor's theorem, we give a condition for a smooth map to be analytic.

Proposition 1.6   Let $O \subseteq \mathbb{R}^m$ be an open set and let $f: O \to \mathbb{R}$ be a smooth (that is C^∞) function. Suppose that for any compact set $K \subseteq O$, there is a number $M > 0$, such that

$$|D^\alpha f(x)| \leq M^{|\alpha|+1}\alpha!$$

for all $x \in K$ and all m-tuples α. Then f is analytic; that is, the Taylor series converges in a neighborhood of every point.

Proof   We estimate the remainder term in the Taylor series as follows:

$$\left| \sum_{|\alpha|=n} \frac{1}{\alpha!} D^\alpha f(\xi)(x-y)^\alpha \right| < n^m M^{n+1} \|x - y\|^n.$$

Now, if $\|x - y\| < 1/M$, then we may rewrite our estimate as

$$n^m M \left(\frac{\|x-y\|}{1/M} \right)^n$$

and see immediately that it goes to zero in n. This proves the convergence of the Taylor series.

It is not difficult to show that the converse of this proposition is also true.

FUNCTIONS WITH PRESCRIBED PROPERTIES; NORMS

Taylor's theorem has many applications, such as showing the completeness of various function spaces. Before considering function spaces, we need to take a hard look at functions from an arbitrary manifold to the real numbers.

Our first goal is to construct a partition of unity, an ingenious device which allows us to localize many properties on manifolds. Then we shall look at the set of zeros of a smooth function, a point where the differences between the smooth and analytic cases become apparent.

We shall build the functions that are involved in a partition of unity from a series of special cases.

Lemma 1.1   There is a C^∞-function $f: \mathbb{R} \to \mathbb{R}$ with $f(x) \geq 0$, $f(0) > 0$, and $f(x) = 0$ for $x > C > 0$. (C is any given positive constant.)

Proof   We set

$$f(x) = \begin{cases} e^{-1/(C-x)} & \text{if} \quad x < C \\ 0 & \text{if} \quad x \geq C. \end{cases}$$

We calculate

$$f'(C) = \lim_{h \to 0} \frac{[f(C+h) - f(C)]}{h} = \lim_{\substack{h \to 0 \\ h < 0}} e^{-1/h}/h = 0.$$

For $x < C$,

$$f'(x) = e^{-1/(C-x)}(1/(C-x)^2).$$

Continuing in this way, one may easily check that f is C^∞.

Lemma 1.2   There is a C^∞-function $f: \mathbb{R} \to \mathbb{R}$ such that $f(x) > 0$ for $a < x < b$, while $f(x) = 0$ for either $x \leq a$ or $x \geq b$.

Proof   This is an elementary algebraic modification of the previous lemma. For example, we may take a product of one function that vanishes precisely when $x \leq a$ with one that vanishes precisely when $x \geq b$.

Lemma 1.3   There is a C^∞-function $g: \mathbb{R}^m \to \mathbb{R}$ with $g(x) \geq 0$, $g(0) > 0$, $g(x) = 0$ for $\|x\| > C > 0$.

Proof   For a suitable f from the previous lemma, we may set

$$g(x) = f(\|x\|^2).$$

Lemma 1.4   Let K be a compact (or even merely closed) subset of $\mathbb{R}^m$ and U an open set with $K \subseteq U$. Then there is a C^∞-function $h: \mathbb{R}^m \to \mathbb{R}$ such that

(a) $h(x) \geq 0$,
(b)   if $x \in K$, $h(x) > 0$,
(c)   if $x \notin U$, $h(x) = 0$.

Proof Let d be the greatest lower bound of all the distances $d(x, y)$ with $x \in K$ and $y \in \mathbb{R}^m - U$. Since K is compact, it is clear that $d > 0$. Taking g as in Lemma 1.3 with $C = 1$, and taking $a \in K$, we put

$$\phi_a(x) = g((1/d)(x - a)).$$

Set $O_a = \{x \in \mathbb{R}^m | \phi_a(x) > 0\}$. Clearly, each $a \in O_a \subseteq U$, so that by compactness, there are finitely many $a_1, \ldots, a_r$ with

$$K \subseteq O_{a_1} \cup \cdots \cup O_{a_2}.$$

We may then set

$$h(x) = \sum_{i=1}^{r} \phi_{a_i}(x).$$

If K is merely closed, one may divide $\mathbb{R}$ into compact pieces and use the above argument to find infinitely many such ϕ_a for which the sum in the definition of $h(x)$ is everywhere just a finite sum.

Definition 1.9 Let $S \subseteq \mathbb{R}^m$ be a subset, and let $\{O_\alpha\}$ be a family of open sets in $\mathbb{R}^m$ that cover S, i.e., $S \subseteq \bigcup_\alpha O_\alpha$.

A family of continuous functions $f_\alpha : \mathbb{R}^m \to \mathbb{R}$, indexed by the same $\{\alpha\}$, is called a *partition of unity* for S, subordinate to $\{O_\alpha\}$, if

(a) for all α and x, $0 \le f_\alpha(x) \le 1$; if $x \notin O_\alpha$, then $f_\alpha(x) = 0$;
(b) for each point $x \in \mathbb{R}^m$, only finitely many of the $f_\alpha(x)$ are different from zero; and
(c) for each $x \in S$, we have

$$\sum_\alpha f_\alpha(x) = 1.$$

Theorem 1.4 Let $W \subseteq \mathbb{R}^k$ be an open set and $\{O_\alpha\}$, $\alpha \in A$, an open cover of W. Then there is a partition of unity for W subordinate to the cover $\{O_\alpha\}$. The functions in this partition of unity may be assumed to be smooth.

Proof We use some standard material from point-set topology (consult the sections on paracompact spaces in the books of Bourbaki [12], Gaal [36], or Kelley [62]). There is an open cover $\{U_\beta\}$, $\beta \in B$, and a map $r : B \to A$ such that $\bar{U}_\beta \subseteq O_{r(\beta)}$ and each $\bar{U}_\beta$ is compact. Furthermore, we may select a cover $\{V_\beta\}$ in such a way that $\bar{V}_\beta \subseteq U_\beta$ and every point of W has an open neighborhood which may be assumed to meet only finitely-many U_β.

Now, using Lemma 1.4, we select for each index $\beta \in B$ a C^∞ function $h_\beta : \mathbb{R}^k \to \mathbb{R}$ such that $h_\beta(x) \ge 0$,

$$h_\beta(x) > 0 \qquad \text{if} \quad x \in V_\beta,$$
$$h_\beta(x) = 0 \qquad \text{if} \quad x \notin U_\beta.$$

By the earlier remarks, in a sufficiently small neighborhood of any point, only finitely many h_β are different from zero.

The function

$$\phi'_\beta(x) = h_\beta(x) \bigg/ \sum_{\gamma \in B} h_\gamma(x)$$

is a C^∞-function that is never negative and is zero outside of U_β. Obviously, $\sum_{\beta \in B} \phi'_\beta(x) = 1$ for all $x \in W$.

In order to modify the $\{\phi'_\beta\}$ to get a partition of unity that is subordinate to our original cover, one simply defines ϕ_α to be the sum of all the functions ϕ'_β for which $r(\beta) = \alpha$, or zero in the case when there are none.

This theorem extends at once to the case where W is an open set in a paracompact manifold. This will include the cases where the manifold is a union of countably many compact subsets.

Corollary 1.7 Let $K \subseteq W \subseteq \mathbb{R}^k$, where K is compact (or just closed) and W is open. Then there is a C^∞-function $f : \mathbb{R}^k \to \mathbb{R}$ such that $f(x) = 1$ if $x \in K$ and $f(x) = 0$ if $x \notin W$.

Proof Select a partition of unity subordinate to the cover of $\mathbb{R}^k$ consisting of the two open sets W and $\mathbb{R}^k - K$. Then there are C^∞-functions ϕ_1 and ϕ_2 with $\phi_1(x) = 0$ if $x \notin W$ and $\phi_2(x) = 0$ if $x \in K$, yet $\phi_1(x) + \phi_2(x) = 1$ for all x. Hence $\phi_1(x) = 1$ if $x \in K$ and $\phi_1(x) = 0$ if $x \notin W$. Choose $f = \phi_1$.

Remark It is clear from this corollary that for any closed set S in $\mathbb{R}^k$ there are many nontrivial, smooth functions that vanish on S. The following theorem, due to Whitney, shows that there is a C^∞-function f such that S is precisely the set of zeros of f.

Theorem 1.5 Let $S \subseteq \mathbb{R}^k$ be closed. Then there is a C^∞-function $f : \mathbb{R}^k \to \mathbb{R}$ such that

$$S \equiv \{x \in \mathbb{R}^k \mid f(x) = 0\}.$$

Proof The open set $\mathbb{R}^k - S$ is a countable union of open balls

$$B_{r_n}(x_n) = \{y \in \mathbb{R}^k \mid d(y, x_n) < r_n\}.$$

By the method of Lemma 1.3, there are C^∞-functions $f_m : \mathbb{R}^k \to \mathbb{R}$ such that $f_m(x) > 0$ if and only if $x \in B_{r_m}(x_m)$.

We define

$$e_j = \text{l.u.b.} \{|D^\beta f_j(x)|\}.$$
$$\underset{\substack{|\beta| \le j \\ x \in \mathbb{R}^k}}{}$$

This least upper bound clearly exists, because there are only a finite number of k-tuples β, $|\beta| \leq j$, and the function f_j is identically zero outside a compact set. Note that each e_j must be positive, because f_j is not identically zero.

Choose a sequence of positive numbers d_j such that

$$0 < d_j < 1/e_j 2^j.$$

We define

$$f(x) = \sum_j d_j f_j(x).$$

The series, as well as any derivative of it, is eventually dominated by the series of constants

$$\sum (1/2^j).$$

Hence it converges uniformly (and the series of derivatives for $D^{\alpha}f(x)$ converges uniformly). Clearly, $f(x)$ is a C^{∞}-function, but it is also clear that $f(x) = 0$ if and only if $f_j(x) = 0$ for all j or $x \notin B_{r_m}(x_m)$ for all m. Since this means $x \in S$, the proof of our theorem is complete.

Remarks (1) This theorem clearly holds for any closed subset of a smooth manifold that satisfies the second axiom of countability, because the f_j's may easily be constructed for small open subsets lying in parts of the manifold that are identical to $\mathbb{R}^k$.

(2) This theorem indicates some of the important differences between various classes of functions on a manifold. The sets of zeros of analytic (or polynomial) functions are the analytic submanifolds (or algebraic varieties), and there are many fewer of these than just all the closed sets. For example, if $f : \mathbb{R} \to \mathbb{R}$ is real analytic (a convergent power series), then the set of zeros of f cannot be a closed interval $[a, b]$, $a < b$.

There are many other fascinating questions concerning smooth functions on $\mathbb{R}^k$, or on a manifold M^k. Some of these deal with approximation problems, and some deal with the differences between C^r, C^{∞}, or analytic functions, either locally or in the large. We can only touch on these here, but we shall be able to state some of the more powerful theorems and give appropriate references. We begin with some norms that generalize the usual absolute value and that offer important ways of measuring the size of functions.

Definition 1.10 Let $S \subseteq \mathbb{R}^k$ be any subset and $f : \mathbb{R}^k \to \mathbb{R}$ a continuous function.

(a)
$$\sup_{x \in S} f(x) = \text{l.u.b.} \{ f(x) | x \in S \},$$

that is, the least upper bound (if it exists) of the set of numbers $f(x)$, $x \in S$. We shall often write this as $\sup_S f$.

(b)   If $f \in C^m(\Omega)$, $S \subseteq \Omega \subseteq \mathbb{R}^k$, with Ω an open set, we define

$$\|f\|_m^S = \sum_{|\alpha| \le m} \frac{1}{\alpha!} \sup_{x \in S} |D^\alpha f(x)|.$$

Remarks   (1)   If f is continuous and S is compact, then clearly $\sup_S f$ exists. If f is also C^m on Ω, $S \subseteq \Omega$, then $\|f\|_m^S$ exists.

(2)   For any real k, $\|kf\|_m^S = |k| \|f\|_m^S$.

(3) $\|f + g\|_m^S \le \|f\|_m^S + \|g\|_m^S$ (check this easy triangle inequality).

(4) $\|fg\|_m^S \le \|f\|_m^S \|g\|_m^S$ (this follows easily from the Leibnitz rule for derivatives of products).

(5)   If $\Omega \subseteq \mathbb{R}^k$ is open, one may define a topology on $C^m(\Omega)$ by specifying a basis for the open sets around $g \in C^m(\Omega)$ to be the sets

$$\{f \in C^m(\Omega) \, | \, \|f - g\|_m^K < \varepsilon\}$$

for all positive ε's and compact sets $K \subseteq S$. Similarly, if the K_i are chosen to be an increasing family of compact subsets of Ω whose union is Ω, one may define a metric on this space by

$$d_m(f, h) = \sum_{i=0}^{\infty} \frac{1}{2^i} \frac{\|f - h\|_m^{K_i}}{1 + \|f - h\|_m^{K_i}}.$$

This metric may be checked to be complete. We shall consider details of such function spaces at several later points.

Definition 1.11   Let $f \in C^m(\Omega)$, $\Omega \subseteq \mathbb{R}^k$ being an open subset.

(a)   If $n \le m$, $x \in \Omega$, we say that f is *n-flat* at x when

$$D^\alpha f(x) = 0, \qquad \text{all} \quad |\alpha| \le n.$$

(b)   If $S \subseteq \Omega$ is a subset, we say that f is *n-flat on S* if f is *n*-flat at every point of S.

As a basic local approximation theorem, we have now

Proposition 1.7   Let $\mathbb{O} \in S \subseteq \mathbb{R}^k$ where $\mathbb{O}$ is the origin and S is open. Given $f \in C^\infty(S)$, with f being *n*-flat at $\mathbb{O}$, and given $\varepsilon > 0$, then there is an open set T, $\mathbb{O} \in T \subseteq S$, and a function $g \in C^\infty(S)$ such that

(a)   if $x \in T$, $g(x) = 0$;

(b) $\|g - f\|_n^S \le \varepsilon$.

Proof   We suppose $0 < \varepsilon < 1$. Choose an open ball

$$B_\varepsilon(\mathbb{O}) = \{x \, | \, d(x, \mathbb{O}) = \|x\| < \varepsilon\}$$

that lies entirely in S. As in Lemma 1.4, there is a C^∞-function η with $\eta(x) \geqq 0$, and $\eta(x) = 1$ if $\|x\| > \varepsilon$, and $\eta(x) = 0$ if $\|x\| \leq \frac{1}{2}\varepsilon$.

For $\delta > 0$, we set

$$g_\delta(x) = \eta(x/\delta)f(x).$$

Obviously, g_δ is a C^∞-function, and if we choose T to be $B_{\delta\varepsilon/2}(\mathbb{O})$, our first condition will be satisfied for g_δ.

It remains to be shown that there is a sufficiently small $\delta > 0$, so that our second condition is satisfied. Referring to the definition of $\| \ \|_n^S$, we need only prove that if $|\alpha| \leq n$ then for a sufficiently small δ, we may make

$$W = \sup_{x \in S} |D^\alpha g_\delta(x) - D^\alpha f(x)|$$

arbitrarily small.

But clearly

$$\sup_{x \in S} |D^\alpha g_\delta(x) - D^\alpha f(x)| = \sup_{\|x\| \leq \delta\varepsilon} |D^\alpha g_\delta(x) - D^\alpha f(x)|,$$

since when $\|x\| > \delta\varepsilon$ the expression in the absolute value sign is zero. But f is n-flat at zero, which means that if $|\alpha| \leq n$,

$$\lim_{\delta \to 0} \left(\sup_{\|x\| \leq \delta} |D^\alpha f(x)| \right) = 0.$$

Therefore, we must concentrate on g_δ.

Note that

$$\sup_{x \in \mathbb{R}^k} |D^\alpha \eta(x)|$$

exists for any α because η is constant outside a suitable ball. Using the Leibnitz formula on $g_\delta(x) = \eta(x/\delta)f(x)$, the chain rule, and the above remark, we see that there must be a positive constant A such that

$$|D^\alpha g_\delta(x)| \leq A \sum_{\beta+\gamma=\alpha} \delta^{-|\gamma|}|D^\beta f(x)|.$$

It is easy to see (using, e.g., Taylor's theorem) that each of these terms will go to zero with δ, $\|x\| < \delta$, because f is n-flat and $|\alpha| \leq n$.

We then see that our original W can be made as small as we wish. Then our second claim, about the norm of $g - f$, follows at once.

This approximation result has now a rather surprising consequence, due originally to Borel (see, e.g., [13]).

Theorem 1.6   We suppose that for each k-tuple

$$\alpha = (\alpha_1, \ldots, \alpha_k)$$

we are given a real number C_α. Then there is a C^∞-function $f: \mathbb{R}^k \to \mathbb{R}$, whose Taylor series about the origin has the numbers C_α as coefficients; that is,

$$D^\alpha f(0)/\alpha! = C_\alpha.$$

(Note that we make no claim that the Taylor series converges at any point other than the origin.)

Proof We begin by choosing polynomials with the numbers C_α as coefficients. Specifically, we set

$$f_m(x) = \sum_{|\alpha| \le m} C_\alpha x^\alpha,$$

where as usual $x^\alpha = x_1^{\alpha_1} \cdots x_k^{\alpha_k}$.

The polynomial $f_{m+1}(x) - f_m(x)$ is visibly m-flat at $\mathbb{O}$, so that we may use the previous proposition to find a C^∞-function g_m such that

$$\left\| f_{m+1} - f_m - g_m \right\|_m^{\mathbb{R}^k} < \rho^m,$$

where ρ is a fixed number, $0 < \rho < 1$, and g_m vanishes in an open set containing the origin.

We then define

$$f(x) = f_0 + \sum_{m \ge 0} (f_{m+1}(x) - f_m(x) - g_m(x)).$$

(Clearly, the series converges uniformly, since it is dominated by $\sum \rho^m$.)

Now because $\left\| f_{m+1} - f_m - g_m \right\|_m^{\mathbb{R}^k} < \rho^m$, we have

$$\frac{1}{\alpha!} \sup_{\mathbb{R}^k} \left| D^\alpha(f_{m+1}(x) - f_m(x) - g_m(x)) \right| < \rho^m$$

whenever $|\alpha|$ is less than m. Thus, we see that the series of derivatives D^α of our original series is also dominated by a convergent series of constants, and therefore converges to $D^\alpha f(x)$. So $f(x)$ is a C^∞-function.

We now see at once that

$$\sum_{i \ge m} (f_{i+1}(x) - f_i(x) - g_i(x))$$

is m-flat, since each term is (recall that every g_i vanishes in a neighborhood of $\mathbb{O}$). Then, for any $m > 0$, we have

$$f(x) = f_0(x) + f_1(x) - f_0(x) + \cdots + f_m(x) - f_{m-1}(x)$$

$$- (g_0(x) + \cdots + g_{m-1}(x)) + \sum_{i \ge m} (f_{i+1}(x) - f_i(x) - g_i(x))$$

$$= f_m(x) - (g_0(x) + \cdots + g_{m-1}(x)) + \sum_{i \ge m} (f_{i+1}(x) - f_i(x) - g_i(x)).$$

Now, the second term, $-(g_0(x) + \cdots + g_{m-1}(x))$, vanishes near $\mathbb{O}$. The rightmost term, the sum, as well as all derivatives of it of order $\leq m$, vanishes at $\mathbb{O}$. We conclude that $f(x)$ and $f_m(x)$ have the same derivatives, of order $\leq m$, at $\mathbb{O}$. They therefore have the same Taylor coefficients, through degree m. Since m was arbitrary, the theorem is completely proved.

Remarks (1) The preceding proposition and theorem represent a first level of approximation theory for C^∞-functions. The matter has been generalized to C^n-functions, and studied in considerable depth by H. Whitney. Whitney (1934; see [84] for references and details) proves that if $\Omega \subseteq \mathbb{R}^k$ is open, $f \in C^n(\Omega)$, and $\varepsilon(x) < 1/n$ is a positive, continuous, real function defined on Ω, then there is a real analytic $g:\Omega \to \mathbb{R}$ such that

$$\left|D^\alpha f(x) - D^\alpha g(x)\right| < \varepsilon(x)$$

for α satisfying $0 \leq |\alpha| \leq n$.

(2) The function in Borel's theorem (Theorem 1.6) need not have a convergent power series when $x \neq \mathbb{O}$. For example, in one variable, we could take $C_n = n!$ The resulting Taylor series has radius of convergence 0.

(3) Whitney has also looked at the generalization to Borel's theorem to the case where the derivatives are prescribed on a compact set.

(4) The theorem of Borel shows the richness of the possible localizations of a smooth function at a point. It thus offers a sort of motivation for the study of such functions in a small neighborhood of a point. The algebra of differentiable functions in a local setting is the theory of germs and jets. This theory, which we now look at briefly, has become quite fashionable; it is, in fact, rather more than merely convenient language.

GERMS AND JETS

If one wishes to study the local behavior of functions near a point in Euclidean space, one might naively look at the function at a single point. This would amount to a forfeiture of all information about derivatives; to understand derivatives one has to look at a function in an open neighborhood of a point. But a big open neighborhood contains too much information to analyze the function near a given point. The idea behind a germ of a function is the just the right compromise; it is a sort of limit over smaller and smaller open sets containing a given point.

Definition 1.12 Let $x \in \mathbb{R}^k$, and $U \subseteq \mathbb{R}^k$ be open, $x \in U$. Let $f \in C^n(U)$ (n possibly ∞). We denote such a function by (U, f). We shall introduce an equivalence relation on such pairs. We write $(U_1, f_1) \sim (U_2, f_2)$ (and say that the pairs are equivalent) if (a) there is an open set W, $x \in W \subseteq U_1 \cap U_2$, and (b) $f_1|W = f_2|W$, that is, when $x \in W$, $f_1(x) = f_2(x)$. An equivalence class of such pairs is called a *germ* of a C^n function at x.

More generally, we may speak of germs of functions who take values in a set, such as $\mathbb{R}^{k_2}$. The germs of C^n-functions $f : \mathbb{R}^{k_1} \to \mathbb{R}^{k_2}$, at $x \in \mathbb{R}^{k_1}$, will be denoted $G_n(\mathbb{R}^{k_1}, \mathbb{R}^{k_2}, x)$.

Remarks   Germs have the obvious nice properties with respect to addition and multiplication. More important, they compose well in the following sense: If we are given $\{(U, f)\} \in G_\infty(\mathbb{R}^{k_1}, \mathbb{R}^{k_2}, x)$, $\{(V, g)\} \in G_\infty(\mathbb{R}^{k_2}, \mathbb{R}^{k_3}, f(x))$, then the composition is defined to be

$$\{(U \cap f^{-1}(V), g \circ f)\} \in G_\infty(\mathbb{R}^{k_1}, \mathbb{R}^{k_3}, x).$$

Here, we use the brackets $\{\ \}$ to represent the equivalence class of what is inside.

By the same token, one sees at once that the germs $G_n(\mathbb{R}^k, \mathbb{R}, x)$ form an algebra [in that there exist addition, multiplication, and scalar multiplication (by constants) of germs of functions at a point satisfying the usual properties]. In this context, one may ask whether a germ has a (multiplicative) inverse. For a given germ $\{(U, f)\} \in G_n$ $(\mathbb{R}^k, \mathbb{R}^k, x)$, $n \geq 1$, one may construct the differential or Jacobian matrix at a point x. We may use Theorem 1.1 to conclude that if the Jacobian is nonsingular the local inverse to f represents the inverse to the germ (U, f) on a suitable small neighborhood of $f(x)$.

Finally, if we let $C_n(\mathbb{R}^k, \mathbb{R})$ denote the algebra of C^n-functions from $\mathbb{R}^k$ to $\mathbb{R}$, we have a natural projection map

$$\phi : C_n(\mathbb{R}^k; \mathbb{R}) \to G_n(\mathbb{R}^k, \mathbb{R}, x),$$

defined by $\phi(f) = \{(\mathbb{R}^k, f)\}$, for any given $x \in \mathbb{R}^k$. ϕ is a ring homomorphism that is easily checked to be onto, because any C^n function in a neighborhood U of x may be multiplied by a C^∞-function that is 1 in a neighborhood (small) of x and that is zero outside of U, giving an extension of our original function to all of $\mathbb{R}^k$, yet having the same germ at x.

Similarly, we might let $\mathbb{R}[[x_1, \ldots, x_k]]$ be the ring of formal power series in k variables, that is to say, power series in k variables, subject to the usual operations, but with no regard for convergence. In this context, Borel's theorem (Theorem 1.6) says that the homomorphism of rings

$$\psi : C_\infty(\mathbb{R}^k; \mathbb{R}) \to \mathbb{R}[[x_1, \ldots, x_k]],$$

defined by $\psi(f) = \sum_\alpha (D^\alpha f(0)/\alpha!)x^\alpha$, is onto.

We will now look briefly at jets, which like germs are a local way of regarding differentiable functions. A jet focuses attention on the lower part of a Taylor polynomial, or alternatively, ignores higher derivatives. We now attempt to make this precise.

Given two germs $\{(U, f)\}$ and $\{(V, g)\}$ in $G_\infty(\mathbb{R}^k, \mathbb{R}, x)$, we say that these germs are n-equivalent if for each α, $|\alpha| \leq n$, $D^\alpha f(x) = D^\alpha g(x)$. This trivially gives a well-defined equivalence relation, independent of the choice of representative for the germs. An n-jet is an equivalence class of n-equivalent germs of C^∞-function at x. We write the n-jets as $J_n(\mathbb{R}^k, \mathbb{R}, x)$, or $J_n(\mathbb{R}^k, x)$ for short. Some of the basic facts about jets, all easy to verify, are the following:

(1)   An n-jet $\{(U, f)\} \in J_n(\mathbb{R}^k, x)$ may be represented by a polynomial. In fact, the terms of order less than or equal to n of the Taylor polynomial of order $n + 1$ will suffice.

(2)   Jets may be added, multiplied, etc. If $\mathbb{O} \in \mathbb{R}^k$ is the origin, the ring $J_n(\mathbb{R}^k, \mathbb{O})$ may be described as follows: Let $\mathbb{R}(x_1, \ldots, x_k)$ be the ring of polynomials in k variables, $(x_1, \ldots, x_k)$ the ideal made up of those polynomials having no constant term. The product of two ideals is the set of finite sums of elements each of which is a product of one element in the first ideal multiplied by one element in the second ideal. The $(n + 1)$-power of an ideal means the product with itself $(n + 1)$ times. In this language, the ring $J_n(\mathbb{R}^k, \mathbb{O})$ is isomorphic to

$$\mathbb{R}(x_1, \ldots, x_k)/(x_1, \ldots, x_k)^{n+1}.$$

In fact, the map which sends f to the terms of degree less than or equal to n, of the $(n + 1)$th Taylor's polynomial, yields the isomorphism.

(3)   A function determines a germ at any point interior to its domain, and thus a jet. The zero-order jets are just the collections of germs having a common value at a point.

There are nontrivial C^∞-functions, whose nth order jet at a point vanishes for all n. In $C_\infty(\mathbb{R}, \mathbb{R})$, for example,

$$f(x) = \begin{cases} e^{-1/x^2}, & x \neq 0 \\ 0, & x = 0 \end{cases}$$

has all derivatives 0 at 0 (check, using l'Hôpital's rule). Hence, the intersection of the kernels of all the homomorphisms

$$\pi_n: C_\infty(\mathbb{R}^k, \mathbb{R}) \to J_n(\mathbb{R}^k, \mathbb{R}, x), \qquad x \in \mathbb{R}^k,$$

is not zero. In our earlier language, a function f may be C^∞ and n-flat for all n at a point without being constant.

Up to now we have spoken of germs and jets from the point of view of localization. But there are times where one wishes to study the local data at each point all at once. For example, one might consider the union of all the germs of C^∞-functions at all points of a space such as $\mathbb{R}^k$, a sort of localization followed by globalization. This is the area of sheaf theory (compare with [53, 106]). The result is bigger than the set of all original functions; in fact, the original functions emerge as sections of the sheaf. These ideas have become very important in complex analysis, algebraic geometry, etc., and will be touched on indirectly when we study vector bundles and differential forms.

PROBLEMS AND PROJECTS

1. There are many points worth examination in connection with the inverse and implicit function theorems.

(a) If $f: U \to V$ is a diffeomorphism from an open set U of $\mathbb{R}^m$ to another open set V of $\mathbb{R}^m$, and $x \in U$, then $J_f(x)$ is nonsingular.

(b) Give examples of such diffeomorphisms f as in (a) for which all higher derivatives vanish, as well as examples where the higher derivatives do not vanish.

(c) If $f: \mathbb{R}^2 \to \mathbb{R}$ is differentiable, $f(a, b) = 0$, but $\partial f(a, b)/\partial y \neq 0$, then one may use the implicit function theorem to solve for $y = g(x)$, with $f(x, g(x)) = 0$ for all x near a. Show that the condition is not necessary—in fact, find $f(x, y)$ for which $\partial f/\partial y = 0$ at $(0, 0)$—but that there is a continuous $g(x)$ for which $f(x, g(x)) = 0$ for all x.

2. The map that converts between rectangular and cylindrical coordinates may be written

$$x = r \cos \theta, \qquad y = r \sin \theta \qquad z = \zeta,$$

while the conversion for spherical coordinates may be written

$$x = r \sin \phi \cos \theta, \qquad y = r \sin \phi \sin \theta, \qquad z = r \cos \phi.$$

In each case, these represent maps from $\mathbb{R}^3$ to itself. Calculate the Jacobians, and discuss the ambiguities in these various systems.

3. Describe specific functions as guaranteed by Theorem 1.5. For example, let S be a submanifold or the union of the coordinate axes.

4. (a) Prove the converse to Corollary 1.3: If N^k is a smooth submanifold of M^n, there is a smooth 1–1 map with each Jacobian injective $g: S^k \to M^n$ (S^k a smooth manifold) such that

$$g(S^k) = N^k.$$

(b)   If M^n is compact and N^k is a submanifold, then N^k has no more than a finite number of connected components.

5.   (a)   If f and g are analytic functions from $\mathbb{R}^k$ to $\mathbb{R}$ that define the same germ at a point, they are equal. (Calculate the Taylor series, treating the point in question as the origin.)

(b)   If $C_\infty(\mathbb{R}^k; \mathbb{R})$ is the ring of smooth functions and $x \in \mathbb{R}^k$, let $E(x) = \{f \in C_\infty(\mathbb{R}^k, \mathbb{R}) \,|\, \text{there is a neighborhood } U \text{ of } x \text{ on which } f \text{ vanishes identically}\}$. Show that $E(x)$ is an ideal in $C_\infty(\mathbb{R}^k; \mathbb{R})$ and that the quotient ring by this ideal is $G_\infty(\mathbb{R}^k, \mathbb{R}, x)$.

For many standard problems in the inverse and implicit function theorems, consult such books as [24, 93].

2

Embeddings and Immersions
of Manifolds

The ideas in this chapter arose in the desire to view manifolds as sub-manifolds of Euclidean space. More generally, one might wish to view one manifold as a submanifold of some bigger (and possibly simpler) manifold. The notion of an embedding is very simple; a manifold M^n may be embedded in a bigger manifold N^s if M^n is identical to a submanifold of N^s (precise wording is given below). Perhaps a more useful point of view would be to say that an embedding is a nice 1–1 map from the smaller manifold M^n to the bigger manifold N^s. A simple example would be the embedding of the circle in the plane given by

$$\theta \to (\cos \theta, \sin \theta)$$

(where the arrow means that the mapping sends θ to the pair on the right in $\mathbb{R}^2$).

An immersion—in the most informal possible language—is a map that would appear, to a nearsighted person, to be an embedding, but that might fail to be 1–1. In other words, after one moves a certain distance from a given point, the mapping might fold back on itself. The simplest, nontrivial example would be a figure eight. If $0 \le \theta \le 2\pi$, we define

$$f(\theta) = \begin{cases} (-\cos 2\theta + 1, \sin 2\theta) & \text{if} \quad 0 \le \theta \le \pi \\ (\cos 2\theta - 1, \sin 2\theta) & \text{if} \quad \pi \le \theta \le 2\pi. \end{cases}$$

Except for the points that are mapped to the origin, $\theta = 0$ and $\theta = \pi$, the mapping is like an embedding.

Of course, to be an immersion is a condition of a local nature. Thus, the 2–1 map

$$\theta \to (\cos 2\theta, \sin 2\theta)$$

is an immersion of S^1 in $\mathbb{R}^2$ whose image happens to be homeomorphic to a circle. This is less frequent; the usual image of an immersion, such as $f(\theta)$ above, is not a manifold.

The importance of putting a manifold into a simpler one (via an embedding or immersion) cannot be overemphasized. Once a manifold is a submanifold of Euclidean space, one may introduce various analytic or differential-geometric concepts that may not have been clear otherwise. Of course much can be done in an intrinsic manner—without regard to embeddings—but other concepts, such as a normal bundle, are fundamentally tied to embeddings. And sometimes an embedding or immersion gives the best intuitive hold on a construction. (The classifying map of the tangent bundle is a good example.) It is a remarkable result of Whitney (Theorem 2.1) that a differentiable manifold of dimension n may always be immersed in $\mathbb{R}^{2n}$ and embedded in $\mathbb{R}^{2n+1}$.

In addition to results on the existence of immersions and embeddings, it is of importance to study approximation questions. For example, in what range of dimensions can one approximate a smooth map by an immersion or embedding? Here too, we shall obtain the basic results of Whitney; we can only refer to some of the important results that have followed in recent years.

SOME IMPORTANT EXAMPLES

To give further motivation beyond the examples that are in the first chapter and to develop some technical tools that we need to prove the basic theorems, we begin with a discussion of some more sophisticated examples of manifolds, such as projective spaces and Grassman manifolds. We shall also need to discuss the tangent space to a manifold. This is a conglomerate construction: by the usual precepts of advanced calculus, a smooth submanifold of Euclidean space has a tangent space at a point, and the tangent space of the entire manifold consists of putting together the tangent spaces at each point. (This is an important special case of the vector bundles of Chapter 4.) Armed with these tools, we shall give complete proofs of the basic theorems of Whitney in the compact case (see [27, 51]).

Definition 2.1   The n-dimensional (real) *projective space* $\mathbb{R}P^n$ is defined as the quotient space of the n-sphere

$$S^n = \{x \in \mathbb{R}^{n+1} \,|\, \|x\| = 1\}$$

by the equivalence relation

$$x \sim y \qquad \text{if and only if} \qquad \text{(i) } x = y \quad \text{or} \quad \text{(ii) } x = -y.$$

(Recall that if $x = (x_1, \ldots, x_{n+1})$, $-x = (-x_1, \ldots, -x_{n+1})$ is the antipode.)

Proposition 2.1   The projective plane has the following alternate descriptions, i.e., $\mathbb{R}P^n$ is homeomorphic to the following spaces:

(a)   the quotient space of the ball

$$\mathbb{B}^n = \{x \in \mathbb{R}^n \,|\, \|x\| \le 1\}$$

by the equivalence relation

$$x \sim_1 y \qquad \text{if and only if} \qquad \text{(i)} \quad x = y \qquad \text{or}$$
$$\text{(ii)} \quad \|x\| = \|y\| = 1 \quad \text{and} \quad x = -y;$$

(b)   the set of all straight lines through the origin in $\mathbb{R}^{n+1}$, topologized by requiring that an open neighborhood of a given line L consists, for an $\varepsilon > 0$, of all lines L' through the origin for which if $x \in L'$, $\|x\| = 1$, there is a $y \in L$, $\|y\| = 1$, with $d(x, y) < \varepsilon$.

Proof   The equivalence of the definition with (b) is quite simple. For each line L there are precisely two points x and $-x$ on L with $\|x\| = \|-x\| = 1$. Define a map from the space in (b) to $\mathbb{R}P^n$ by $\phi(L) = \{x\}$. Conversely, given $\{x\} \in \mathbb{R}P^n$, let $\psi(x)$ be the line in $\mathbb{R}^{n+1}$ containing the points x, $\mathbb{O}$, and $-x$. It is trivial to check that both these maps are continuous and inverse to one another (do it!).

To prove the equivalence of the space in (a) with our definition, note that the upper hemisphere of S^n,

$$S^n_+ = \{x \in \mathbb{R}^{n+1} \,|\, \|x\| = 1, \quad x = (x_1, \ldots, x_{n+1}), \quad x_{n+1} \ge 0\},$$

is homeomorphic to $\mathbb{B}^n$, where the desired homeomorphism is given by

$$h((x_1, \ldots, x_{n+1})) = (x_1, \ldots, x_n) \in \mathbb{R}^n.$$

(Here we identify $\mathbb{R}^n$ with the subspace of $\mathbb{R}^{n+1}$ for which $x_{n+1} = 0$, and we observe that there is exactly one point on $S^n \subseteq \mathbb{R}^{n+1}$ with a given first n coordinates and nonnegative last coordinate.) It then suffices to show that

$\mathbb{R}P^n$, as in our Definition 2.1, is in fact the quotient space of the subset S^n_+ of S^n by the equivalence relation:

$$(x_1, \ldots, x_{n+1}) \sim_2 (y_1, \ldots, y_{n+1})$$

if and only if they are equal or both x_{n+1} and y_{n+1} are 0, and $x_i = -y_i$, for $1 \leq i \leq n$. Denoting this quotient space $S^n_+/\sim_2$, define

$$k: S^n_+/\sim_2 \to \mathbb{R}P^n$$

by $k(\{x_1, \ldots, x_{n+1}\}) = \{x_1, \ldots, x_{n+1}\}$, the brackets referring to the appropriate equivalence class. This map is visibly well defined and easily checked to be continuous. It is $1-1$, for points are equivalent, according to Definition 2.1, only if they are the same or antipodal, just as in the equivalence relation $\sim_2$. Finally, we note that any point in S^n is either in S^n_+ or antipodal to a point in S^n_+ (i.e., $-x \in S^n_+$), which shows at once that k is onto.

Both $S^n_+/\sim_2$ and $\mathbb{R}P^n$ are compact and easily checked to be Hausdorff spaces. It follows, therefore, from basic point-set topology, that k is a homeomorphism.

Remarks These various descriptions of $\mathbb{R}P^n$ are convenient for different circumstances. Proposition 2.1b is often the most useful for embedding and immersion problems, while Definition 2.1 itself shows immediately that the map $p: S^n \to \mathbb{R}P^n$ given by $p(x) = \{x\}$ is a 2–1 covering map. Proposition 2.1a is a concrete form, which in low dimensions (such as $\mathbb{R}P^2$) is easier to visualize.

Proposition 2.2 $\mathbb{R}P^n$ is a smooth or C^∞-manifold.

Proof The proof rests heavily on the relationship between $\mathbb{R}P^n$ and the n-sphere from which it was obtained (and that covers it in a 2–1 fashion.)

The map $\pi: S^n \to \mathbb{R}P^n$, which associates to each point its equivalence class, consisting of the point and its antipode, is onto. Clearly if $y \neq x$ and $y \neq -x$, then $\pi(y) \neq \pi(x)$. It follows that any subset of S^n that does not contain a pair of antipodal points is mapped in 1–1 fashion by π.

In particular, let U_x be the set of points $y \in S^n$, so that the angle, at the origin $\mathbb{O}$, of the segments from $\mathbb{O}$ to x and from $\mathbb{O}$ to y is less than $\pi/2$; it is mapped by π, in a 1–1 fashion, onto a subset of $\mathbb{R}P^n$. It may be trivially checked that the sets $\pi(U_x)$ cover $\mathbb{R}P^n$, and because $\pi^{-1}(\pi(U_x)) = U_x \cup U_{-x}$ (check!), it is clear—from the definition of the quotient topology—that the sets $\pi(U_x)$ are open. By the same token, π takes open subsets of U_x to open sets. Given a homeomorphism $\phi_x: U_x \to \mathbb{R}^n$, defined in terms of the manifold structure on S^n, we may define an obvious homeomorphism

$$\psi_x: \pi(U_x) \xrightarrow{\pi^{-1}} U_x \xrightarrow{\phi_x} \mathbb{R}^n,$$

where π^{-1} selects the point (out of the two antipodal points $\pi^{-1}(x)$) that is in the same connected component of $\pi^{-1}(\pi(U_x))$ as x.

Now, we note that if $z \in \pi(U_x) \cap \pi(U_y)$ we may look at the map

$$\psi_x(\pi(U_x) \cap \pi(U_y)) \xrightarrow{\psi_x^{-1}} \pi(U_x) \cap \pi(U_y) \xrightarrow{\psi_y} \mathbb{R}^n.$$

This map is easily calculated to be $(\phi_y \circ \pi^{-1}) \circ (\pi \circ \phi_x^{-1}) = \phi_y \circ \phi_x^{-1}$. Therefore because we know that S^n is a smooth manifold, it follows that $\mathbb{R}P^n$ is also a smooth manifold, having C^∞-functions for changes in coordinates (compare with Definition 1.2).

One may show in fact with little more difficulty that $\mathbb{R}P^n$, just as S^n, is an analytic manifold.

Definition 2.2   A differentiable manifold (C^r-manifold, $r \geq 1$) is called *orientable* if one may choose coordinate charts for the manifold, say (U_α, ϕ_α), in such a way that all the coordinate change functions $\phi_\beta \circ \phi_\alpha^{-1}$ have Jacobian matrices with positive determinant at every point for which they are defined.

Note that if a manifold is orientable and $\{O_\alpha, \psi_\alpha\}$ is another cover by coordinate charts, then one may choose the homeomorphisms ψ_α so that the Jacobians have positive determinant. One simply has to choose ψ_α so that, for some $x \in O_\alpha \cap U_\beta$, $\phi_\beta \circ \psi_\alpha^{-1}$ has Jacobian of positive determinant, and our claim follows by an easy calculation.

Examples   1.   All S^n are orientable. Cover S^n by

$$U = \{x \in S^n \mid x \neq (0, \ldots, 0, -1)\}$$

and

$$V = \{x \in S^n \mid x \neq (0, \ldots, 0, +1)\}.$$

Choose any suitable homeomorphisms $\phi_U: U \to \mathbb{R}^n$ and $\phi_V: V \to \mathbb{R}^n$. If $\phi_V \circ \phi_U^{-1}$ has Jacobian of negative determinant, replace ϕ_V by $h \circ \phi_V$, where

$$h(x_1, \ldots, x_n) = (-x_1, x_2, \ldots, x_n).$$

This verifies that S^n is orientable. [Note that if the Jacobian of $\phi_V \circ \phi_U^{-1}$ has positive (resp. negative) determinant at a single point, then it has positive (resp. negative) determinant at every point in its domain of definition, when this domain of definition is connected.]

  2. $\mathbb{R}P^1$ is homeomorphic to S^1 (check!) and hence is orientable.

The interesting cases arise when the manifold requires three or more co-ordinate charts, so that no simple modification could possibly make the Jacobians of all the possible coordinate changes have positive determinant.

3. The real projective plane $\mathbb{R}P^2$ is *not* orientable. We use the description of Proposition 2.1a that represents $\mathbb{R}P^2$ as a quotient of a closed disk by the equivalence relation that identifies pairs of diametrically opposed points on the boundary. We shall cover $\mathbb{R}P^2$ by three open sets U_1, U_2, U_3, and in each open set we use coordinate axes to show the homeomorphisms ϕ_1, ϕ_2, ϕ_3 with $\mathbb{R}^2$ at some point (see Fig. 2.1). In each case, the open set is the shaded region. The points $A, \ldots, F$ (along with their antipodes, with which they are to be identified) are placed in the pictures simply to show how segments are to be identified. The choice of the homeomorphisms with $\mathbb{R}^2$ is a matter of convenience and does not affect the argument that the determinants of the Jacobians of the coordinate change maps cannot all be positive.

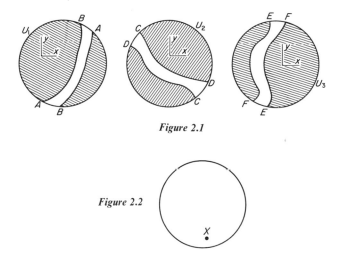

Figure 2.1

Figure 2.2

Now consider the point X, in both U_1 and U_3 (Fig. 2.2). Because the lower region of U_1 is identified with the boundary of the upper region antipodally, the coordinate axes, were they placed in the lower region, would be (we place the letter x on the positive segment of the x-axes, etc.) Since the axes in this region of U_3 are

$$\begin{array}{c} y \\ \llcorner\!\!\!\!\!\!\_\_\_\_\_ \; x, \end{array}$$

one sees at once that the determinant of the Jacobian of the coordinate transformation from U_1 to U_3, i.e.,

$$\phi_1(U_1) \xrightarrow{\phi_1^{-1}} U_1 \cap U_3 \xrightarrow{\phi_3} \phi_3(U_3),$$

must be negative.

On the other hand, at a point Y, in $U_1 \cap U_2$ (Fig. 2.3a), or a point Z in $U_2 \cap U_3$ (Fig. 2.3b) one easily checks that the determinant of the Jacobian would have to be positive (the coordinate axes have the same orientation at each of these points). In particular, the determinants of the Jacobians for the coordinate change maps for any pair of overlapping sets cannot all be positive simultaneously.

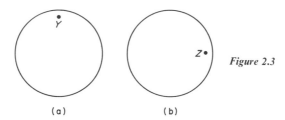

(a)   (b)   *Figure 2.3*

Remarks The more general results on orientability of manifolds best follow from the methods of algebraic topology. M^n is orientable if and only if $H_n(M^n; Z)$, the nth integral homology group, is the integers (compare with [25, 33], or the material in this text following Definition 8.5 for a brief survey). In the case of the real projective spaces, the conclusion is then that $\mathbb{R}P^n$ is orientable exactly when n is odd.

Now, Proposition 2.1b describes the projective space $\mathbb{R}P^n$ in terms of lines through the origin in $\mathbb{R}^{n+1}$ with a suitable topology. This immediately suggests two directions in which one might wish to generalize this construction. First, one may start with a vector space over another topologized field, for example $\mathbb{C}^{n+1}$, where $\mathbb{C}$ represents the complex numbers. The results are generally important, thought not related to our immediate goals. We include this as Definition 2.3 below. Second, one may replace the lines through the origin by linear subspaces of some fixed dimension. This construction, of the famous Grassman manifolds, is of great importance in what follows. We shall introduce it as Definition 2.4.

Definition 2.3   Let $\mathbb{C}^{n+1}$ be the complex vector space consisting of ordered $(n + 1)$-tuples of complex numbers $(z_1, \ldots, z_{n+1})$. A complex line through the origin consists of a given complex $(n + 1)$-tuple $(z_1, \ldots, z_{n+1})$, with not all z_i's equal to zero, and the multiples $(\lambda z_1, \ldots, \lambda z_{n+1})$ for all complex $\lambda \neq 0$. The set of all (complex) lines in $\mathbb{C}^{n+1}$ is clearly the quotient of $\mathbb{C}^{n+1}$ by the equivalence relation, which identifies $(n + 1)$-tuples that may be obtained from one another by a nonzero complex multiple.

The quotient space of $\mathbb{C}^{n+1}$ by this relation is defined to be the complex projective space $\mathbb{C}P^n$.

Remarks   (1)   As before, one may verify that $\mathbb{C}P^n$ is a compact C^∞-manifold. Using material from algebraic topology (see [25]), one may verify that $\mathbb{C}P^n$ is always orientable.

(2) $\mathbb{C}P^1$ is the sphere S^2. The points all have representatives of the form $(1, z_2)$ expect for $(0, z_2)$, which is equivalent to $(0, 1)$. Comparing this to S^2, as the one-point compactification of the complex plane, the point $(1, z_2)$ goes to the point z_2 and $(0, 1)$ goes to the point at infinity. ($(1, z_2)$ is equivalent for $z_2 \neq 0$ to $(1/z_2, 1)$, which clearly approaches $(0, 1)$ as $\|z_2\| \to \infty$).

Definition 2.4   The *Grassman manifold* $G_{n,m}$ of (unoriented) n-dimensional planes, through the origin in $\mathbb{R}^{n+m}$, consists—as a set—of all n-dimensional subspaces of the vector space $\mathbb{R}^{n+m}$. In other words, a point of $G_{n,m}$ is an n-dimensional subspace $\mathbb{R}^{n+m}$. The topology on $G_{n,m}$ is specified by describing an open neighborhood of $X \in G_{n,m}$. This consists, for any $\varepsilon > 0$, of all elements $Y \in G_{n,m}$, so that if u is any point of the subspace X, with $\|u\| \leq 1$, there is a point v of Y, with $\|v\| \leq 1$, such that $d(u, v) = \|u - v\| < \varepsilon$.

Remarks   (1)   Clearly $G_{1,n}$ is the same as $\mathbb{R}P^n$.

(2)   The operation of passing from a subspace to its orthogonal complement shows at once that that $G_{n,k}$ and $G_{k,n}$ are homeomorphic.

(3)   A similar construction can be made using *oriented* n-planes in $\mathbb{R}^{n+m}$. That is, each n-dimensional subspace X is endowed with a specific homeomorphism $\phi_X : X \to \mathbb{R}^n$, so that $\phi_Y \circ \phi_X^{-1}$ always has a Jacobian whose determinant is positive. The insistance on oriented planes changes the situation seriously; for example, the manifold of one-dimensional oriented subspaces of $\mathbb{R}^{n+1}$ is easily checked to be homeomorphic to S^n, not $\mathbb{R}P^n$. These oriented Grassman manifolds appear naturally in connection with the theory of homogeneous spaces (mentioned in the end of Chapter 9).

Proposition 2.3 $G_{n,m}$ is a C^∞-manifold of dimension nm.

Proof We shall describe the coordinate neighborhoods of each point and leave to the reader the (easy) task of verifying that the change of coordinate transformations are C^∞. To begin we fix an n-dimensional subspace L_0, and let $M = L_0^+$ be the orthogonal complement of L_0 in $\mathbb{R}^{n+m}$, dim $M = m$. Let U denote the set of all n-dimensional subspaces of $\mathbb{R}^{n+m}$ that intersect M at precisely the origin. We shall give U the structure of a coordinate neighborhood or chart.

Select a basis for the subspace M, say $z_1, \ldots, z_m$, and a basis for the subspace L_0, say $y_1, \ldots, y_n$. We claim that there is a homeomorphism from the matrices (x_{ij}), $i \le i \le n$, $1 \le j \le m$, that form $\mathbb{R}^{nm}$, to the n-dimensional subspaces, belonging to U, that have as basis the vectors

$$y_i + \sum_{j=1}^{m} x_{ij} z_j,$$

with $1 \le i \le n$.

Clearly, these vectors are linearly independent for a given matrix, and no linear combination of them (other than zero) can lie in M, so they determine an element of U. This gives a continuous map from $\mathbb{R}^{nm}$ to U.

Conversely, given $L \in U$, we wish to construct a matrix (x_{ij}). The vectors in L have the form

$$\sum_{i=1}^{n} \alpha_i y_i + \sum_{j=1}^{m} \beta_j z_j,$$

where not all α_i are zero. It is clear from adding multiples of such vectors together, as in the standard reduction of a matrix to diagonal form, that the vectors

$$y_i + \sum_{j=1}^{m} \bar{\beta}_j z_j$$

must belong to L and for each i, we call the numbers $\bar{\beta}_j$ by the name x_{ij}. I claim that these x_{ij} are uniquely determined. For suppose $y_i + \sum_j x_{ij} z_j$ and $y_i + \sum_j x'_{ij} z_i$ belong to L. Since L is a vector subspace, their difference

$$\sum_j (x_{ij} - x'_{ij}) z_i$$

belongs to L. But this vector also belongs, visibly, to M, so that it must be zero vector, since $L \cap M = \mathbb{O}$. Since the set $z_1, \ldots, z_m$ is a linearly independent set, we conclude $x_{ij} = x'_{ij}$. Thus, the x_{ij} are uniquely determined.

The two maps that we have described, from $\mathbb{R}^{nm}$ to U and vica versa, are trivially inverse to one another and continuous, so that we have the desired homeomorphism from U to $\mathbb{R}^{nm}$.

This completes the important examples that we need to use in immersion and embedding, etc.

THE TANGENT SPACE

We now need to study the tangent space to a differentiable manifold. If M^n is a submanifold of $\mathbb{R}^m$, $n < m$, at any given point x one may construct, using standard ideas from advanced calculus, an n-dimensional linear subspace in $\mathbb{R}^m$ that is tangent to M^n at x. [Here, we call a linear subspace the translation, by a fixed vector, of a vector subspace through $\bigcirc$. If L is a (vector) subspace, $x \in \mathbb{R}^m$, the set $x + L = \{x + y \mid y \in L\}$ is a linear subspace through x.] This would give the tangent space at x.

But, more generally, we wish to discuss the total tangent space, which will be the union of all the tangent spaces at each point of M^n. This will turn out to be a manifold of dimension $2n$, say TM^n, endowed with a projection map

$$\pi: TM^n \to M^n$$

with the additional property that if $x \in M^n$, $\pi^{-1}(x) \subseteq TM^n$ will be—in a reasonable way—homeomorphic to $\mathbb{R}^n$. Each tangent space at $x \in M^n$, as indicated in the preceding paragraph, will be identified with $\pi^{-1}(x)$. (Details will follow in a moment.)

It often appears that the various concepts and examples in the study of manifolds are only loosely related. The notions of tangent space and Grassman manifold, for example, are closely tied together. Let $M^n \subseteq \mathbb{R}^m$ be a submanifold whose tangent space is TM^n and whose tangent space at the point x we shall denote $\pi^{-1}(x) = (TM^n)_x$. Then there is an important and natural map, often called the Gauss map

$$\rho: M^n \to G_{n,m-n},$$

defined by sending each $x \in M^n$ into $\rho(x)$, which is the result of translating the n-plane $(TM^n)_x$ down to the origin and then viewing the result as an n-dimensional vector subspace of $\mathbb{R}^m$: In vector notation, we would have

$$\rho(x) = (TM^n)_x - x.$$

Apart from its remarkably simple and elegant definition, the map ρ has a deeper meaning. It is a *classifying map* in the sense of vector bundles (as in Chapter 4), and one may recover the space TM^n entirely from M^n and the map ρ.

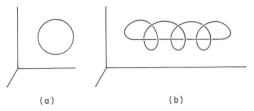

(a)   (b)

Figure 2.4

Unfortunately, the point of view that we have just outlined also has its disadvantages. For one thing, it is true (but far from trivial) that any differentiable manifold is diffeomorphic to a submanifold of Euclidean space. There might even be many different ways in which it is diffeomorphic to a submanifold of Euclidean space. For example, Fig. 2.4a,b shows submanifolds of $\mathbb{R}^3$ that are diffeomorphic to S^1. It needs some work to show that the tangent space, as in the intuitive perception given above, is independent of the way in which the manifold is embedded in Euclidean space.

All this points to the need for an abstract definition of the tangent space of a differentiable manifold. Such a definition may be given in various ways, for example, using equivalence classes of smooth curves or differentials on the algebra of smooth functions (an approach that we shall utilize in Chapter 9, especially in Proposition 9.1). We have chosen the definition below at this time because it clearly exhibits the tangent space to a manifold as a linear approximation and because it fits naturally into the framework of vector bundles (Chapter 4).

If we accept the idea, from the motivation above, that the tangent space over a point $\pi^{-1}(x) = (TM^n)_x$ should be identical to $\mathbb{R}^n$ and that the tangent space to $\mathbb{R}^n$ should simply be $\mathbb{R}^n \times \mathbb{R}^n$ (one $\mathbb{R}^n$ for each point in $\mathbb{R}^n$), then the problem of defining the tangent space to an arbitrary n-manifold is simply that of deciding how to glue together the tangent spaces over different coordinate charts. This idea is what is behind the following definition.

Definition 2.5   Let M^n be a differentiable manifold (of class C^r, $r \geq 1$). Let $\{O_\alpha\}$ be a family of coordinate charts, i.e., the $\{O_\alpha\}$ are an open cover, each O_α is endowed with a homeomorphism $\phi_\alpha: O_\alpha \to \mathbb{R}^n$, so that the maps $\phi_\beta \circ \phi_\alpha^{-1}$ are differentiable when defined, just as in Definition 1.2.

We consider the sets

$$T_\alpha = O_\alpha \times \mathbb{R}^n$$

for each α, with the usual product topology. Form the disjoint union of all these sets,

$$T' = \bigcup_{\text{disjoint}} O_\alpha \times \mathbb{R}^n.$$

We introduce an equivalence relation in T' as follows:

Given $(x, y) \in O_\alpha \times \mathbb{R}^n$, $(x', y') \in O_{\alpha'} \times \mathbb{R}^n$, we write $(x, y) \sim (x', y')$, if and only if $x = x'$ and $J_{\phi_{\alpha'} \circ \phi_{\bar{\alpha}}^{-1}}(y) = y'$. (Here $x = x'$ means that they are the same point of M^n, and J refers to the Jacobian.)

The relation is clearly reflexive and symmetric, and the transitivity is an immediate consequence of the rule of composition for Jacobians (Proposition 1.2b).

Define the tangent space to be the quotient space

$$TM^n = T'/\sim$$

and the projection map to be

$$\pi: TM^n \to M^n \qquad \text{by} \qquad \pi(\{x, \ y)\}) = x.$$

Proposition 2.4 TM^n is a noncompact, $2n$-dimensional manifold. If M^n is a differentiable manifold of class C^r, then TM^n is a C^{r-1}-manifold. $\pi: TM^n \to M^n$ is continuous, and if M^n is a C^r-manifold, π is a C^{r-1}-map. $\pi^{-1}(x) = (TM^n)_x$ is homeomorphic with $\mathbb{R}^n$.

Proof We set

$$U_\alpha = \{O_\alpha \times \mathbb{R}^n\},$$

where the brackets refer to the equivalence classes, of the set of elements inside, in the quotient topology. Define the obvious quotient map

$$\rho: O_\alpha \times \mathbb{R}^n \to U_\alpha,$$

which associates to each element its equivalence class. Define a map

$$i: U_\alpha \to O_\alpha \times \mathbb{R}^n$$

by observing that no different points of $O_\alpha \times \mathbb{R}^n \subseteq T'$ are equivalent (consult Definition 2.5 of the equivalence relation). In other words, in the diagram where (1 means the identity) no pair of distinct points has the same image under ρ, so that the quotient map i is well defined, with $i \circ \rho = 1$,

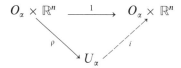

It follows immediately that ρ and i are inverses of one another, so that U_α is homeomorphic to $\mathbb{R}^n \times \mathbb{R}^n = \mathbb{R}^{2n}$. It follows, without difficulty, that TM^n is Hausdorff and has a countable base.

To check that TM^n is C^{r-1}, observe that the coordinate change maps for the sets U_α are given by the original coordinate change maps for M^n in the first n coordinates and the Jacobian of such maps in the last n coordinates. The functions which make up the entries of the Jacobian of a C^r-map are obviously C^{r-1}, and the remainder of the proposition follows at once.

Remarks   (1)   It is not difficult to work out some examples. For instance, $TS^1 = S^1 \times \mathbb{R}$. To see this, define a map

$$\phi: S^1 \times \mathbb{R} \to TS^1$$

by setting $\phi(x, y)$ to be that point in the tangent line to S^1 at x that is y units from x moving counterclockwise for positive y, along that line. (Recall that there is no difficulty defining a metric on the entire tangent space TS^1, and in the small the tangent space looks like a tangent line to S^1 in $\mathbb{R}^2$, so that it makes sense to speak of counterclockwise.)

   ϕ is clearly continuous and onto. If $\phi((x, y)) = \phi((\bar{x}, \bar{y}))$, then $x = \bar{x}$ (they must be over the same point in S^1) and y and $\bar{y}$, being the distances, in $\pi^{-1}(x)$, from x, must therefore be the same. Hence, ϕ is 1–1.

   To check that ϕ is open, it suffices to check that $\phi(O_\alpha \times V)$ is open, where O_α is a proper open subset of S^1, and V is an open interval of $\mathbb{R}^1$. But $\phi(O_\alpha \times V)$ lies in $\pi^{-1}(O_\alpha) = O_\alpha \times \mathbb{R}$ (by Definition 2.5), and it consists of all points, on sets $\pi^{-1}(x)$, $x \in O_\alpha$, whose distances from x lie in the open interval V. It is trival to check that any point near enough to such a point will also be in $\phi(O_\alpha \times V)$, and hence this set is open. We conclude that ϕ, being continuous, 1–1, onto, and open, is a homeomorphism.

   The key to this construction is that there is a global notion of counterclockwise, at every point on a circle.

   (2) $T(M^n \times N^k) \equiv TM^n \times TN^k$. In this case, the projection map, on a pair, is simply the projection map on each factor.

   (3)   Regardless of whether M^n is orientable, if M^n is a C^r-manifold, $r > 1$, TM^n is an *orientable* C^{r-1}-manifold. To show this, one need only calculate the matrix of the change of coordinate map for

$$\{(z, y)\} \in (O_\alpha \times \mathbb{R}^n) \cap (O_\beta \times \mathbb{R}^n).$$

Looking at the proof of Proposition 2.4, TM^n is covered by open neighborhoods $O_\alpha \times \mathbb{R}^n$ that are homeomorphic to $\mathbb{R}^n \times \mathbb{R}^n$, and the suitable change of coordinate function is

$$(\phi_\beta \circ \phi_\alpha^{-1}) \times J_{\phi_\beta \circ \phi_\alpha^{-1}}: \mathbb{R}^n \times \mathbb{R}^n \to \mathbb{R}^n \times \mathbb{R}^n.$$

In particular, every smooth curve whose first coordinate is constant is carried—by this map—to a smooth curve whose first coordinate is constant.

The usual result, from advanced calculus, is that the desired Jacobian has the form

$$J_{(\phi_\beta \circ \phi_\alpha^{-1}) \times J\phi_\beta \circ \phi_\alpha^{-1}} = \left[\begin{array}{c|c} J_{\phi_\beta \circ \phi_\alpha^{-1}} & 0 \\ \hline * & J_{\phi_\beta \circ \phi_\alpha^{-1}} \end{array} \right].$$

Clearly, the determinant of this matrix is

$$(\det(J_{\phi_\beta \circ \phi_\alpha^{-1}})^2 > 0.$$

(4) If M^n is nonorientable and N^k is arbitrary, then $M^n \times N^k$ is nonorientable. (In fact, one may check that for one open set $U \subseteq N^k$ it is not possible to find an open cover of M^n so as to make even $M^n \times U$ orientable.)

(5) If M^n is nonorientable, then (3) and (4) above imply that TM^n is not homeomorphic to $M^n \times \mathbb{R}^n$.

In particular, we may conclude that $T(\mathbb{R}P^2)$ is not homeomorphic to $\mathbb{R}P^2 \times \mathbb{R}^2$.

We note that the case when TM^n is homeomorphic to $M^n \times \mathbb{R}^n$ is a sort of trival extreme. In such a case, one says either that the manifold has a trivial tangent space (or sometimes a trivial tangent bundle) or that M^n is parallelizable. This last term has a historical origin that is best clarified in terms of the following definition.

Definition 2.6 A *tangent vector field*, to a C^r-manifold, $r > 1$, is a continuous map

$$v: M^n \to TM^n$$

such that $\pi \circ v = 1_{M^n}$ (that is, $\pi(v(x)) = x$ for each $x \in M^n$).

In other words, a tangent vector field consists of a continuous assignment of a tangent vector to each point such that $v(x)$ lies in $\pi^{-1}(x)$ for all $x \in M^n$.

Remarks (1) Every manifold has various tangent vector fields. The most trivial is given by $v(x) \equiv 0 \in \pi^{-1}(x)$. (Note that the equivalence relation used in Definition 2.5 preserves the presence of a zero in the second coordinate, so that this formula is well defined.)

Given an open set U whose closure lies in a coordinate chart $O_\alpha \subseteq M^n$, select a smooth function $\phi: M^n \to \mathbb{R}$ with $\phi(x) = 1$, for some $x \in U$, and $\phi(y) = 0$, if $y \notin U$ (compare Lemma 1.4). Define a constant tangent vector field over O_α by setting

$$v(x) = (x, z)$$

for some fixed $z \neq 0$ and all $x \in O_\alpha$. (Recall $\pi^{-1}(O_\alpha) \equiv O_\alpha \times \mathbb{R}^n$.) Extend v

to a vector field on all of M^n by setting

$$\bar{v}(x) = \begin{cases} \phi(x)v(x) & \text{if } x \in O_\alpha \\ 0 & \text{if } x \notin O_\alpha. \end{cases}$$

This constructs a tangent vector field that is zero outside of O_α but not everywhere zero.

(2)   The question of the existence of an everywhere nonvanishing tangent vector field is deeper and relates to the invariants which are studied in algebraic topology. Some results which may be so proved are

(i) S^n has a nowhere vanishing tangent vector field, if and only if n is odd;

(ii)   a compact, odd-dimensional manifold always has a nowhere vanishing vector field;

(iii)   a differentiable manifold that is a topological group has a rich structure in nowhere vanishing tangent vector fields (Chapter 9). For some details, consult [20, 54].

(3)   The term *parallelizable*, mentioned before Definition 2.6, means $TM^n \equiv M^n \times \mathbb{R}^n$. Such a manifold has n (nowhere vanishing) tangent vector fields, namely,

$$v_i(x) = (x, (0, \ldots, 1, \ldots, 0)),$$

with the 1 in the ith place, for all $x \in M^n$. At every point, the n vectors $v_i(x)$ are a basis for $\pi^{-1}(x) \subseteq TM^n$. Thus, a nonzero vector $v \in \pi^{-1}(x)$ may be expressed as

$$v = \sum \alpha_i v_i(x),$$

and then transported, parallel to itself, over the entire manifold to give a nowhere vanishing vector field by setting

$$v(x) = \sum \alpha_i v_i(x)$$

for all $x \in M^n$. There is this global notion of *parallel motion* in such a manifold, and hence the term.

Next, we shall investigate the effect on the tangent spaces of a differentiable map between differentiable manifolds.

Definition 2.7   Given a differentiable map ($C^r, r \geq 1$, or smooth)

$$f : M^n \to N^k$$

between C^r-manifolds, we define

$$df : TM^n \to TN^k$$

as follows: For $x \in M^n$, select O_α, $x \in O_\alpha \subseteq M^n$, $\phi_\alpha : O_\alpha \xrightarrow{\approx} \mathbb{R}^n$, and U_β, $f(x) \in U_\beta \subseteq N^k$, $\psi_\beta : U_\beta \quad \mathbb{R}^k$. Let $(x, y) \in O_\alpha \times \mathbb{R}^n$. We define

$$df\{(x, y)\} = \{(f(x), J_{\psi_\beta \circ f \circ \phi_\alpha^{-1}}(\phi_\alpha(x))(y))\},$$

where $\{\ \}$ refers to equivalence and the Jacobian matrix is to be evaluated at $\phi_\alpha(x)$ and then multiplied by the vector y.

I claim that df is indeed well defined. Using a prime to denote a similar coordinate chart, one may calculate, for example,

$$\begin{aligned} df\{(x, J_{\phi_{\alpha'} \circ \phi_\alpha^{-1}}(y))\} &= \{(f(x), J_{\psi_\beta \circ f \circ \phi_{\alpha'}^{-1}} \circ J_{\phi_{\alpha'} \circ \phi_\alpha^{-1}}(y))\} \\ &= \{(f(x), J_{\psi_\beta \circ f \circ \phi_{\alpha'}^{-1}\phi_{\alpha'} \circ \phi_\alpha^{-1}}(y))\} \\ &= \{(f(x), J_{\psi_\beta \circ f \circ \phi_\alpha^{-1}}(y))\}, \end{aligned}$$

where we have surpressed the point at which the Jacobians are evaluated in order to keep the notation under control. The reader should have no difficulty completing the rest of the verification that our definition is independent of the choices.

Remarks (1) With all the given data in Definition 2.7, one checks at once, that df is a C^{r-1}-map. If f, M^n, N^k are all smooth, then so is df.

(2) The construction df respects composition of differentiable maps, i.e., $d(g \circ f) = dg \circ df$. In fact, the construction, which associates with each C^r-manifold M^n, the C^{r-1}-manifold TM^n, and each C^r-map f, the C^{r-1}-map df, is a functor from the category of C^r-manifolds and maps to that of C^{r-1}-manifolds and maps.

(3) The map df is a sophisticated version of a linear approximation. Trivially, for each fixed x, the map $df\{(x, y)\}$, as a function of y, is a linear transformation from $\pi^{-1}(x) = \mathbb{R}^n$ to $\pi^{-1}(y) \equiv \mathbb{R}^k$ that, in one representation, is $J_{\psi_\beta \circ f \circ \phi_\alpha^{-1}}(\phi_\alpha(x))$.

It is clear that we may speak unambiguously of the rank of df at a point. In particular, it makes sense to say that df is the zero map, or df is an isomorphism, at a point, regardless of the choice of coordinate charts.

EXISTENCE OF EMBEDDINGS
AND IMMERSIONS

The definition of df now enables us to tackle the basic concepts of embedding and immersion, the principal concerns of this chapter.

Definition 2.8   Given smooth manifolds M^n and N^k, with $n \le k$,

(a)   a smooth map $f: M^n \to N^k$ is called an *immersion* if df has maximal possible rank at every point in M^n. That is, $df\{(x, y)\}$, as a function of y, is a linear transformation of rank n (occasionally, we say f is regular at x to mean df has maximal rank);

(b)   an immersion is called an *embedding* if it is $1-1$ and open (that is, takes open sets to relatively open sets in its image).

Remarks   (1)   The inclusion map of a submanifold is a standard example of an imbedding. The map of S^1 into $\mathbb{R}^2$ as a figure-eight is an immersion. The $2-1$ covering map $\pi: S^2 \to \mathbb{R}P^2$ is an immersion.

(2)   A $1-1$ immersion can fail to be an embedding if the image doubles up on itself infinitely often. Figure 2.5 shows a $1-1$ smooth map of maximal rank from $\mathbb{R}^1$ to $\mathbb{R}^2$ (which may be formalized using functions like $\sin 1/x$). An open neighborhood of x is sent to a subset of the image that is not open in the relative topology, since a relatively open set would contain a part of the oscillatory region of the curve.

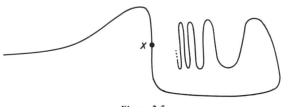

Figure 2.5

However, if the domain M^n is compact, then we may use Corollary 1.3 fo see that a $1-1$ immersion must necessarily be an embedding.

If one does not care about specific dimensions, then the existence of embeddings is not a difficult matter.

Proposition 2.5   Let M^n be a compact, smooth manifold. Then there is a $k > n$ and an embedding (smooth)

$$i: M^n \to \mathbb{R}^k.$$

Proof   Let $O_1, \ldots, O_s$ be a finite cover of M^n by open sets, each homeomorphic to $\mathbb{R}^n$ by $\phi_\alpha: O_\alpha \to \mathbb{R}^n$, defining the differentiable structure on M^n. We can assume, as in Chapter 1, that the cover may be refined to a cover by U_i's as well as V_i's, so that $\overline{U_i} \subseteq V_i \subseteq \overline{V_i} \subseteq O_i$, each i. As in Lemma 1.4

select smooth function $g_i: M^n \to [0,1] \subseteq \mathbb{R}$ so that

$$g_i(x) = \begin{cases} 1 & \text{if} \quad x \in U_i \\ 0 & \text{if} \quad x \notin V_i. \end{cases}$$

Define $\tilde{g}_{ij}: O_i \to \mathbb{R}$ to the jth coordinate of the homeomorphism function $\phi_i: O_i \overset{\approx}{\to} \mathbb{R}^n$. Define $g_{ij}: M_n \to \mathbb{R}$ to be the product of $\tilde{g}_{ij}$ and g_i, i.e., $g_{ij}(x) = g_i(x) \cdot \tilde{g}_{ij}(x)$. It is trivial to check that the g_{ij} are smooth functions defined on all M^n.

We define $i: M^n \to \mathbb{R}^{n+sn}$ to be the function whose coordinates are—in order—the functions g_i and then g_{ij} (ordered as $g_{11}, \ldots, g_{s1}, g_{12}, \ldots, g_{s2}, \ldots$). i is obviously a smooth map. We check that i is 1–1. Suppose $i(x) = i(y)$. Clearly x and y belong to a common O_i, for $g_i(x)$ must be 1 for some i, and then $g_i(y) = 1$, which means $y \in V_i \subseteq O_i$. But when $g_i(x) = g_i(y) = 1$, we may conclude that $\tilde{g}_{ij}(x) = \tilde{g}_{ij}(y)$, for all j, so that x and y have the same coordinates under the homeomorphism $\phi_i: O_i \to \mathbb{R}^n$. Hence $x = y$.

Lastly, we must show that di has maximal rank n, or in other words, di is injective 1–1 for all $x \in M^n$ (since $n < n + sn$). Given $x \in M^n$, $x \in U_i$ for some i, in which case $g_{ij}(x)$ is just the jth coordinate function of the homeomorphism ϕ_i. But then the n tangent vectors, which are tangent to the coordinate axes in $O_i = \mathbb{R}^n$ are mapped into n linearly independent vectors by these functions g_{ij}. Therefore, the dimension of the image of di is at least n, proving that the rank is at least n. Of course, since $\dim (TM^n)_x = n$, for any x, the rank of di is also at most n, proving that it is exactly n.

Our goal is now to improve this result, to get embeddings and immersions in lower dimensions that are independent of the number of coordinate charts necessary to cover the manifold. We need several preliminary propositions.

Proposition 2.6 Let M^n be a smooth manifold and $f: M^n \to \mathbb{R}^k$ be a smooth map. We assume that at the point $a \in M^n$, df is 1–1 (which implies that $n \leq k$). Furthermore, we let $\pi: \mathbb{R}^k \to \mathbb{R}^{k-1}$ be the projection $\pi(x_1, \ldots, x_k) = (x_1, \ldots, x_{k-1})$.

Then a necessary and sufficient condition for the differential of $\pi \circ f$ to be 1–1 at the point a is that the line $f(a) + L$, where $L = \{(0, \ldots, 0, t)\}$, is not contained in the image of df at a.

Proof Since df is 1–1 (or a monomorphism) at a, $d(\pi \circ f)$ will be 1–1 (a monomorphism) if and only if no vector of the linear subspace $\operatorname{Im} df(a)$ (the image of df at a) of the linear space $(T\mathbb{R}^k)_{f(a)}$ is annihilated under projection onto the first $k - 1$ coordinates. If the line $f(a) + L$ is contained in the image, then any nonzero vector lying in L will be annihilated under the projection π. Conversely, if a (nonzero) vector is annihilated by the projection, it must be parallel to the last coordinate axis; if it lies in Im $df(a)$, it will lie in the line $f(a) + L$, showing that that line lies in the image.

I suggest drawing a sketch in the plane, using for f an embedding of the circle in the plane.

Remark   By rotation, one easily sees that Proposition 2.6 holds for any line L when π is replaced by the projection to the orthogonal complement of L.

Proposition 2.7   Now, suppose that M^n and N^k are smooth, that M^n is compact, and that $n < k$. Let $f : M^n \to N^k$ be a smooth (or even just C^r, $r \geq 1$) map. Then $f(M^n) \subseteq N^k$ is a nowhere dense subset of N^k; i.e., for any open $O \subseteq N^k$, the set $O \cap f(M^n)$ is a proper subset of O. (In case M^n is not compact, it will follow that $f(M^n)$ is a set of the first category.)

Proof   For each $x \in M^n$, we choose an open coordinate chart O_α, $x \in O_\alpha$, $\phi_\alpha : O_\alpha \xrightarrow{\approx} \mathbb{R}^n$, as well as an open subset $U_\beta \subseteq N^k$, $f(x) \in U_\beta$, $\psi_\beta : U_\beta \xrightarrow{\approx} \mathbb{R}^k$. By passing to a possibly smaller O_α, we see that it is no loss of generality to assume that $f(O_\alpha) \subseteq U_\beta$. Let B be an open ball in $\mathbb{R}^n$ whose center is $\phi_\alpha(x)$, and set $O_x = \phi_\alpha^{-1}(B)$, so that clearly $\overline{O_x} \subseteq O_\alpha$. Since M^n is compact, only a finite number of O_x's are needed to cover M^n, so that we need only show that $f(\overline{O_x})$ is nowhere dense in U_β.

But because we are then in the world of individual coordinate charts, it is easily checked that we need only prove that $f(K)$ is nowhere dense in $\mathbb{R}^k$, where $K \subseteq \mathbb{R}^n$ is compact and $f : \mathbb{R}^n \to \mathbb{R}^k$ is differentiable, $n < k$. For this, let C be a cube in $\mathbb{R}^n$, $K \subseteq C$, where the diagonal of C has length d. We shall repeatedly divide C into identical cubes (or subcubes), so that after l subdivisions, each cube will have diagonal of length $d/2^l$.

Because of our assumptions, the usual material from advanced calculus (compare with Chapter 1) implies the existence of $A > 0$ with

$$\|f(x) - f(y)\| \leq A\|x - y\|,$$

for all $x, y \in K$. Now, if two points lie in a cube of the l-fold subdivision of C mentioned above, then after applying f their distances will be no more than $Ad/2^l$.

Now, in the range space $\mathbb{R}^k$, let C' be a cube containing the image under f of the intersection of K and a cube of the l-fold subdivision of C. Our above remarks show that we may assume that the side of C' is at most $2Ad/2^l$. Since the number of cubes in the l-fold subdivision of C is precisely 2^{nl}, we may estimate the total volume of all possible such C' cubes in $\mathbb{R}^k$ as at most

$$2^{nl}(2Ad/2^l)^k \leq (2Ad)^k/2^l.$$

Naturally, as l grows, the total volume of all these cubes in $\mathbb{R}^k$, which contain $f(K)$, may be made arbitrarily small. Since $f(K)$ is already closed, it is clear that $f(K)$ cannot be dense in any open cube, and hence it is nowhere dense.

Because any manifold M^n that we consider is a countable union of compact subsets, we conclude at once that in the general case $f(K)$ is a set of the first category (i.e., a countable union of nowhere dense subsets).

Now, in order to obtain Whitney's basic theorems on embedding and immersion, one must apply some reduction in dimension scheme to the basic result of Proposition 2.5. For this purpose, it is convenient to study a certain quotient of the tangent space to a manifold; this is the space of directions or rays in TM^n, defined as follows.

Definition 2.9 If M^n is a differentiable manifold, we let $(TM^n)^0$ denote the subspace of TM^n that is the complement to the set of all $\{(x,0)\}$; i.e., $(TM^n)^0$ is all points $\{(x,y)\} \in TM^n$, $y \neq 0$. We introduce an equivalence relation in $(TM^n)^0$ as follows: $\{(x,y)\} \sim \{(x',y')\}$, if and only if $x = x'$ and there is a real $\alpha > 0$, so that $\alpha y = y'$. (Recall that each $(TM^n)_x$ is a real vector space, isomorphic to $\mathbb{R}^n$, so that scalar multiplication is defined.)

The *space of rays* $R(M^n)$ is the quotient space of $(TM^n)^0$ by this equivalence relation. Note that equivalent points in $(TN^n)^0$ have the same image under $\pi: TM^n \to M^n$, so that we have a well-defined continuous projection map

$$\pi_R : R(M^n) \to M^n.$$

Remarks Suppose we endow TM^n with some metric, and let UM^n be the subset of all points $\{(x,y)\}$ for which the distance between $\{(x,y)\}$ and $\{(x,0)\}$ is 1. Clearly in each equivalence class of $(TM^n)^0$ there is exactly one element of UM^n. This means that the composition

$$UM^n \to UM^n/\sim \; \equiv (TM^n)^0/\sim \; \equiv R(M^n)$$

is a homeomorphism. $R(M^n)$ or UM^n occasionally go by the alternate name of the *unitangent bundle* of M^n.

Clearly $UM^n (= R(M^n))$ is a smooth manifold if M^n is smooth, because locally it is $O_\alpha \times S^{n-1}$. The dimension of $R(M^n)$ must then be $2n - 1$. One may think of the various α in the equivalence relation in Definition 2.9 as accounting for the additional dimension to get $2n = \dim TM^n$. (Recall that we shall often abbreviate the dimension of X by $\dim X$.)

The next two propositions make use of the techniques that we have now developed, and they are the last remaining steps before reaching Whitney's basic theorem. For simplicity, we shall now take our manifolds and maps to be smooth (although usually merely a little differentiability will suffice).

Proposition 2.8   Let $f: M^n \to \mathbb{R}^m$ be a smooth immersion, $2n < m$. Suppose M^n is compact. Let l be a line through the origin in $\mathbb{R}^m$, i.e., $l \in G_{1,m-1} \equiv \mathbb{R}P^{m-1}$. Finally, let

$$P_l: \mathbb{R}^m \to \mathbb{R}^{m-1}$$

be the projection of $\mathbb{R}^m$ onto the orthogonal complement to l at the origin (identified with $\mathbb{R}^{m-1}$). (For example, if l is the last coordinate axis, P_l is precisely the map π in Proposition 2.6.)

Then the set of all such $l \in \mathbb{R}P^{m-1}$ for which $P_l \circ f$ is not regular (i.e., $d(P_l \circ f)$ does not have maximal rank) is nowhere dense in $\mathbb{R}P^{m-1}$.

Proof   We propose first to define a map

$$\delta_f: R(M^n) \to \mathbb{R}P^{m-1}.$$

Because f is regular, for each $x \in M^n$ and ray $\{(x, y)\}$, df, being a linear map, yields a line in $\mathbb{R}^m$ through the point $f(x)$. We translate that line parallel to itself down to the origin and call $\delta_f(\{(x, y)\})$ the resulting element in $\mathbb{R}P^{m-1}$.

Now, we note that dim $R(M^n) = 2n - 1$ (1 less than dim TM^n), and thus, by our hypothesis,

$$\dim R(M^n) < \dim \mathbb{R}P^{m-1}.$$

Also, $R(M^n)$ is compact, because it is convered by a finite number of sets of the form $\overline{O}_\alpha \times S^{n-1}$, O_α a chart in M^n, and every $\overline{O}_\alpha \times S^{n-1}$ is compact. We may thus apply Proposition 2.7 to $\delta_f: R(M^n) \to \mathbb{R}P^{m-1}$ and conclude that the image of δ_f is nowhere dense in $\mathbb{R}P^{m-1}$.

Finally, we claim that the set of all $l \in \mathbb{R}P^{m-1}$ such that $P_l \circ f$ is not regular somehwere is precisely the image of δ_f. If l belongs to this set, we use a rotated Proposition 2.6 to find a direction such that the map df applied to this direction at some point is parallel to l. But this means by definition that l is in the image of δ_f. Conversely, if l is in the image of δ_f, there is some ray, at some point x, such that df applied to that ray yields a line parallel to l. Once again, Proposition 2.6—rotated so that l may be taken for L—shows that $P_l \circ f$ is not regular at every point, showing that l is an element of the set described in the proposition.

Putting all this together, the set of l such that $P_l \circ f$ fails to be regular at some point is nowhere dense in $\mathbb{R}P^{m-1}$.

Proposition 2.9   If $f: M^n \to \mathbb{R}^m$ is a smooth map that is 1–1, and if $2n + 1 < m$, the set of those lines $l \in \mathbb{R}P^{m-1}$ such that $P_l \circ f$ is not 1–1 is a subset of the first category in $\mathbb{R}P^{m-1}$. (As before, P_l is the projection onto the orthogonal complement of l.)

Proof $M^n \times M^n$ is a compact, smooth manifold of dimension $2n$. The diagonal

$$\Delta = \{(x, x) \mid x \in M^n\}$$

is a closed subset of $M^n \times M^n$, so that $M^n \times M^n - \Delta$ is a (noncompact) smooth manifold of dimension $2n$. It is occasionally called the space of secants in M^n, because a secant is determined by two distinct points, and we denote it $S(M^n)$.

Given $(x, y) \in S(M^n)$, we define $d^f(x, y)$ to be the unique line in $\mathbb{R}^m$ through $f(x)$ and $f(y)$ (which are distinct, since f is 1–1) translated parallel to itself so as to pass through the origin. It is routine to check that

$$d^f : S(M^n) \to \mathbb{R}P^{m-1}$$

is a smooth map.

Now, our assumptions about dimensions assure us that

$$\dim S(M^n) < \dim \mathbb{R}P^{m-1}.$$

Therefore, by Proposition 2.7, the image of d^f is a subset of the first category of $\mathbb{R}P^{m-1}$.

As before, we claim that our sets are the same. To be specific, we claim that the image of d^f, $\operatorname{Im} d^f$, is precisely those lines $l \in \mathbb{R}P^{m-1}$ such that $P_l \circ f$ is not 1–1. If $l \in \operatorname{Im} d^f$, there are distinct $x, y \in M^n$ with l parallel to the line segment through $f(x)$ and $f(y)$. But then $P_l \circ f$ is surely not 1–1. Conversely, if $P_l \circ f(x) = P_l \circ f(y)$, but $f(x) \neq f(y)$ (equivalently $x \neq y$), then the segment from $f(x)$ to $f(y)$ is parallel to l, so l is in the image of d^f.

Note that we can only claim that our set is of the first category, rather than nowhere dense, because our domain is not compact. Nevertheless, our set is a proper subset of $\mathbb{R}P^{m-1}$, because in a complete metric space, such as the compact $\mathbb{R}P^{m-1}$, any open set—such as all of $\mathbb{R}P^{m-1}$ itself—would be of the second category. (This is the basic Baire category theorem, from point-set topology.)

We are now ready for Whitney's results (see [27, 51]).

Theorem 2.1 (Whitney) Let M^n be a compact, smooth manifold. Then there is an immersion $f : M^n \to \mathbb{R}^{2n}$ and an embedding $g : M^n \to \mathbb{R}^{2n+1}$.

Proof Construct by means of Proposition 2.5 an embedding

$$h : M^n \to \mathbb{R}^k$$

for k sufficiently large. With no loss of generality, we shall assume that $k > 2n + 1$. The two preceding propositions show that there is a line $l \in \mathbb{R}P^{k-1}$

such that $P_l \circ h$ is both regular everywhere (Jacobian has maximal possible rank) and 1–1. (The union of a nowhere dense set and a set of the first category is still a set of the first category, and in a complete metric space the complement of such a set must be nonempty.) Because the image of P_l is identical to $\mathbb{R}^{k-1}$, we may assume that we now have an embedding $h_1 : M^n \to \mathbb{R}^{k-1}$. If $k - 1 = 2n + 1$, we have found our g, as required. If $k - 1 > 2n + 1$, repeat the above argument with h_1 and find $h_2 : M_n \to \mathbb{R}^{k-2}$, etc. In this way, we may eventually find the desired embedding $g : M^n \to \mathbb{R}^{2n+1}$.

Once we have an embedding $g : M^n \to \mathbb{R}^{2n+1}$, we apply Proposition 2.8 to select a direction l with

$$P_l \circ g : M^n \to \mathbb{R}^{2n}$$

regular at each point. This is possible because there are points in $\mathbb{R}P^{m-1}$, indeed most points, outside the nowhere dense set where regularity breaks down. $P_l \circ g$ is then our desired immersion $f : M^n \to \mathbb{R}^{2n}$.

Remarks This theorem is not the best possible result. First of all, compactness is not really necessary for the proof, although to do without it would draw us into some cumbersome details. Roughly, to handle the embedding problem in the noncompact case one must be careful to place the different compact pieces of M^n in different parts of $\mathbb{R}^{2n+1}$. Second, the dimensions for the ranges of f and g are not optimal; excluding certain obvious irregular, low-dimensional cases, one may improve the dimensions in Theorem 8.1 by 1 (see [114, 115]). There is also a vast literature on special types of embedding—analytic, metric preserving, etc.—that may be traced down through our general references (such as [27, 51]).

One should also note that there are some very beautiful theories of immersion (see Smale [100] or Hirsch [50]) and embedding (see Haeffliger [43]) that reduce the problem of in what Euclidean space the manifold may be immersed or embedded to other problems that may be attacked by the methods of algebraic topology. Unfortunately, these ideas do not bear directly on global analysis, and even a statement of the results would lead us too far a field.

APPROXIMATION OF SMOOTH MAPPINGS

As a final topic in this chapter, we study the question of approximation of smooth mappings by immersions and embeddings. Obviously, this requires that dimension of the range be sufficiently large (for example, if M^n is nontrivial, no map $f : M^n \to \mathbb{R}^0$ could ever be approximated by a regular map).

To make sense of approximation, we must first introduce a topology on the set of smooth maps. There are various ways of doing this: for the purpose at hand, where the domain manifold is compact, the following definition suffices.

We shall formulate our approximation results in the C^k case, $k \geq 1$. A C^k-map is an immersion (or embedding) if everywhere regular (or everywhere regular, 1–1, and takes open sets to relatively open sets). The smooth case, $k = \infty$, is not difficult to include in this somewhat more general framework.

Definition 2.10   Let M^n be a compact, smooth manifold and $O_1, \ldots, O_s$ a collection of open sets with homeomorphisms $\phi_i : O_i \Rightarrow \mathbb{R}^n$, which define the differentiable structure. Let $f : M^n \to \mathbb{R}^l$ be a C^k-map. Let $\varepsilon > 0$.
We define

$$S(f, k, \varepsilon) = \{g : M^k \to \mathbb{R}^l \,|\, g \text{ is } C^k \quad \text{and whenever } x \in O_i$$
$$\|D^\beta f \circ \phi_i^{-1}(x) - D^\beta g \circ \phi_i^{-1}(x)\| < \varepsilon \quad \text{for all } |\beta| \leq k\}.$$

It is easy to check that the sets $S(f, k, \varepsilon)$ form a subbase for a topology on the set of maps (C^k-maps)

$$M^n \to \mathbb{R}^k.$$

This topological space of functions is denoted $C^k(M^n; \mathbb{R}^l)$. The topology is easily verified to be independent of the choice of cover.

Theorem 2.2   Let M^n be a smooth, compact n-manifold. We write

$$\text{Imm}(M^n; \mathbb{R}^l) = \text{set of } C^k\text{-immersions } M^n \to \mathbb{R}^l,$$
$$\text{Emb}(M^n; \mathbb{R}^l) = \text{set of } C^k\text{-embeddings } M^n \to \mathbb{R}^l.$$

Suppose $k > 1$. Then we have

(a) $\text{Imm}(M^n; \mathbb{R}^l)$ is an open subset of $C^k(M^n; \mathbb{R}^l)$. If $2n \leq l$, it is also a dense subset.

(b) $\text{Emb}(M^n; \mathbb{R}^l)$ is an open subset of $C^k(M^n; \mathbb{R}^l)$. If $2n + 1 \leq l$, it is also a dense subset.

Proof   The proof that immersions are open is virtually trivial. We may assume that M^n is covered by U_i, $U_i \subseteq O_i$, and $\bar{U}_i$ compact. For each $\bar{U}_i$ one can find an $\varepsilon_i > 0$ such that every map that is closer to a given regular map is automatically regular, because the determinant in the Jacobian is continuous, and thus assumes a minimum, in absolute value, on $\bar{U}_i$. Simply choose $\varepsilon > 0$ to be the minimum of all these ε_i (check details here).

Now, suppose $l \geq 2n$. Let $f: M^n \to \mathbb{R}^l$ be a C^k map, and $i: M^n \to \mathbb{R}^p$ be an embedding, by Proposition 2.5. Let

$$h: M^n \to \mathbb{R}^l \times \mathbb{R}^p = \mathbb{R}^{l+p}$$

be the product map $h(x) = (f(x), i(x))$. Let

$$\pi: \mathbb{R}^{l+p} = \mathbb{R}^l \times \mathbb{R}^p \to \mathbb{R}^l$$

be the projection. Note that $\pi \circ h = f$ and that h is an embedding, because different $x_1, x_2 \in M^n$ will have different second coordinates under the map h.

Now our earlier argument from Proposition 2.8 showed that if $l \geq 2n$ there is a projection $\rho: \mathbb{R}^{l+p} \to \mathbb{R}^l$ giving an immersion. This argument was based on the fact that for projections that drop the dimension by 1, those not everywhere regular are nowhere dense. Using the fact that we are in the complement of a nowhere dense set, it follows at once that we can find a projection ρ as close as we wish to π for which $\rho \circ h$ is an immersion. But $\pi \circ h = f$, and $\rho \circ h$ is an immersion, for ρ as close as we wish to π; it is therefore clear that we have an immersion as close as we wish in our topology to the original C^k map.

For the second part of the theorem, it is best to deal with the density first. Proposition 2.9 showed that for embeddings the set of projections down to $\mathbb{R}^l$, $l \geq 2n + 1$, that are not 1–1 maps is a set of the first category. Since a set of the first category does not contain an open set in any of the projective spaces, which describe the directions, we may find a projection $\rho': \mathbb{R}^{l+p} \to \mathbb{R}^l$ arbitrarily near

$$\pi: \mathbb{R}^{l+p} \to \mathbb{R}^l$$

for which $\rho' \circ h(h(x) = (f(x), i(x))$, i an embedding) is both regular and 1–1. Since $\pi h = f$, and a 1–1 and regular map with compact domain is an embedding, we have produced an embedding as near as we wish to a given f, proving the density.

We now must show that the embeddings are open in $C^k(M^n; \mathbb{R}^l)$. We shall do this by contradiction. Suppose $i: M^n \to \mathbb{R}^l$ is a C^k embedding but for any $\varepsilon > 0$ there is a map in $S(i, k, \varepsilon)$ that is not 1–1. Then for each positive integer j we may find a C^k map $f_j: M^n \to \mathbb{R}^l$ lying in $S(i, k, 1/j)$ and a pair of points $x_j, y_j \in M^n$, $x_j \neq y_j$, such that $f_j(x_j) = f_j(y_j)$.

By the compactness of M^n, we may use the result that every sequence contains a convergent subsequence. Therefore, it is no loss of generality if we assume that the x_i's and the y_i's converge, $x_i \to x$, $y_i \to y$. The ideas will perhaps be more clear if we separate two cases.

Case i: $x \neq y$.   For j large, we may assume that $i(x_j)$ is as near as we wish to $i(x)$ and $i(y_j)$ is as near as we wish to $i(y)$. But $i(x_j)$ is arbitrarily near

$f_j(x_j)$ for j large, and $i(y_k)$ is arbitrarily near $f_k(y_k)$ for k large. Since $f_q(x_q) = f_q(y_q)$, for any q, we conclude that $i(x)$ and $i(y)$ are closer than any positive ε, whence we must have $i(x) = i(y)$, which implies $x = y$. This reduces the proof to the more difficult second case.

Case ii: $x = y$. Note that if $f_j(x_j) = f_j(y_j)$ and x_j and y_j are near enough so that we may take them in a single coordinate chart, then we may think of f_j as a map $\mathbb{R}^n \to \mathbb{R}^l$ near the point $x = y$ that takes the line segment from x_j to y_j into a closed loop beginning and ending at $f_j(x_j) = f_j(y_j)$.

We recall that the space of rays in $\mathbb{R}^l$ is compact, in fact homeomorphic to S^{l-1} (a ray is a half line emanating from the origin). Thus, a sequence of rays always has a convergent subsequence. The elements

$$di(\{(u, (y_j - x_j)/\|y_j - x_j\|)\}),$$

where the elements in the bracket are elements of the tangent space to $\mathbb{R}^n$ at u, are nonzero vectors at $i(u) \in \mathbb{R}^l$, because i is regular. Translating to the origin, they determine a sequence of rays in $\mathbb{R}^l$, and we select a subsequence j_n of the j's for which these rays converge to a given ray λ. For large enough m, we see that

$$di(\{(u, (y_{j_m} - x_{j_m})/\|y_{j_m} - x_{j_m}\|)\})$$

has positive component in the direction λ. Since x_{j_m} and y_{j_m} approach $x = y$, we may also assume with no loss of generality that these vectors

$$di(\{(u, (y_{j_m} - x_{j_m})/\|y_{j_m} - x_{j_m}\|)\})$$

have a positive component in the direction λ for every point u on the line segment from x_{j_m} to y_{j_m}.

Because f_j may be taken arbitrarily near i in our topology and the segment from y_{j_m} to x_{j_m} is compact, we see that for large enough m

$$df_{j_m}(\{(u, (y_{j_m} - x_{j_m})/\|y_{j_m} - x_{j_m}\|)\})$$

has a positive component in the direction λ, for all u on the segment in question.

For clarity, recoordinatize $\mathbb{R}^l$ so that λ is an axis. Then, for large m, the projection onto this axis of the map f_{j_m}, applied to the segment from x_{j_m} to y_{j_m}, is strictly increasing in x as x runs from x_{j_m} to y_{j_m}. Thus, for large m, it is not possible that $f_{j_m}(x_{j_m}) = f_{j_m}(y_{j_m})$. Thus, $x = y$ is not possible either.

In conclusion, the assumption that for any $j > 0$ there is $f_j \in S(i, k, 1/j)$ and distinct x_j, y_j with $f_j(x_j) = f_j(y_j)$ leads to a contradiction. Therefore, for some large enough j, all the maps in $S(i, k, 1/j)$ are 1–1. It follows at once from this fact and the openness of the immersions that the embeddings are also open.

We note that the heart of the argument was in the second case.

I will now consider some important examples of embeddings and immersions.

Examples ***1.*** Most of the examples that we have presented so far come equipped with a natural example of an embedding in some Euclidean space. For example,

(a) S^n is actually presented as embedded in $\mathbb{R}^{n+1}$,
(b) $\mathrm{Gl}(n;\mathbb{R})$ is presented as embedded in $\mathbb{R}^{n^2}$,
(c) the torus, $S^1 \times S^1$, is often presented as embedded in $\mathbb{R}^3$.

On the other hand, many examples of manifolds that we have studied have no special embedding in any Euclidean space. Examples would be $\mathbb{R}P^n$, $G_{n,k}$, and $\mathrm{SO}(n;\mathbb{R})$, the group of rotations of n-space (that is, the orthogonal matrices of positive determinant).

In most cases, the smallest r such that a given manifold can be embedded (or immersed) in $\mathbb{R}^r$ is not known, and there is by now a vast literature dealing with this fascinating question. Closely related manifolds may have different behavior, as far as is known, with respect to embeddings and immersions.

2. Occasionally one knows specific immersions for a manifold. For instance, we have the Klein bottle, which is the quotient space of the closed square of Fig. 2.6 under an equivalence relation which identifies the side AB to DC, preserving directions, and the side DA to the side BC, reversing directions. The Klein bottle (Fig. 2.7) is often shown in terms of the image of an immersion in $\mathbb{R}^3$. The points at which the neck cuts through the side are points where the map is not 1–1. In fact it is known from basic algebraic topology (see [25, 102]) that no nonorientable, two-dimensional manifold may be embedded in $\mathbb{R}^3$.

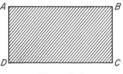
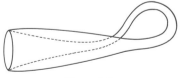

Figure 2.6   *Figure 2.7*

We note that an entirely different view of the Klein bottle is obtained, if one takes two different copies of $\mathbb{R}P^2$, cuts a hole out of each, and connects them with a hollow tube, with each end attached to a hole. This is shown schematically in Fig. 2.8. This tube construction is called the connected

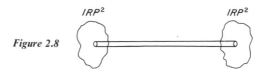

Figure 2.8

sum. One can then infer that if $\mathbb{R}P^2$ may be embedded in $\mathbb{R}^k$, then so may the Klein bottle. Such a k, not guaranteed from out theorem, would be 4.

There is a famous immersion of $\mathbb{R}P^2$ in $\mathbb{R}^3$, called Boy's surface (see e.g. [48]).

A covering map (see [75, 102]) such as $f: \mathbb{R} \to S^1$, $f(t) = e^{2\pi i t}$, is an immersion. Thus, if we are given an embedding

$$f: M^n \to \mathbb{R}^k$$

and a covering map

$$\pi: \tilde{M}^n \to M^n,$$

the composition $f \circ \pi$ is always an immersion.

In general, the theory of immersions is quite advanced. For example, Hirsch [50] has shown that any Lie group, or more generally any parallelizable manifold, may be immersed in a Euclidean space of one higher dimension. The best possible dimensions for embedding such manifolds are not known in general.

3. Along with the idea of an immersion or embedding there is a corresponding notion of equivalence. Two immersions are equivalent if they may be deformed into one another in a smooth way, so that the deformation map (which has domain $M^n \times I$, I the closed unit interval) has maximal rank, in the obvious sense, at each point. Two embeddings are equivalent if, in addition, the deformation consists, at each time, of an embedding. The classification of the appropriate equivalence classes has been studied, and there are many interesting results for immersions. For suitable dimensions, the theory has been worked out for embeddings, too. All this would take us too far afield; the interested reader should look at the papers of Smale [100], Hirsch [50], and Haeffliger [43].

PROBLEMS AND PROJECTS

1. One may show, using simple low-dimensional examples, that Whitney's basic theorems are close to optimal (compare with the remarks after Theorem 2.1). Using basic facts from advanced calculus, prove that

(a) there is no smooth immersion of S^1 in $\mathbb{R}^1$ (check regularity at a maximum or minimum value);

(b)   A smooth map of S^1 to $\mathbb{R}^1$ may be found, which then by (a) cannot be approximated by an immersion;

(c)   the figure-eight map (from the beginning of this chapter) may not be approximated, arbitrarily, by an embedding of S^1 in $\mathbb{R}^2$; and

(d)   there are immersions of S^2 in $\mathbb{R}^3$, similar to (c), that may not be be approximated closely by embeddings.

2.   Since there are canonical embeddings $i: \mathbb{R}^p \to \mathbb{R}^{p+q}$, $i(x_1, \ldots, x_p) = (x_1, \ldots, x_p, 0, \ldots, 0)$, there is a natural map which associates to an n-plane through the origin in $\mathbb{R}^{n+m}$ an n-plane through the origin in $\mathbb{R}^{n+m+k}$, for $k > 0$. Define an embedding

$$G_{n,m} \to G_{n,m+k}.$$

Check the conditions of Definition 2.8b.

In the case $n = 1$, this amounts to an embedding $\mathbb{R}P^m \to \mathbb{R}P^{m+k}$. Recover this map from a natural map $S^m \to S^{m+k}$.

3. (a)   If M^n is an open subset of N^n and N^n is parallelizable, i.e., TN^n is a Cartesian product of N^n and $\mathbb{R}^n$, prove that the same is true for M^n.

(b)   Construct an embedding of $S^n \times \mathbb{R}$ in $\mathbb{R}^{n+1}$ (view $\mathbb{R}$ as diffeomorphic to $(-\varepsilon, \varepsilon)$, for a small positive ε, and embed, for each $x \in S^n$, $(-\varepsilon, \varepsilon)$ along a vector perpendicular to S^n at x).

(c)   Show, therefore, that $S^n \times \mathbb{R}$ is parallelizable (recall $\mathbb{R}^{n+1}$ is parallelizable).

(d)   Using the remark following Definition 2.6, show that S^2 is not parallelizable ($TS^2 \neq S^2 \times \mathbb{R}^2$) but $S^2 \times \mathbb{R}$ is parallelizable.

4.   In our examples, following Theorem 2.2, we sketch a standard immersion f of the Klein bottle K. In this map there is a circle in the image where the map is 2–1. Let $j: \mathbb{R}^3 \to \mathbb{R}^4$ be $j(x_1, x_2, x_3) = (x_1, x_2, x_3, 0)$, and consider the map $j \circ f: K \to \mathbb{R}^4$.

(a)   Show that $(j \circ f)^{-1}$ applied to the circle in question is two disjoint circles.

(b)   Construct a smooth function $\phi: K \to \mathbb{R}$ that is unity on one of these circles and zero outside a small open set around the other circle.

(c)   Show that the map $g: K \to \mathbb{R}^4$, defined by $g(x) = (f(x), \phi(x)) \in \mathbb{R}^3 \times \mathbb{R} = \mathbb{R}^4$, is an embedding.

(d)   In the topology of Definition 2.10, is there an embedding of K in $\mathbb{R}^4$ arbitrarily close to $(j \circ f)$? (That is, given any open neighborhood of $j \circ f$, is there an embedding in this neighborhood?)

5.   Construct an immersion of $\mathbb{R}P^2$ in $\mathbb{R}^3$. (Either attack it directly or observe that the Klein bottle K is the union of two copies of $\mathbb{R}P^2$ – hole, sewed along the hole. One may then use the immersion of Problem 4 to get an immersion of $\mathbb{R}P^2$ – hole and add a disk to fill out the hole in a nice enough way to have an immersion of $\mathbb{R}P^2$.)

6. Let $D^n = \{x \in \mathbb{R}^n \,|\, \|x\| \le 1\}$. An n-sphere is then homeomorphic to a union of two copies of D^n, attached along the boundary. Ignoring the boundary, there is an embedding of $D^n \times S^m$ in $\mathbb{R}^{n+m+1}$ (obtained trivially from $S^m \subseteq \mathbb{R}^{m+1}$). Attach two nearby copies of $D^n \times S^m$ in $\mathbb{R}^{n+m+1}$ to get a smooth embedding of $S^n \times S^m$ in $\mathbb{R}^{n+m+1}$.

7. D^n is the most simple example of a manifold with boundary. Much of our theory generalizes to these. Consult $[51, 15, 84]$ to learn about these extensions of our results.

8. There are various refined notions, such as that of a *clean immersion*, which means that

(a) there are no three distinct points with the same image under f and

(b) If $a \ne b$, $f(a) = f(b)$, then $df(TM^n_a)$ and $df(TM^n_b)$ have a zero-dimensional intersection at $f(a)$.

Draw clean and not-clean immersions of S^1 in $\mathbb{R}^2$. Consult $[27]$ for a brief but interesting discussion of this theory.

3

Critical Values, Sard's Theorem, and Transversality

Just as in freshman calculus, there is substantial interest in points for which the differential (or Jacobian) of a function vanishes, or collapses to something less than what it could be. We mean, of course, those points where the differential fails to be onto; the case where the differential is actually zero (the zero transformation) is a rather extreme example of the phenomena. These points, usually called critical points, are connected with the geometry of the domain and range, as related to each other by the map. The greatest information for geometry comes when the critical points are isolated (as in Morse theory, Chapter 8).

Of course, it is possible for every point to be a critical point, for example, when the function is a constant. It is interesting that the situation is different for the critical values, the points in the range that are images of critical points. In a sense to be made precise shortly, these are rather rare. This is the famous theorem of Sard [96], which we treat as our first major topic in this chapter. After that we shall look at transversality, a notion that refers to a map and a submanifold in the range of the map. Roughly speaking, it means that the image of the differential, which is a subspace of the tangent space at a point in the image, and the tangent space to the submanifold should have an intersection of the lowest possible dimension. We shall prove a version of the Thom transversality lemma, a tool that is useful in many areas.

This chapter brings us up to the Lebesgue measure theory. Here we only need to know the notion of a set of measure zero, so that we give a definition of this concept alone. But the reader might be advised to read some general measure theory here (see [44, 95]) to get a general perspective and because we shall need some of this material in Chapter 6, when we study differential operators on manifolds. Of course, this is related to sets of the first category, and various authors have discussed slightly weaker versions of Sard's theorem that do not mention measure (see [27]). But the theorem is both important and useful, so that I feel that it is worth an exposition in general, at least in the case of a smooth (or C^∞) map.

CRITICAL POINTS AND VALUES

Definition 3.1 Let $f: M^n \to N^k$ be a differentiable map of differentiable manifolds ($C^r, r \geq 1$).

(a) A point $a \in M^n$ is called a *critical point* for f, if df is not onto at a, i.e., the matrix representing df has rank less than k.

(b) $b \in N^k$ is called a *critical value*, if $b = f(a)$ for $a \in M^n$ a critical point. We say that b is a *regular value* if $f^{-1}(b)$ contains no critical points. Thus, f maps the set of critical points onto the set of critical values. Note that the use of the word regular here is not quite the same as in the previous chapter. There a map was regular at a point if df was 1–1 at that point. Here we speak of a regular value as one for which df is onto at all points in the pre–image.

For example, if $M^n = \mathbb{R}^1 = N^k$, a critical point is a point where the derivative vanishes. A function on an interval without critical points is either an increasing or a decreasing function. If $M^n = (0, \infty) \subseteq \mathbb{R}^1$ and $N^k = \mathbb{R}^1$, we may set $f(x) = (1/x) \sin x$. The set of critical values is then an infinite subset of $\mathbb{R}$ that is not closed.

The interest in critical values in either calculus or advanced calculus centers on the search for extrema. While this is important in applications, it is not the sole interest of the concept. In fact, there is great geometrical interest in regular values, as the following proposition shows.

Proposition 3.1 Let $f: M^n \to N^k$ be a differentiable map and $b \in N^k$ be a *regular* value. Then $f^{-1}(b)$ is a submanifold of M^n whose dimension is $(n - k)$.

Proof Since b is a regular value, if $a \in f^{-1}(b)$, then df is onto at a. Clearly, this means that the rank of df at a is k. In matrix language, a matrix representing df at a has k linearly independent columns. Obviously k-vectors near k-independent vectors are independent. Thus, df has maximal rank k in a neighborhood of a, say O_α. By possibly passing to a smaller neighborhood, we may assume that O_α is homeomorphic to $\mathbb{R}^n$ by some homeomorphism ϕ_α. Suppose $b = f(a) \in U_\beta$ with appropriate homeomorphism ψ_β: $U_\beta \to \mathbb{R}^k$. Then $\psi_\beta \circ f \circ \phi_\alpha^{-1}$ is defined and has maximal rank k in a neighborhood of a.

Choose a k-dimensional subspace of $\mathbb{R}^n$ passing through $\phi_\alpha(a)$ such $\psi_\beta \circ f \circ \phi_\alpha^{-1}$ has maximal rank on this subspace at $\phi_\alpha(a)$ and also in an open neighborhood of $\phi_\alpha(a)$, say W. By a suitable change of coordinates, it is clearly no loss of generality to assume that $\phi_\alpha(a)$ is the origin and our k-dimensional subspace is given as $\{(x_1, \ldots, x_k, 0, \ldots, 0)\}$. We define

$$g: W \to \mathbb{R}^k \times \mathbb{R}^{n-k}$$

by $g(x_1, \ldots, x_n) = (\psi_\beta \circ f \circ \phi_\alpha^{-1}(x_1, \ldots, x_n), x_{k+1}, \ldots, x_n)$ (recall that the range of $\psi_\beta \circ f \circ \phi_\alpha^{-1}$ is $\mathbb{R}^k$).

It is trivial to calculate the Jacobian of g at $\phi_\alpha(a)$. In the lower-right corner, $(n - k) \times (n - k)$-dimensional, we have the identity. In the upper-left corner, one has the Jacobian of the inclusion of the k-dimensional subspace followed by $\psi_\beta \circ f \circ \phi_\alpha^{-1}$. Because the range of $\psi_\beta \circ f \circ \phi_\alpha^{-1}$ is $\mathbb{R}^k$, the lower-left box must be zero. It follows at once that at $\phi_\alpha(a)$, and hence in a neighborhood of $\phi_\alpha(a)$, the Jacobian of g is nonsingular.

We may then invoke the inverse function theorem (Theorem 1.1) to conclude that g is a diffeomorphism on a suitably small open neighborhood of $\phi_\alpha(a)$. Our procedure then has had the net advantage of expressing f as the first k-coordinates of a diffeomorphism. One checks immediately that $g^{-1}(z, \{x_{k+1}, \ldots, x_n\})$, with z fixed, is the same as $(\psi_\beta \circ f \circ \phi_\alpha^{-1})^{-1}(z)$ in the neighborhood in question. (In fact, $\psi_\beta \circ f \circ \phi_\alpha^{-1}$ equals the projection of g on the first k coordinates.)

In particular, up to our homeomorphism ϕ_α, we have that $f^{-1}(b)$ and

$$g^{-1}(\psi_\beta \circ f(a), \{x_{k+1}, \ldots, x_n\}),$$

with the $x_{k+1}, \ldots$ varying, are the same sets. But g is a diffeomorphism, and $\phi_\alpha(a)$ is assumed to be the origin of our coordinate neighborhood. We then have our desired coordinate neighborhood of the form required by Definition 1.5 (since $\mathbb{R}^k \times \mathbb{R}^{n-k} \equiv \mathbb{R}^{n-k} \times \mathbb{R}^k$).

It is easy to give examples. For example, if $f: \mathbb{R}^n \to \mathbb{R}$ is defined by $f(x) = \|x\|^2$, then when $b > 0$, $f^{-1}(b) \equiv S^{n-1}$. If $g: \mathbb{R}^2 \to \mathbb{R}$ is given by $g(x, y) = x^2 - y^2$, then the only critical value is 0. If $b \neq 0$, $g^{-1}(b)$ is the hyperbola $x^2 = y^2 + b$, clearly a submanifold of $\mathbb{R}^2$.

On the other hand, the last example above shows that when b is a critical value $f^{-1}(b)$ need not be a submanifold (taking b to be the origin, $f: \mathbb{R}^2 \to \mathbb{R}$ by $f(x, y) = x^2 - y^2$).

SARD'S THEOREM; APPLICATIONS

Our goal is now the basic theorem of Sard [96]. See also A. Morse [81] and Pontrjagin [91] to the effect that the set of critical values of a smooth map forms a set of measure zero.

Definition 3.2   Let W_i denote a cube or a parallelopiped in $\mathbb{R}^n$. Denote the volume of W_i by $\mu(W_i)$.

We say that a set $S \subseteq \mathbb{R}^n$ has measure zero, written $\mu(S) = 0$, if given any $\varepsilon > 0$ there is a countable family of cubes $\{W_i\}$ such that

(a) $S \subseteq \bigcup_i W_i,$

(b) $\sum_i \mu(W_i) < \varepsilon.$

(Here, of course, we mean that first the series converges, and second, its sum is less than ε.)

Remarks   (1)   Of course, this is a special case of the theory of Lebesgue measure, where sets of any measure are studied. But, in this particular case, the definition is remarkably simple.

(2)   If $T \subseteq S \subseteq \mathbb{R}^n$ and $\mu(S) = 0$, then it is obvious that $\mu(T) = 0$. Note that in the area of sets of measure zero there is no question of whether a subset of a measurable set is measurable.

(3)   If S_i is a countable family of sets, $\mu(S_i) = 0$ for each i, then

$$\mu(\bigcup_i S_i) = 0$$

(Choose $\varepsilon > 0$, cover S_1 with cubes $\{W_1^n\}$, $\mu(\bigcup W_1^n) < \varepsilon/2$, cover S_2 with cubes $\{W_2^n\}$ $\mu(\bigcup W_2^n) < \varepsilon/2^2$, etc. Then the union is covered by the cubes W_k^n, for all n and k, and their total volume is less than $\sum_{i=1}^{\infty} \varepsilon/2^i$.)

(4)   It is unfortunately possible for the continuous image of a set of measure zero to have positive measure. (In fact, the famous space-filling curves of Peano fill entire boxes in the plane.) However, the following proposition excludes this possibility when the maps are differentiable. It should be compared with Proposition 2.7.

Proposition 3.2   Let $C \subseteq U \subseteq \mathbb{R}^n$, where $\mu(C) = 0$ and U is open. Let $f : U \to \mathbb{R}^m$ be a differentiable map $(C^r, r \geq 1)$. Then $\mu(f(C)) = 0$.

Proof   Examining the definition, it suffices to show that there is a constant $A > 0$ such that if D is a cube in $\mathbb{R}^n$, $\mu(D) < \varepsilon$, then $f(D)$ is contained in a family of cubes in $\mathbb{R}^m$, of total volume less that $A\varepsilon$. In fact, as a matter of convenience, we may just as well prove that if $B \subseteq U$ is a closed ball, $B = \{x \in \mathbb{R}^n \,|\, d(x, y) \leq a$, fixed $y \in U\}$, and the volume of B is less than $\varepsilon > 0$, then $f(B)$ lies in a ball of volume less than $A\varepsilon$.

Such as assertion will follow at once if one can show that there are numbers A' and δ, both positive such that if $d(x, y) < \delta$ and

$$\{tx + (1 - t)y\} \subseteq U, \qquad \text{for} \quad 0 \leq t \leq 1,$$

then

$$\|f(x) - f(y)\| < A' \|x - y\|.$$

But this follows at once from the finite increments lemma (Proposition 1.3), because if f is C^r, $r \geq 1$, and B is compact, then the $\|J_f(x)\|$, $x \in B$, are bounded. (Check this!)

Our next proposition is merely a reformulation, in the special case of sets of measure zero, of the famous theorem of Fubini (change of order in multiple integration) in Lebesgue integration theory.

Proposition 3.3   Let $\mathbb{R}_t^{n-1} = \{(x, t) \in \mathbb{R}^{n-1} \times \mathbb{R} = \mathbb{R}^n \,|\, \text{where } t \in \mathbb{R}$, is fixed, $x \in \mathbb{R}^{n-1}\}$. We let $\mu_i(\ )$ refer to the measure of a subset of $\mathbb{R}^i$, which we use only in the case of measure 0. We note that $\mathbb{R}_t^{n-1}$ is identical to $\mathbb{R}^{n-1}$, via the map $i(x) = (x, t)$, so we may speak of μ_{n-1} in $\mathbb{R}_t^{n-1}$.

Suppose that C is compact and for any $t \in \mathbb{R}$ we have $\mu_{n-1}(C \cap \mathbb{R}_t^{n-1}) = 0$. Then $\mu_n(C) = 0$.

Proof   Because C is compact it would be no loss of generality to assume that

$$C \cap \mathbb{R}_t^{n-1} = \phi,$$

unless $0 \leq t \leq 1$. We shall abreviate

$$C_t = C \cap \mathbb{R}_t^{n-1}.$$

If $\mu(C_t) = 0$, we cover C_t by countably many cubes D_t^i in $\mathbb{R}_t^{n-1}$, $\sum_i \mu(D_t^i) < \varepsilon$, with $\varepsilon > 0$. We set $D_t = U_i \text{ Int } D_n^i$ (Int abbreviates Interior), and we see at once that we may assume that $C_t \subseteq D_t$.

The continuous real function $f: C \to \mathbb{R}$ defined by

$$f(x_1, \ldots, x_n) = |x_n - t|$$

vanishes precisely on $C_t \times \{t\}$. On each closed set

$$C - \text{Int}(D_t \times [0, 1]),$$

f achieves a positive minimum by compactness; we denote this minimum by m.

We then see that

$$C \cap \{(x_1, \ldots, x_n) | f(x_1, \ldots, x_n) < m\} \subseteq D_t \times (t - m, t + m).$$

Focusing on the last coordinate, we see that the intervals $(t - m, t + m)$, where naturally m depends on t, form a cover of the closed interval $[0, 1]$ by open intervals. By compactness, a finite number are needed to cover $[0, 1]$; we shall need to estimate their length. (Note that we utilize the usual conventions whereby if $t - m < 0$ we look at $[0, t + m) = \{r \in \mathbb{R} | 0 \leq r < t + m\}$ instead of $(t - m, t + m)$, etc.)

We now need a momentary digression (a sort of easy lemma) that if $\{O_\alpha\}$ is a cover of $[0, 1]$ by open intervals, there is a finite subcover whose total length is less that or equal to 2. For this, we choose a finite cover by open intervals (p_i, q_i), $p_i < q_i$, $i = 1, \ldots, l$, that is assumed to be minimal. Then we may take $p_i < p_{i+1} < q_i$ and $q_i < q_{i+1}$. If it were the case that $q_i > p_{i+2}$, the interval (p_{i+1}, q_{i+1}) would be totally superfluous, so that we have $q_i \leq p_{i+2}$. (Note again that if $p_1 \leq 0$, (p_1, p_2) means $[0, p_2)$, etc.) We calculate

$$2 \geq \sum_i (p_{i+2} - p_{i+1}) + \sum_i (p_{i+1} - p_i)$$
$$= \sum_i (p_{i+2} - p_{i+1} + p_{i+1} - p_i) \geq \sum_i (q_i - p_i),$$

as desired.

Returning to our proof, the sets $D_t \times (t - m, t + m)$ offer a cover of C, because each of them covers a set of the form

$$C \cap \{(x_1, \ldots, x_n) | f(x_1, \ldots, x_n) = |x_n - t| < m\},$$

and hence there is a finite subcover. Since we now know that we may assume that the total length of the intervals, for all $t \in [0, 1]$, is no bigger than 2 and that the total volume of the set D_t, a union of cubes, is arbitrarily small, we have $\mu_n(C) = 0$, as was to be proved.

We are now in a position to tackle our basic theorem.

Theorem 3.1 (Sard) Let $f : M^n \to N^k$ be a smooth function between smooth manifolds, $k > 0$. Then the set of critical values of f has measure 0 in N^k, in the sense that in any coordinate chart or neighborhood the set of critical values in that neighborhood has measure zero. (Note that Proposition 3.2 assures us that this statement is independent of the choice of coordinate neighborhood.)

Proof Naturally, it will suffice to prove this locally, that is, to assume $U \subseteq M^n$ is open and $f : U \to \mathbb{R}^k$ is smooth (C^∞). Let $C_0 \subseteq U$ be the set of critical points, i.e., those points in the domain for which df is not onto, i.e., rank $df < k$. Then we shall show that $\mu(f(C_0)) = 0$.

The proof shall go by induction on n, it being totally obvious for $n = 0$. We define for each positive integer j the set C_j to be those points $x \in U$ for which all partial derivatives of the coordinate functions of order less than or equal to j vanish. If C_0 is the set of critical points, we obviously have a decreasing sequence of subsets

$$C_0 \supseteq C_1 \supseteq C_2 \supseteq \cdots .$$

Our proof will then be based on showing that

(i) for some sufficiently large p, $\mu(f(C_p)) = 0$, and
(ii) for all $i \geq 0$, $\mu(f(C_i - C_{i+1})) = 0$.

First, to prove (i), we shall in fact show that $\mu(f(C_p)) = 0$ whenever $p \geq n/k$. But it will then be sufficient to show that

$$\mu(f(W \cap C_p)) = 0,$$

for any cube W in $U \subseteq \mathbb{R}^n$, for U is clearly a countable union of such cubes.

Since a cube in the domain is convex, we may use Taylor's theorem (Proposition 1.5) to get

$$f(y) = f(x) + r, \ |r| \leq C \|y - x\|^{p+1},$$

when $x \in W \cap C_p$, $y \in W$, and where C is a constant.

If $l(W)$ is the length of a side of W, we divide W into s^n congruent subcubes W' such that $l(W') = l(W)/s$. Supposing W' is one such cube, with $x \in W' \cap C_p$, we realize immediately that every point y of W' is closer to x than $l(W) \cdot \sqrt{n}/s$, so that by the above estimate from Taylor's theorem, we have

$$\|f(y_1) - f(y_2)\| \leq 2C(l(W)\sqrt{n}/s)^{p+1}$$

for any two points $y_1, y_2 \in W'$.

The total volume of a set of cubes, in the range, containing $f(W \cap C_p)$, which would be no more in number than s^n, would then be

$$V = s^n 2^k C^k (l(W)\sqrt{n})^{k(p+1)} s^{-k(p+1)}.$$

Now recall that we are assuming that $p \geq n/k$, or alternatively, $k(p + 1) > n$. Thus, we may arrange our expression V,

$$V = 2^k C^k (l(W)\sqrt{n})^{k(p+1)} / s^{k(p+1)-n},$$

so that the exponent in the denominator is positive. Allowing the number of cubes in the subdivision of W go to infinity, this converges to 0. Hence (i) is proved.

To prove (ii), we note that for $k = 1$, $C_0 = C_1$, so that we may freely assume $k > 1$ and check $\mu(f(C_0 - C_1)) = 0$. The assertion (ii), for $i > 0$, will be checked at the end. We let $x \in C_0 - C_1$; clearly, some partial derivative of some coordinate function is nonzero. Up to a possible change of order, we may define

$$g_1(x_1, \ldots, x_n) = (f_1(x_1, \ldots, x_n), x_2, \ldots, x_n),$$

with f_1 the first coordinate function of f and with g_1 having nonsingular Jacobian at $x = (x_1, \ldots, x_n)$. Invoke the inverse function theorem (Theorem 1.1) to see that locally g_1 must be a diffeomorphism. Writing the inverse of g_1 as simply g_1^{-1} and examining the definition of g_1 we see that, for suitable functions $p_2(x), \ldots, p_k(x)$, we must have

$$f \circ g_1^{-1}(x_1, \ldots, x_n) = (x_1, p_2(x), \ldots, p_k(x)).$$

(g_1^{-1} takes a point whose first coordinate is $f_1(x)$ to one whose first coordinate is x_1; f_1 does the opposite. Hence, $f \circ g_1^{-1}$ does not alter the first coordinate.)

We may then prepare to apply Proposition 3.3 by observing that locally the map $f \circ g_1^{-1}$ sends the hyperplane $t \times \mathbb{R}^{n-1}$ into $t \times \mathbb{R}^{k-1}$. We note also that $f \circ g_1^{-1}$ has the following Jacobian:

$$
\begin{bmatrix}
1 & 0 & \cdots & 0 \\[2mm]
  & \dfrac{\partial p_2}{\partial x_2} & \dfrac{\partial p_3}{\partial x_3} & \cdots \\[3mm]
* & \dfrac{\partial p_3}{\partial x_2} & & \\
  & \vdots & &
\end{bmatrix}
$$

This means that $(t, z) \in \mathbb{R} \times \mathbb{R}^{n-1} = \mathbb{R}^n$ is a critical point of $f \circ g_1^{-1}$ precisely when (t, z) is a critical point of the restricted mapping $f \circ g_1^{-1} | t \times \mathbb{R}^{n-1}$.

We are now able to use induction on n as a variable. The case $n = 0$ being obvious, our above remarks, combined with Proposition 3.3, show that whenever part (ii) of the theorem is true for the dimension of the domain being $n - 1$, it is also true when the dimension is n. Thus, claim (ii) is proved for $i = 0$.

Finally, we must argue that $\mu(f(C_i - C_{i+1})) = 0$, $i > 0$. But if $x_0 \in C_i - C_{i+1}$ for some coordinate of f, say the first, there is an $(i + 1)$-th order derivative that does not vanish at x. That is, for some multi-index α, $|\alpha| = i + 1$,

$$D^\alpha f_1(x_0) \neq 0.$$

We may therefore assume $\alpha_1 \geq 1$, and write

$$\alpha' = \alpha - (1, 0, \ldots, 0).$$

Call $w(x) = D^{\alpha'} f_1(x)$, so that with no loss of generality $\partial w(x)/\partial x_1 \neq 0$ but $w(x_0) = 0$.

Examine the map

$$g(x) = (w(x), x_2, \ldots, x_n),$$

which is clearly a local diffeomorphism. Consider $f \circ g^{-1}$ locally. Restricting attention to $O \times \mathbb{R}^{n-1}$ and reasoning by induction on n, just as before, we assume that the critical values of $f \circ g^{-1} | O \times \mathbb{R}^{n-1}$ have measure 0.

But locally, every point in $g(C_i)$ is a critical point for $f \circ g^{-1}$ (because all partials of order $\leq i$ vanish and $i \geq 1$). When $x_0 \in C_i - C_{i+1}$ we see that $g(x_0) \in O \times \mathbb{R}^{n-1}$; by induction, as remarked in the previous paragraph, $f \circ g^{-1} | O \times \mathbb{R}^{n-1}$ has, as a set of critical values, a set of measure zero. Thus locally the set

$$f \circ g^{-1}(g(C_i - C_{i+1})) = f(C_i - C_{i+1})$$

must have measure zero. Of course, an easy countability argument shows that if the set has measure zero locally it has measure zero. (Every open ball contains a point whose coordinates are all rational, so it suffices to take a countable collection of open balls; a countable union of sets of measure zero has, as we remarked, measure zero.)

Finally, we note that a similar argument covers the case where $\alpha_1 = 0$ but some other $\alpha_j \geq 1$. Taking the (finite) union of all these sets of measure zero, we get the complete result (ii).

There are many striking applications of this theorem, some of which do not appear—on the surface—to have any relation to sets of measure zero. I plan to discuss several corollaries, beginning with some obvious ones.

Corollary 3.1   Let $f: M^n \to N^k$ be a smooth map, $k \geq 1$, that is onto. Then, except for a subset of N^k of measure zero, for all $y \in N^k$, $f^{-1}(y)$ is a submanifold of M^n. In particular, there is always some $y \in N^k$ such that $f^{-1}(y)$ is a (proper) submanifold.

Proof   Direct use of Theorem 3.1 and Proposition 3.1.

Corollary 3.2   Let $M^n \subseteq \mathbb{R}^k$ be a smooth submanifold. Let $S \subseteq \mathbb{R}^k$ be a $(k - 1)$-dimensional linear subspace through the origin. Then there is a vector $v \in \mathbb{R}^k$ such that $(v + S) \cap M^n$ is a submanifold of M^n.

Proof Let l be a line through the origin perpendicular to S. For $x \in M^n$, let $p(x)$ be the perpendicular projection of x onto l. This is clearly a smooth map $p: M^n \to \mathbb{R}$.

If $p(M^n)$ is merely a single point, p^{-1} of some point is all of M^n, which is trivially a submanifold. If M^n is a union of several connected components, each of which is mapped to a point, p^{-1} of some point is simply a connected component and thus a submanifold. But if $p(M^n)$ contains an interval, then $p(M^n)$ is surely not a set of measure zero. We conclude that there must be at least one value in this interval which is regular, so that p^{-1} applied to this value is a submanifold.

To complete the proof, observe that

$$p^{-1}(y) = (y + S) \cap M^n,$$

because y lies in the normal line to the plane (hyperplane) S.

Corollary 3.3 (Lemma to the Browner Fixed Point Theorem) Let $D^n = \{x \in \mathbb{R}^n \mid \|x\| \leq 1\}$ and $S^{n-1} = \partial D^n$, the boundary of D^n, which is an $(n-1)$-dimensional sphere. Let $i: S^{n-1} \to D^n$ be the inclusion map.

Then there is *no* continuous map $\rho: D^n \to S^{n-1}$ such that $\rho \circ i = 1_{S^{n-1}}$, the identity map. In other words, there is no continuous ρ such that for each $x \in S^{n-1}$, $\rho(i(x)) = x$. (Such a map ρ, if it existed, would be called a retraction.)

Proof We suppose ρ exists. Let S^{n-1}_{++} be the subset of D^n consisting of all $x \in D^n$ such that $\frac{1}{2} \leq \|x\| \leq 1$. We claim, first of all, that we may modify ρ so as to be the outward projection onto S^{n-1} for all points of S^{n-1}_{++}. To be specific, we define

$$\rho'(x) = \begin{cases} x/\|x\| & \text{if } \|x\| \geq \frac{1}{2} \\ \rho(2x) & \text{if } \|x\| \leq \frac{1}{2}. \end{cases}$$

It is clear that ρ' is continuous and $\rho' \circ i = 1_{S^{n-1}}$ for $\rho'(i(x)) = \rho'(x) = x/\|x\| = x/1 = x$ when $x \in S^{n-1}$.

Second, I claim that we can find a smooth approximation to ρ' that actually equals ρ' on

$$S^{n-1}_+ = \{x \mid \tfrac{3}{4} \leq \|x\| \leq 1\}.$$

Choose $0 < \varepsilon < \frac{1}{8}$, and find, using standard approximation theorems (as in [51]), a smooth map f' that differs in distance from ρ' by no more than ε.

Let $\phi: D^n \to [0,1]$ be a smooth map, selected so that if $x \in S^{n-1}_+$, $\phi(x) = 0$, and if $\|x\| \leq \frac{5}{8}$, $\phi(x) = 1$. (See Lemma 1.4 and Corollary 1.7.) Set $\tilde{f}: D^n \to D^n - \{0\}$ by

$$\tilde{f}(x) = (1 - \phi(x))\rho'(x) + \phi(x)f'(x).$$

The second summand is a smooth map, and ρ' is smooth for all values where $1 - \phi(x) \neq 0$. Hence, we see, at once that $\tilde{f}(x)$ is smooth. Note that $\tilde{f}(x)$ is never 0, because of the size of ε and the fact the range of ρ' is $S^{n-1} = \partial D^n$.

Third, we set

$$f(x) = \pi(\tilde{f}(x)),$$

where $\pi(y) = y/\|y\|$. Now f and ρ' agree on $\{x \in D^n \,|\, \|x\| > \frac{3}{4}\}$, so $f \circ i = \rho' \circ i = \rho \circ i = 1_{S^{n-1}}$; in other words, such an f will also be a retraction.

The purpose of this construction is to find a smooth map with the same basic properties as ρ. Having done so, we may use Theorem 3.1.

To this end, we extend f to a map

$$\overline{f} : S^n \to S^{n-1}$$

by using f on each hemisphere of S^n. (The hemispheres are essentially the same as D^n. Check this here!) By Theorem 3.1 (Sard's theorem), there is a regular value y in the image of $\overline{f}$, because $\overline{f}$ is onto (as f was), and a set of 0 measure in S^{n-1} is a proper subset. By Proposition 3.1, $\overline{f}^{-1}(y)$ is a 1-dimensional submanifold of S^n. Since $\overline{f}$ is the identity on the equator $S^{n-1} \subseteq S^n$, which is the boundary of each hemisphere, $\overline{f}^{-1}(y)$ meets S^{n-1} at precisely one point, namely y.

Therefore, we have a 1-dimensional submanifold of S^n that is closed, and hence compact, and that therefore must be S^1, yet that has points in both hemispheres and meets the equator at a single point. But such a space cannot be a submanifold at the point on the equator, for the portion of $\overline{f}^{-1}(y)$ that enters the northern hemisphere must return to get to the southern hemisphere, crossing at y, and ditto the portion in the southern hemisphere. In the case of S^2, this is illustrated by Fig. 3.1; y cannot possibly have a neighborhood homeomorphic to $\mathbb{R}$.

We conclude that no such map ρ could have existed.

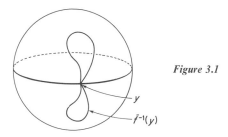

Figure 3.1

Corollary 3.4 (Brouwer Fixed Point Theorem) As before, we write $D^n = \{x \in \mathbb{R}^n \,|\, \|x\| \leq 1\}$. Let $f : D^n \to D^n$ be a continuous map. Then f has a fixed point; that is, there is some $x_0 \in D^n$ such that $f(x_0) = x_0$.

Proof   Suppose not. Then for each $x \in D^n$, $x \neq f(x)$, let l_x be the half line from $f(x)$ through x, and let $\psi(x)$ be the point where l_x meets the boundary of D^n, which is S^{n-1}. ψ is clearly a retraction, in that it is visibly continuous and if $x \in S^{n-1}$, $\psi(x)$ must be x. This contradicts the previous corollary, and therefore we see that $f(x) \neq x$ is impossible for all x.

Corollary 3.5   Let $f: M^n \to N^k$ be a smooth map of smooth manifolds. Then the set of critical values of f is a set of the first category in N^k, i.e., a countable union of nowhere dense subsets ($k > 0$).

Proof   It is easy to check that there is no loss of generality in assuming that $N^k = \mathbb{R}^k$. Also, since M^k is assumed to satisfy the second axiom of countability, we see that M^k is a countable union of compact subsets. Hence, we lose no generality in assuming that M^n is compact. But then $f(M^n)$ is compact and hence closed. The set of critical values is also closed, because if b is a limit of critical values b_n then, in the set of all $f^{-1}(b_n)$, we may choose $a_n \in f^{-1}(b_n)$ and (by compactness) a convergent subsequence of the a_n's whose limit must be a critical point in $f^{-1}(b)$.

We are thus reduced to showing that a compact subset of $\mathbb{R}^k$, of measure zero, must be of the first category. *First*, we might note that if $S \subseteq \mathbb{R}^k$ is closed, then S is of the first category if and only if it is nowhere dense. If it is nowhere dense, it is of the first category trivially. If it is not nowhere dense, and it is closed, it must contain some nonempty open set. But a nonempty open set in $\mathbb{R}^k$ is of the second category (Baire's theorem), and the same automatically holds for a possibly bigger set. *Second*, if S is closed, of measure zero, but fails to be nowhere dense, then S is dense in an open ball O. Being closed, it would have to contain O. Since the measure of an open ball (nonempty) is positive, we would get a contradiction, proving that S is nowhere dense.

THOM'S TRANSVERSALITY LEMMA

There are, of course, many other applications of Sard's theorem. Many of the arguments that we used in the previous chapter with respect to sets of the first category, such as those relating to Whitney's theorems, can be reformulated, and even on occasion made more precise, by the use of Sard's theorem. But our final application will be to the very important concept of transversality. Given a smooth map of smooth manifolds

$$f: M^n \to N^k$$

and a submanifold (not necessarily a sphere)

$$S^p \subseteq N^k,$$

the concept that f is transversal to S^p at x means roughly that the intersection in N^k of $f(M^n)$ and S^p has lowest possible dimension in a neighborhood of $f(x)$, yet the sum of the dimensions of $f(M^n)$ and S^p is at least k. Of course, it is not easy to give precision to the dimension of a set like $f(M^n)$, so that our formal definition, which now follows, expresses this all in terms of tangent spaces.

Definition 3.3   Given a smooth map $f : M^n \to N^k$ between smooth manifolds and a submanifold $S^p \subseteq N^k$, we say that f is *transversal* (or transverse) to S^p if for each $x \in M^n$, $f(x) \in S^p$, we have

$$df(TM^n)_x) + (TS^p)_{f(x)} = (TN^k)_{f(x)}.$$

Here the symbol $+$ means that we take the set of all vectors in $(TN^k)_{f(x)}$ that are sums of a vector in the image of df and a tangent vector to the submanifold to S^p. The left side is always contained in the right, and the concept *transversal* means that we have equality for all $x \in f^{-1}(S^p)$.

Remarks   (1)   If $f(M^n) \cap S^p = \phi$, the map f is automatically transversal to S^p.
(2)   If $n + p \le k$, then the dimension of the left side of the equation is less than or equal to k (with equality in the case where the intersection of the two summands is the zero element). Thus, the left side can never equal the right side, which has dimension k, whenever $n + p < k$.
(3)   Let $M^n = S^1$, $N^k = \mathbb{R}^2$, and $S^p = $ the x-axis in $\mathbb{R}^2$. Figure 3.2 shows transversality (Fig. 3.2a) and nontransversality (Fig. 3.2b), respectively. The concept requires that intersections be in their most general possible positions.

One of the important aspects of transversality, which surely led to the recognition of the concept, is the following proposition concerning inverse images. It is clearly an extension of the earlier Proposition 3.1.

Proposition 3.4   Let M^n, N^k be smooth, and let $f : M^n \to N^k$ be a smooth map. Given a submanifold $S^p \subseteq N^k$. We suppose that f is transversal to S^p. Then $f^{-1}(S^p)$ is a submanifold of M^n, whose codimension in M^n is $k - p$; that is, its dimension as a submanifold, and hence a manifold, is $n - k + p$.

Proof   Choose a coordinate chart $\psi_\beta : U_\beta \xrightarrow{\equiv} \mathbb{R}^k$, $f(x) \in U_\beta$, so that the last p coordinates represent S^p in this neighborhood of $f(x)$. By the definition

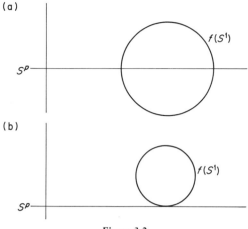

Figure 3.2

of transversality, $df((TM^n)_x)$ contains the subspace

$$\{(y_1, \ldots, y_{k-p}, 0, \ldots, 0)\},$$

spanned by the first $k - p$ coordinates in $(TN^k)_{f(x)}$. Define

$$\pi(y_1, \ldots, y_k) = (y_1, \ldots, y_{k-p}),$$

and carry this map over to U_β via ψ_β^{-1}, that is,

$$\tilde{\pi} = \psi_\beta^{-1} \circ \pi \circ \psi_\beta.$$

Locally, then, S^p is the set of points that are sent to $f(x)$ by $\tilde{\pi}$. And so locally $f^{-1}(S^p)$ consists of those points in M^n that are sent to $f(x)$ by the map $\tilde{\pi} \circ f$.

Our assumption of transversality assures us that $f(x)$ is a regular value of $\tilde{\pi} \circ f$ in this neighborhood U_β, so that we infer, from Proposition 3.1, that $f^{-1}(S^p)$ has the structure of a submanifold. (Note that the neighborhood of x lies entirely within $f^{-1}(U_\beta)$.)

To verify our claim about dimensions, observe that the map $\tilde{\pi} \circ f$ has an image that is $k - p$ dimensional. By Proposition 3.1, $\dim(\tilde{\pi} \circ f)^{-1}(f(x))$ must be $n - (k - p)$, completing the proof. We assume $f^{-1}(S^p)$ nonempty.

Corollary 3.6 If M^n and S^p are both submanifolds of N^k and for each $x \in S^p \cap M^n$

$$(TS^p)_x + (TM^n)_x = (TN^k)_x,$$

then $S^p \cap M^n$ is a submanifold of N^k.

Proof   Let $i: M^n \to N^k$ be the inclusion. Then i is transversal to S^p and $i^{-1}(S^p) = M^n \cap S^p$, as desired.

Before we get to the basic transversality result of Thom, we need refinement of our earlier study of distances and topologies on the set of differentiable maps between differentiable manifolds (Definitions 1.9 and 2.10). The topology that we shall now construct will, in the noncompact case, actually be finer than the obvious generalizations of the previous constructions.

Definition 3.4   (a)   Let $C^\infty(\mathbb{R}^n; \mathbb{R}^k)$ be the set of C^∞-functions (that is, smooth functions) $f: \mathbb{R}^n \to \mathbb{R}^k$. We shall topologize this set as follows:

If $\varepsilon(x)$ is a positive, continuous function defined on $\mathbb{R}^n$, and $p > 0$ is any integer, we set all

$$B(0, \varepsilon(x), p) = \{f \in C^\infty(\mathbb{R}^n; \mathbb{R}^k) \mid |D^\alpha f_j(x)| < \varepsilon(x) \quad \text{for all } |\alpha| \le p \text{ and } j\},$$

where f_j is the jth coordinate of f. This gives a basis for the neighborhoods of the constant function 0. To define a similar basis neighborhood for $g \in C^\infty(\mathbb{R}^n, \mathbb{R}^k)$, set

$$B(g, \varepsilon(x), p) = \{f \in C^\infty(\mathbb{R}^n, \mathbb{R}^k) \mid (f - g) \in B(0, \varepsilon(x), p)\}.$$

Note that the integer p is allowed to vary in this definition of basis set; that is, we consider sets $B(g, \varepsilon(x), p)$ for all possible p.

(b)   If M^n and N^k are two smooth manifolds, then to topologize $C^\infty(M^n, N^k)$ one may proceed as in (a) by choosing a locally finite cover of N^k by coordinate charts and demanding that condition of (a) hold in all the coordinate charts, near any point. Alternatively, one can reduce this question to (a) by using embeddings in Euclidean space, as in Chapter 2.

Remarks   (a)   This is, in general, a rather fine topology to put on $C^\infty(M^n, N^k)$, because functions such as $\varepsilon(x)$ may decrease to zero rapidly, even though $\varepsilon(x) > 0$ for all x.

(b)   In the compact case, we easily see that the same topology results if we choose all the functions $\varepsilon(x)$ to be (positive) constants. Thus, Definition 3.4 is similar to the earlier definitions. (If we were to use constants $\varepsilon(x)$ in the noncompact case, we would get a rather coarse topology.)

Proposition 3.5   Let M^n, N^k be smooth manifolds and $S^p \subseteq N^k$ a submanifold. Then the set $C^\infty_{S^p}(M^n, N^k)$, consisting of all maps in $C^\infty(M^n, N^k)$ that are transversal to S^p, is an open subset of $C^\infty(M^n, N^k)$.

Proof   Referring to Definition 3.4, we see that the problem is local. That is, if we can show that in a neighborhood of a point $x \in M^n$ all functions

$g \in C^\infty(M^n, N^k)$ that are sufficiently near $f \in C^\infty_{S^p}(M^n, N^k)$, say nearer than $\varepsilon > 0$, for derivatives of order $\leq r$, must also be transverse to S^p, for points x in this neighborhood, then we can put together different ε's to build a suitable $\varepsilon(x)$, so that if $g \in B(f, \varepsilon(x), r)$ then g is transversal to S^p. (Check that if $\{O_\alpha\}$ is a locally finite cover of M^n and for each O_α we have $\varepsilon_\alpha > 0$, then there is a continuous, positive $\varepsilon(x)$ such that when $x \in O_\alpha$, $\varepsilon(x) \leq \varepsilon_\alpha$.)

But from Definition 3.2, if $f(x) \in S^p$, we need to verify

$$df((TM^n)_x) + (TS^p)_{f(x)} = (TN^k)_{f(x)},$$

which is obviously equivalent to verifying that the composition

$$(TM^n)_x \xrightarrow{df} (TN^k)_{f(x)} \xrightarrow{\pi} (TN^k)_{f(x)}/(TS^p)_{f(x)},$$

with π being the projection onto quotient space, is onto. If we express this in terms of matrices, this means that the linear transformation $\pi \circ (df)$ is expressed by a matrix of maximal rank.

To rephrase what we have done, in order to prove that near x any map sufficiently near a transversal map is transversal, we need only check that the matrix representation of $\pi \circ (df)$ has maximal rank. But if f is transversal, so that $\pi \circ df$ has maximal rank, then in some matrix representation $\pi \circ df$ has a $(k - p) \times (k - p)$ submatrix with nonvanishing determinant. Clearly, for g near enough to f, and all points in a neighborhood of x, that same determinant will be nonzero (because the topology of Definition 3.3 requires $p > 0$, and thus *any* small neighborhood of a given function consists of functions whose first partials are close to those of the given function.) Of course, one may argue alternatively with $k - p$ linearly independent columns of the matrix.

If on the other hand $f(x) \notin S^p$, since S^p is a submanifold $f(x)$ is at a positive distance from S^p. Clearly, all functions g near enough to f in our topology will have $g(y)$ at a positive distance from S^p for y near x.

This resolves the problem locally, and by our initial remarks, proves the proposition.

Theorem 3.2 (Thom's Transversality Lemma) Keeping the notations of Proposition 3.5, $C^\infty_{S^p}(M^n, N^k)$ is dense in $C^\infty(M^n, N^k)$.

Proof The proof will be broken into two steps, the first a local version, to be followed by a globalization. $f: M^n \to N^k$ will denote a given smooth function.

(a) Let U_β be a coordinate chart or neighborhood in N^k containing $y \in S^p \subseteq N^k$, with a given homeomorphism $\psi_\beta : U_\beta \to \mathbb{R}^k$. We assume ψ_β maps $U_\beta \cap S^p$ to $\{(0, \ldots, 0, y_{k-p+1}, \ldots, y_k)\}$.

We look at

$$\tilde{f} = \psi_\beta \circ f \,|\, f^{-1}(U_\beta) : f^{-1}(U_\beta) \to \mathbb{R}^k.$$

We let $p(y_1, \ldots, y_k) = (y_1, \ldots, y_{k-p})$, and we set, for any $\lambda = (\lambda_1, \ldots, \lambda_{k-p})$,

$$g_\lambda(x) = \tilde{f}(x) - (\lambda_1, \ldots, \lambda_{k-p}, 0, \ldots, 0).$$

Naturally for small λ, $g_\lambda(x)$ is close to $\tilde{f}(x)$. Because of Sard's theorem, the set of critical values of $p \circ \tilde{f}$ has measure zero. We may thus find a small λ for which $p \circ g_\lambda(x)$ has the origin as a regular value. But this means that the mapping

$$(dp)_{g_\lambda(x)} \circ (dg_\lambda)_x : Tf^{-1}(U_\beta) \to T\mathbb{R}^k / T(S^p \cap U_\beta)$$

is onto, when $g_\lambda(x) \in S^p$. Just as in the preceding proposition, this amounts to transversality at the point in question (and hence in a neighborhood of that point).

(b)   We choose a cover of N^k by $\{U_\beta\}$, which are coordinate charts for which we have selected a λ, as in (a), which we now name $\lambda(\beta)$ for this U_β. If $\tilde{f}$ is replaced by the nearby $g_{\lambda(\beta)}$, then we have transverse regularity in a neighborhood, which we write as U_β as well.

By standard point-set topology (as in [36] or [62]), we may assume that $\{U_\beta\}$ is a countable cover of N^k that for each $f^{-1}(U_\beta)$ there are compact sets $K_{\beta_i} \subseteq f^{-1}(U_\beta)$, and that the interiors of the countably many sets K_{β_i} cover M^n. (For the second part, simply note that if $f^{-1}(U_\beta)$ is not compact, at least it is covered by a countable number of compact sets. One assumes, with no loss of generality, that the $\{K_{\beta_i}\}$ are locally finite.)

The idea of the proof is to modify f on each $f^{-1}(U_\beta)$ so that the map becomes transversal on bigger and bigger unions of the sets K_{β_i}. To this end, we note the following:

(i)   If the transversality condition is met on a compact subset of the domain, that is, for each x in the compact set either $f(x) \notin S^p$ or the composition $(TM^n)_x \to (TN^k)_{f(x)} \to (TN^k)_{f(x)}/(TS^p)_{f(x)}$ is onto, then all sufficiently close approximations to f will meet the condition on the compact set. (We may even take $\varepsilon(x)$ constant, due to compactness. See Proposition 3.5.)

(ii)   The $\lambda(\beta)$ in (a) may be chosen to have arbitrarily small norm, for any β. Our modifications may be constructed as follows:

We write

$$M^n = \bigcup_j K_j,$$

where the j's are a numbering of the β_j. We then refer to $\lambda(j)$ from (a). By standard point-set topology [36], there is no loss of generality in assuming that there is an open cover $\{V_j\}$ of M^n, locally finite, with $K_j \subseteq V_j \subseteq f^{-1}(U_j)$ for each j, U_j meaning that U_β with $K_j \subseteq f^{-1}(U_\beta)$. We then define smooth $\phi_j : M^n \to \mathbb{R}$ such that

$$\phi_j(x) \in [0, 1], \qquad \phi_j(K_j) = 1, \qquad \phi_j(M^n - V_j) = 0.$$

We successively replace f by f_j, with $f = f_0$, according to the inductive definition

$$f_j(x) = \begin{cases} \phi_j(x)g_{\lambda(j)}(x) + (1 - \phi_j(x))f_{j-1}(x) & \text{if } x \in V_j \\ f_{j-1}(x) & \text{if } x \notin V_j, \end{cases}$$

where $\lambda(j)$ is chosen by (a) to assure transversality on K_j and yet small enough so as *not* to destroy it on

$$\bigcup_{i=0}^{j-1} K_j.$$

It is clear that each f_j is smooth if each V_j lies in a chart.

The cover $\{K_j\}$ of the domain being locally finite, in a sufficiently small neighborhood of a given point the sequence of functions $\{f_j\}$ will eventually become constant, so that the sequence f_j will converge to a smooth function. Near any point, the last f_j to have changed will be transversal, so that the limit map will be transversal everywhere.

Remarks (1) There are many variants of the basic transversality lemma. For one, it is useful to have a relative version where the map is not changed on a certain prescribed nice subset of the domain where it is already well behaved (see Milnor and Stasheff [78]). It is also possible to consider versions where less differentiability is required. This involves some difficulties with Sard's theorem, in the case of C^r-map, r finite. The transversality lemma is used crucially by Thom in his basic constructions of cobordism theory (see [105, 107]).

(2) Thom has found applications of transversal maps to other problems. For example, if $f : \mathbb{R}^n \to \mathbb{R}^k$ is differentiable and $A : \mathbb{R}^n \to \mathbb{R}^k$ is linear, $f + A$ is called a linear perturbation of f. Then it is known that for almost every A the set of $x \in \mathbb{R}^n$ such that $f + A$ has a differential of a given rank at x is a submanifold of $\mathbb{R}^n$.

PROBLEMS AND PROJECTS

1. The situation with critical points is radically different from that with critical values. Let $A > 0$. Construct a smooth $f : \mathbb{R}^n \to \mathbb{R}^k$ such that if C is the set of critical points of f then $\mu(C) = A$.

2.   Is there a smooth $f : \mathbb{R} \to \mathbb{R}$ where the set of critical values is not countable? (Consider smooth positive functions that vanish precisely on the Cantor set and their integrals.)

3.   Consider Corollary 3.6 in the case where $n + p < k$. Show by example that it is possible to have $M^n \cap S^p$ be a submanifold or not be a submanifold, according to position.

4.   Consult [107, 105, 109] to learn the basic notion of cobordant manifolds and its relation to transversality.

5.   Consult [105] for a full proof of a relative transversality theorem (in Appendix 2).

6.   Consult [2] for relations to flows (dynamical systems).

7.   Consult Smale [101] for an infinite dimensional generalization (see Chapter 7 for the basic theory).

4

Tangent Bundles, Vector Bundles, and Classification

The tangent space to a manifold M^n is built locally from copies of $O_\alpha \times \mathbb{R}^n$, where O_α is a coordinate chart in M^n. If a point lies both in $O_\alpha \times \mathbb{R}^n$ and $O_\beta \times \mathbb{R}^n$, a double representation is avoided (see Definition 2.5) by an identification or equivalence relation that is linear in the second variable. This situation arises in other, rather different situations. For example, we may consider a submanifold $S^p \subseteq \mathbb{R}^k$. For each $x \in S^p$, we then let N_x^{k-p} be the $(k-p)$-dimensional space of vectors that begin at x and are perpendicular to the tangent space to S^p at x, i.e., perpendicular to every vector in $(TS^p)_x$. One may easily topologize the union

$$\bigcup_{x \in S^p} N_x^{k-p},$$

in a manner similar to the definition of the tangent space, and we write this space, often called the normal bundle, with the notation $(NS^p)^{k-p}$. Or, for another example, if $S^p \subseteq N^k$, one may restrict attention, to the tangent space $(TN^k)_x$, for $x \in S^p$. Over each point of the p-dimensional S^p, one then has a k-dimensional vector space $(TN^k)_x$.

The gist of all these constructions, whose geometric basis is really quite well motivated, is the local procedure of taking a bit of a manifold, forming a Cartesian product with some Euclidean space, and then performing some identifications that are linear in the second variable at each point. Such constructions are now called *vector bundles*, I presume because near any point they appear to be a bunch of vector spaces. They have played an increasingly important role in modern mathematics, culminating in the sophisticated subject of *K*-theory, which studies equivalence classes of vector bundles over a given space.

Our goals in this chapter are the following:

(i)   analyze the structure of the tangent space to a differentiable manifold from this somewhat deeper point of view;

(ii)   define vector bundles in general, and develop all the basic properties; and

(iii)   prove a classification theorem for vector bundles over a compact Hausdorff space.

Hardly anything we do in this chapter cannot be done in some greater degree of generality—usually with much more work. We shall include at the end a brief survey of where one may find the more striking generalizations. It will be convenient at many places to discuss topological groups and their actions on spaces. This will be our starting point here.

GROUPS ACTING ON SPACES

Definition 4.1   Let G be a group that is also endowed with the structure of a topological space. We call G a *topological group* if the map

$$\phi: G \times G \to G$$

defined by $\phi(x, y) = xy^{-1}$ is continuous.

Remarks   (1)   It is clearly equivalent to require that the multiplication and inverse maps, respectively

$$(x, y) \to xy \qquad \text{and} \qquad x \to x^{-1},$$

be both continuous.

(2)   Naturally, any group that is endowed with the discrete topology is a topological group.

(3) $Gl(n; \mathbb{R})$, the group of invertible $n \times n$ real matrices, is clearly a topological group.

(4)   Products and subgroups afford easy ways of generating various examples of topological groups.

Definition 4.2 Let G be a topological group and X a topological space. We say that G *acts* on X, or that there is an *action* of G on X, if we are given a continuous map

$$\mu: G \times X \to X$$

such that

(a) $\mu(g_1, \mu(g_2, x)) = \mu(g_1 g_2, x)$ for any $g_1, g_2 \in G$ and $x \in X$; and
(b) if $e \in G$ is the identity element in G, and $x \in X$,

$$\mu(e, x) = x.$$

Sometimes one calls G a *transformation group* (or a left transformation group acting on X).

Remarks (1) A right action, or right transformation group, would be defined by a map $\mu: X \times G \to X$ with the appropriate properties.

(2) If G is a topological group, $\mu: G \times G \to G$, defined by $\mu(x, y) = xy$ defines an action of the topological group G on the topological space G (check this.)

(3) If $G \subseteq \mathrm{Gl}(n; \mathbb{R})$ is any subgroup, there is an action of G on $\mathbb{R}^n$, defined by left matrix multiplication. Specifically, if $M \in G$, $X \in \mathbb{R}^n$, we set

$$\mu(M, X) = MX \in \mathbb{R}^n.$$

(4) If $\mu(g, x) = y$, $\mu(g^{-1}, y) = x$ (for $\mu(e, x) = x$ and $gg^{-1} = g^{-1}g = e$, the identity element.)

(5) A matrix $M \in \mathrm{Gl}(n; \mathbb{R})$ is orthogonal if $M^{-1} = M^t$, where M^t, the transpose of M, interchanges the rows and columns of M, If $(\ ,\ )$ denotes the inner product and $X, Y \in \mathbb{R}^n$, then

$$(MX, MX) = (M^t M X, X) = (M^{-1} M X, X) = (X, X).$$

Thus, if M is orthogonal $\|MX\| = \|X\|$, so that left multiplication by an orthogonal matrix sends S^{n-1} to S^{n-1}. If the orthogonal matrices are denoted $O(n)$, clearly $O(n)$ is a closed subgroup of $\mathrm{Gl}(n; \mathbb{R})$ which acts, as a (left) transformation group, on S^{n-1}.

There are several kinds of transformation groups that are noteworthy. We discuss this more in Chapter 10, but two kinds of action are relevant to our needs here.

Definition 4.3 Given an action $\mu: G \times X \to X$ and a set $S \subseteq X$, we write

$$G_S = \{g \in G | \mu(g, y) = y, \quad \text{for all} \quad y \in S\}.$$

(a) We say that the action is *effective* if $G_X = \{e\}$, that is, the only element leaving all of X fixed is e, the identity.

(b)   We say that the action is *transitive* if given $x, y \in X$ there is some $g \in G$ such that $\mu(g, x) = y$.

Remarks   (1)   In Remark (2) before Definition 4.3., the action is visibly effective and transitive.

(2)   A trivial example of a transitive action that is not effective occurs when G is a nontrivial group but X is a single point.

Similarly, if $G = \{e\}$, the action of G on any X is effective.

(3)   The action of Gl$(n; \mathbb{R})$ on $\mathbb{R}^n$, as described in Remark (3) preceding Definition 4.3., is visibly both effective and transitive, omitting the origin.

(4)   Given an action $\mu: G \times X \to X$ and a subgroup $H \subseteq G$, there is an action

$$\mu_H: H \times X \to X,$$

defined by $\mu(g, x)$, when $g \in H$.

If μ is effective, so is μ_H. On the other hand, if μ_H is transitive, so is μ.

The following proposition clarifies the importance of transitive actions:

Proposition 4.1   Let X be a Hausdorff space and $\mu: G \times X \to X$ a transitive action. Let $x_0 \in X$ and recall

$$G_{x_0} = \{g \in G \,|\, \mu(g, x_0) = x_0\}.$$

Note that G_{x_0} is closed. Define G/G_{x_0} to be the quotient space of G by the equivalence relation $g_1 \sim g_2$ if and only if $g_1^{-1}g_2 \in G_{x_0}$. (G/G_{x_0} is called a coset space; in this case it is easily checked to be Hausdorff.)

If G/G_{x_0} is compact, then there is a natural homeomorphism

$$\phi: G/G_{x_0} \to X.$$

Proof   Define $\phi(\{g\}) = \mu(g, x_0)$. Note that if $g \sim g'$, so that $(g')^{-1}g \in G_{x_0}$, then $\phi(\{g'\}) = \mu(g', x_0) = \mu(g', \mu((g')^{-1}g, x_0)) = \mu(g, x_0) = \phi(\{g\})$, so that ϕ is well defined. Continuity is immediate. Because G acts transitively, ϕ is immediately checked to be onto.

Now if $\phi(\{g_1\}) = \phi(\{g_2\})$, then

$$\mu(g_1, x_0) = \mu(g_2, x_0),$$

or acting by g_1^{-1} we get

$$x_0 = \mu(g_1^{-1}g_1, x_0) = \mu(g_1^{-1}g_2, x_0).$$

But this implies, by definition, that $g_1^{-1}g_2 \in G_{x_0}$ (equivalently; $g_2^{-1}g_1 \in G_{x_0}$, because G_{x_0} is a subgroup). But this means that if $\phi(\{g_1\}) = \phi(\{g_2\})$, then $g_1 \sim g_2$. Therefore, ϕ is 1–1.

Naturally a 1–1 map from a compact Hausdorff space onto another is a homeomorphism.

Remarks (1) A standard case to which this situation applies is the action of O(n), the group of $n \times n$ orthogonal matrices, on S^{n-1} [see Remark (5) preceding Definition 4.3]. The subgroup of O(n) that leaves a unit vector, such as $x_0 = (1, 0, \ldots, 0)$ fixed, is easily checked to isomorphic to O($n-1$) in the form

$$\begin{pmatrix} 1 & 0 & \cdots & 0 \\ \hline 0 & & & \\ \vdots & & * & \\ 0 & & & \end{pmatrix}$$

with $*$ an $(n-1) \times (n-1)$ orthogonal matrix. In this case one occasionally sees written O(n)/O($n-1$) = S^{n-1}.

(2) In case G_{x_0} is a normal closed subgroup of G, G/G_{x_0} and hence X inherits the structure of a topological group. For example, let $G_0 = Z$, the group of integers, and $G = \mathbb{R}$, the additive group of real numbers, in the case of $\mathbb{R}$ acting on the circle S^1 via the prescription that a acts on a point by a counterclockwise rotation of $2\pi a$. Then Proposition 4.1 tells us $\mathbb{R}/Z \equiv S^1$.

THE TANGENT BUNDLE

We next will utilize the action of Gl(n; $\mathbb{R}$) on $\mathbb{R}^n$ to exhibit some richer structure in the tangent space to a manifold. This will be a prototype for our vector bundles. Consider

$$T_\alpha = O_\alpha \times \mathbb{R}^n$$

with $\phi_\alpha : O_x \to \mathbb{R}^n$ some coordinate chart or neighborhood in some manifold. T_α is a building block in our construction of TM^n. We note that we may define an action of Gl(n; $\mathbb{R}$) on T_α by

$$\mu(M, (x, y)) = (x, My)$$

with $M \in$ Gl(n; $\mathbb{R}$) and My left matrix multiplication of $y \in \mathbb{R}^n$. More generally, the equivalence relation that we used in the definition of TM^n (Definition 2.5) places (x, y) and $(x, J_{\phi_\beta \circ \phi_\alpha^{-1}}(\phi_\alpha(x))y)$ in the same equivalence class whenever $x \in O_\alpha \cap O_\beta$. We now rewrite this in the language of actions as follows: there is a continuous map, $J_{\phi_\beta \circ \phi_\alpha^{-1}}$ from $\phi_\alpha(O_\alpha \cap O_\beta)$ to Gl(n; $\mathbb{R}$) such that whenever $x \in O_\alpha \cap O_\beta$,

$$(x, y) \sim (x, J_{\phi_\beta \circ \phi_\alpha^{-1}}(\phi_\alpha(x))y).$$

To put it differently, the equivalence relation in Definition 2.5 is obtained in terms of a continuous map J from $O_\alpha \cap O_\beta$ to $Gl(n; \mathbb{R})$ that identifies two values of the second coordinate whenever the action J at the point takes the one second coordinate into the other. Using these ideas, we rewrite the definition of TM^n is a way that remarkably close to the original definition of a manifold.

Definition 4.4   (Rewording of TM^n according to Definition 2.5)   Let M^n be a C^r-manifold, $r \geq 1$, with a cover by coordinate charts $\{O_\alpha\}$, each with a given homeomorphism $\phi_\alpha : O_\alpha \overset{\equiv}{\to} \mathbb{R}^n$. By the tangent space to M^n we mean the space TM^n endowed with a map $\pi : TM^n \to M^n$, specified by the following structure: Every point $z \in TM^n$ has a neighborhood U_α that is homeomorphic to $O_\alpha \times \mathbb{R}^n$, where O_α is a coordinate chart for M^n. We write the homeomorphism $\psi_\alpha : U_\alpha \to O_\alpha \times \mathbb{R}^n$. If $z \in U_\alpha \cap U_\beta$, then in a neighborhood of $\psi_\alpha(z)$ the map $\psi_\beta \circ \psi_\alpha^{-1}$ is

$$(x, y) \to (x, J_{\phi_\beta \circ \phi_\alpha^{-1}}(\phi_\alpha(x)))y).$$

The map $\pi : TM^n \to M^n$ is defined in any U_α by $\pi(x, y) = x$.

The passage from tangent bundle to arbitrary vector bundle is clear. M^n is generalized to a topological space covered by $\{O_\alpha\}$. The J's are no longer Jacobians in general, but (to assure that we have an equivalence relation) they are functions (continuous) to $Gl(n; \mathbb{R})$ that satisfy $J_{\phi_\alpha \circ \phi_\alpha^{-1}} = $ identity, as well as $J_{\phi_\gamma \circ \phi_\beta^{-1}} J_{\phi_\beta \circ \phi_\alpha^{-1}} = J_{\phi_\gamma \circ \phi_\alpha^{-1}}$, where defined.

VECTOR BUNDLES

Definition 4.5   An n-dimensional real *vector bundle* over a topological space X, written (E, X, p) consists of a space E and a continuous map $p : E \to X$ such that for each $x \in X$, $p^{-1}(x)$ is homeomorphic to $\mathbb{R}^n$, and for each x there is an open set O_α, $x \in O_\alpha$, and a homeomorphism

$$\psi_\alpha : p^{-1}(O_\alpha) \to O_\alpha \times \mathbb{R}^n$$

such that

(a)   if $\pi_\alpha : O_\alpha \times \mathbb{R}^n \to O_\alpha$ is projection onto the first coordinate and $y \in p^{-1}(O_\alpha)$, then $\pi_\alpha(\psi_\alpha(y)) = p(y)$.
(b)   Whenever $O_\alpha \cap O_\beta \neq \phi$, there is a continuous map $g_{\alpha\beta} : O_\alpha \cap O_\beta \to Gl(n; \mathbb{R})$ such that

   (i) $g_{\alpha\alpha} = 1$, the identity,
   (ii) $g_{\alpha\beta} g_{\beta\gamma} = g_{\alpha\gamma}$, where defined, and

(iii)   if $y \in p^{-1}(O_\alpha) \cap p^{-1}(O_\beta)$ and $\psi_\alpha(y) = (x, z)$,
$\psi_\beta \circ \psi_\alpha^{-1}(x, z) = (x, g_{\alpha\beta}(x)z)$.

Remarks   (1)   The definition of a vector bundle has much in common with the definition of a manifold; it is specified in terms of local structure, with the overlap transformation being the identity in the first variable and a (continuous family of) linear transformations in the second variable. Note that if O_γ is a smaller open set than O_α one may define ψ_γ and the $g_{\gamma\beta}$ by restriction.

(2)   If M^n is an n-manifold, TM^n arises by choosing the O_α to be coordinate neighborhoods and setting $g_{\alpha\beta}(x) = J_{\phi_\beta \circ \phi_\alpha^{-1}}(\phi_\alpha(x))$.

If $X \subseteq M^n$ is a subspace, we set $E = \pi^{-1}(X)$ and $p = \pi|E$ to get a vector bundle over X, (E, X, p). This is the restriction of TM^n to X.

For any space Y, a trivial n-dimensional vector bundle may be constructed by setting $Y \times \mathbb{R}^n = E$.

(3)   A similar definition may be given over any topologized field (or even skew field). $p^{-1}(x)$ would then be an n-dimensional vector space over that field, and $\mathrm{Gl}(n; \mathbb{R})$ would be replaced by the general linear group over that field.

For example, in the case of a complex vector bundle, we would have

$$\psi_\alpha^{-1}(O_\alpha) \equiv O_\alpha \times \mathbb{C}^n,$$

where $\mathbb{C}$ is the complex numbers.

(4)   Neither E nor X need be connected, but since the map p is onto (obviously) when E is connected, then so is X. But the converse is also true, because each set $p^{-1}(x)$ is connected, so any two points in E lie in connected sets $p^{-1}(x)$ and $p^{-1}(y)$, both of which would meet the connected set $\{(x, 0) \in E | x \in X\}$ (this latter being homeomorphic to X by $p(x) = \{(x, 0)\}$).

(5)   If $X = M^k$, a k-dimensional manifold, and (E, X, p) is an n-dimensional (real) vector bundle over X, then E is an $(n + k)$-dimensional manifold. Depending on how much differentiability one assumes on X and the functions $g_{\alpha\beta}$, one can prove differentiability on E. In particular, if X and all $g_{\alpha\beta}$ are smooth, so is E.

(6)   The general philosophy will be to fix X and examine various possible vector bundles over X. For example, in the introduction to this chapter we looked at a submanifold $M^n \subseteq \mathbb{R}^k$ and considered the tangent bundle to M^n, the normal bundle to M^n, namely $(NM^n)^{k-n}$, and $(T\mathbb{R}^k)|M^n$, which is the restriction of $T\mathbb{R}^k$ to the subspace M^n as in (2) above. Over any point, the bundle $(T\mathbb{R}^k)|M^n$ clearly splits into perpendicular pieces, one in TM^n and the other in $(NM^n)^{k-n}$. This is the operation of taking a direct sum of two vector subspaces that intersect only at the zero vector. In fact, we shall

see that one can do almost anything with vector bundles that one could have done with vector spaces. First, however, it is important to look at the notion of a section, which is the obvious generalization of a tangent vector field.

Definition 4.6 If (E, X, p) is a vector bundle, a *section* of the bundle means a continuous map

$$s: X \to E$$

such that $p \circ s = 1_X$, the identity on X (i.e., $p(s(x)) = x$ for all x).

For example, one may always set $s(x) = \{(x, 0)\}$, getting the so-called zero section. There may not necessarily exist a section s that is everywhere nonzero; this depends on certain topological invariants (see [25, 102]). But in a simple example, such as TS^1, we may put $s(x)$ to be the unique unit vector in the tangent space to a point that points in, say, the counterclockwise direction. Because each $p^{-1}(x)$ is a vector space, we may form sums and scalar multiples of sections; the sections, therefore, form a vector space, which is usually infinite dimensional.

CONSTRUCTIONS WITH VECTOR BUNDLES

We now look at the basic constructions that are possible with vector bundles.

Definition 4.7 Given two vector bundles (E_1, X, p_1) and (E_2, X, p_2), we define their *Whitney* or *direct sum* as follows:

For each x, choose neighborhoods U_α^1 and U_α^2 and suitable homeomorphisms

$$\psi_\alpha^1: p_1^{-1}(U_\alpha^1) \xrightarrow{\equiv} U_\alpha^1 \times \mathbb{R}^n, \qquad \psi_\alpha^2: p_2^{-1}(U_\alpha^2) \to U_\alpha^2 \times \mathbb{R}^m,$$

where n and m are the dimensions of E_1 and E_2, respectively. Let $O_x = U_\alpha^1 \cap U_\alpha^2$.

Consider the disjoint union

$$S' = \bigcup_{x \in X} O_x \times \mathbb{R}^n \times \mathbb{R}^m.$$

Introduce the equivalence relation

$$(z_1, u_1, v_1) \sim (z_2, u_2, v_2)$$

$$\text{if and only if} \qquad z_1 = z_2, \quad g_{x_1 x_2}^{(1)}(u_1) = u_2, \quad \text{and } g_{x_1 x_2}^{(2)}(v_1) = v_2.$$

Here, we write $g^{(i)}_{x_1 x_2}$ for the function from the bundle E_i concerning the overlap of the sets O_{x_1} and O_{x_2}, containing the points z_1 and z_2.

The quotient space $S'/\sim$ is defined to be $E_1 \oplus E_2$, the Whitney sum. Define $p_1 \oplus p_2 : E_1 \oplus E_2 \to X$ by $(p_1 \oplus p_2)\{(z, u, v)\} = z$. It is trivial to check that $E_1 \oplus E_2$ is a vector bundle of dimension $(n + m)$, over X. We observe that the coordinate change functions in the second variables, namely the $g_{\alpha\beta}$, for the Whitney sum bundle $E_1 \oplus E_2$, are the following matrices:

$$g_{\alpha\beta}(x) = \begin{pmatrix} g^{(1)}_{\alpha\beta}(x) & 0 \\ 0 & g^{(2)}_{\alpha\beta}(x) \end{pmatrix}.$$

When we clarify the notion of isomorphism of bundle, we shall show that if $M^n \subseteq \mathbb{R}^k$ is a submanifold

$$(T\mathbb{R}^k)|M^n \approx (TM^n) \oplus (NM^n)^{k-n}.$$

Formalizing our earlier remarks about restriction, we have

Definition 4.8 Let (E, X, p) be a vector bundle and $A \subseteq X$ a subspace; one defines the vector bundle restricted to A to be (E_A, A, p_A), where $E_A = p^{-1}(A)$ and $p_A = p \,|\, E_A$, that is, p applied to elements of E_A. There is no difficulty in showing that when A is nonempty (E_A, A, p_A) is a vector bundle of the same dimension as E. On occasion, we write $E|A$ for E_A.

Definition 4.9 An n-dimensional vector bundle (E, X, p) is called *trivial* if there is a homeomorphism $\phi : E \overset{\equiv}{\to} X \times \mathbb{R}^n$, such that $\pi \circ \phi = p$, i.e., for any $e \in E$, $\pi(\phi(e)) = p(e)$, where π is the projection of $X \times \mathbb{R}^n$ onto the first factor and where ϕ is linear on each $p^{-1}(x)$.

Notice that for any X and n there is always a trivial bundle $E = X \times \mathbb{R}^n$ with exactly one set $U_\alpha = E$ and the obvious homeomorphism. One quickly verifies that a sum of trivial bundles is trivial, and a restriction of a trivial bundle to a subset is still trivial. Since $T\mathbb{R}^k$ is trivial (because $\mathbb{R}^k$ has a single coordinate chart), $T\mathbb{R}^k|M^n$ is trivial for any submanifold $M^n \subset \mathbb{R}^k$. Since

$$T\mathbb{R}^k|M^n = TM^n \oplus (NM^n)^{k-n}$$

if TM^n is nontrivial, one can see that a nontrivial bundle may be a summand of a trivial bundle.

Note that a trivial bundle has a wealth of nonzero sections. For any $v \in \mathbb{R}^n, v \neq O$, set

$$s(x) = \phi^{-1}(x, v).$$

Then s is clearly a nonzero section (check!) But, it is well known from algebraic topology (see [25]) that the tangent bundle to an even-dimensional sphere is nontrivial; in fact, there is no nowhere vanishing tangent vector field. Hence, TS^{2i} is a nontrivial vector bundle.

I believe it will be helpful to explore in full detail an elementary example. Let $S \subseteq \mathbb{R}^3$ be a surface, that is, a 2-dimensional submanifold. For each $x \in S$, let N_x be the line through x perpendicular to S. Let N be the (disjoint) union of all the lines N_x. If $z \in N_x \subseteq N$, we define $p(z) = x$. N is topologized so that locally it looks like $O \times \mathbb{R}$, where O is a (small) open set in S. Then it is easy to check that N has the structure of a 1-dimensional vector bundle over S whose functions $g_{\alpha\beta}$ take values in $\mathrm{Gl}(1, \mathbb{R})$, which is the multiplicative group of nonzero real numbers. Such a 1-dimensional vector bundle (N, S, p) is often called a *line bundle*.

If S is an open band $(S = S^1 \times (0, 1))$, then this normal line bundle (N, S, p) is trivial because S has two sides, say an inside and an outside, and we define $\phi: N \to S \times \mathbb{R}$ by $\phi(z) = (p(x), r(x))$, where $r(x)$ is the positive distance from z to S along N_x in case z lies on outside, and otherwise the negative of the distance from z to S along N_x. Such a ϕ easily meets all the conditions of Definition 4.9.

But if S is the (open) Möbius band, that is, the interior of the quotient space obtained by gluing the two vertical sides AB of a closed box (Fig. 4.1) in reverse direction, then N is not trivial. To see this, observe that a trivial bundle would have many nonzero sections s. But if you follow a continuous family of nonzero vectors all normal to the Möbius band once around the circle in the center of the band, you will find that the vector at the original point points in the opposite direction to that when you started. Hence, a supposed nonzero normal vector field cannot be well defined. We conclude that a normal line bundle to a Möbius band embedded in $\mathbb{R}^3$ is nontrivial.

Figure 4.1

One may now easily verify that if S is a surface in $\mathbb{R}^3$, there is a homeomorphism

$$(TS) \oplus (NS) \equiv T\mathbb{R}^3 \big| S.$$

The right-hand side is always trivial, being the restriction of a trivial bundle to a subspace. We have just showed that N may be nontrivial. For completeness, consider finally a compact, orientable surface. It is well known that such surfaces are homeomorphic to S^2, the torus, or to a generalized torus with m holes, $m > 1$. It is shown in algebraic topology (see [25, 103]) that only the torus has a nonvanishing tangent vector field, or section of the tangent bundle. In all other cases the tangent bundle must be nontrivial. On the other hand, because each such surface has a well-defined inside and outside (see [25]), N clearly has a nowhere vanishing section, consisting of all unit vectors in each N_x that point, for example, outside. Then just as we had remarked for the open band above, one sees that N is a trivial line bundle. Lastly, we note that the compact 2-dimensional manifolds that are not orientable, namely the Klein bottle and the real projective plane, do not embed in $\mathbb{R}^3$, so that we cannot ask similar basic questions about line bundles.

We must now focus on the proper notion of a map between vector bundles. This concept combines the ideas of continuous maps of spaces, with linear transformations of vector spaces.

Definition 4.10 A *map* or *homomorphism* of vector bundles

$$f : (E_1, X_1, p_1) \to (E_2, X_2, p_2)$$

consists of a pair of continuous maps

$$f_E : E_1 \to E_2, \qquad f_X : X_1 \to X_2$$

such that $p_2 \circ f_E = f_X \circ p_1$ (which implies that f_E maps $p_1^{-1}(x)$ to $p_2^{-1}(f_X(x))$) and such that

$$f_E | p_1^{-1}(x) : p_1^{-1}(x) \to p_2^{-1}(f_X(x))$$

is a linear transformation of vector spaces.

We may rephrase this latter condition by selecting $z \in p_1^{-1}(O_x) \overset{\equiv}{\underset{\psi}{\to}} O_x \times \mathbb{R}^n$, $f(z) \in p_2^{-1}(U_y) \overset{\equiv}{\underset{\chi}{\to}} U_y \times \mathbb{R}^m$, with $\dim E_1 = n$, $\dim E_2 = m$, and specifying that in a neighborhood of $\psi(z)$ the map $\chi \circ f \circ \psi^{-1}$ is

$$(u, v) \to (f_X(u), \gamma(u) \cdot v)$$

for some $\gamma(u) \in L(\mathbb{R}^n; \mathbb{R}^m)$, the set of linear transformations from $\mathbb{R}^n$ to $\mathbb{R}^m$.

Remarks (1) The function $\gamma(u)$ from an open neighborhood of $\psi(z)$ to $L(\mathbb{R}^n, \mathbb{R}^m)$, the set of $m \times n$ real-valued matrices, is easily checked to be continuous because the maps f_X and f_E are continuous.

(2) Trivial examples of maps of vector bundles are the identity map and the inclusion of (E_A, A, p_A) in (E, X, p) when $A \subseteq X$ is a subspace. Slightly less trivial examples come from choosing f_X to be the identity, $E_1 = E_2$, but the maps $\gamma(u)$ all to be the negative of the identity matrix. The obvious inclusions $E_1 \to E_1 \oplus E_2$ or projections $E_1 \oplus E_2 \to E_2$ form further examples.

Definition 4.11 (a)   The most frequently encountered case of a map of vector bundles is the case where $X_1 = X_2$ and f_X is the identity. The set of all bundle maps from (E_1, X, p_1) to (E_2, X, p_2), for which $f_X = 1_X$, the identity, will be denoted

$$\text{Hom}(E_1, E_2).$$

(b)   The subset of the set $\text{Hom}(E_1, E_2)$ consisting of all maps f having the property that for each $x \in X$ the restriction $f \,|\, p_1^{-1}(x)$ is an isomorphism of vector spaces is denoted

$$\text{Iso}(E_1, E_2).$$

Note that $\text{Iso}(E_1, E_2)$ is empty unless $\dim E_1 = \dim E_2$. If there is an element in $\text{Iso}(E_1, E_2)$, we occasionally will write $E_1 \approx E_2$.

Remarks (1)   The above definitions may be somewhat clarified by the following elegant construction. Suppose we are given two vector bundles over the same space X. We shall form a vector bundle that at each point consists of the vector space of linear transformations from $p_1^{-1}(x)$ to $p_2^{-1}(x)$. Choose an open neighborhood O_x of x small enough that $p_1^{-1}(O_x)$ and $p_2^{-1}(O_x)$ are homeomorphic to $O_x \times \mathbb{R}^n$ and $O_x \times \mathbb{R}^m$ by homeomorphisms ψ_x and χ_x, respectively. Our new bundle is formed locally from the spaces

$$O_x \times L(\mathbb{R}^n; \mathbb{R}^m),$$

where $L(\mathbb{R}^n; \mathbb{R}^m)$ is the vector space of linear transformations from $\mathbb{R}^n$ to $\mathbb{R}^m$. If $x \in O_y$ for a similar O_y, then (x, f) is defined to be equivalent to (x, f'), where f' is

$$\chi_y \circ \chi_x^{-1}(f)\psi_x \circ \psi_y^{-1}$$

all over the point x.

We define $p(\{(x, f)\}) = x$. A tedious but elementary verification shows that this is a $m \times n$-dimensional vector bundle over X, which we denote

$$\text{Lin}(E_1, E_2).$$

(2)   In the language of (1), $\text{Hom}(E_1, E_2)$ consists of the sections of $\text{Lin}(E_1, E_2)$, because a section will be a continuous assignment of a linear map from $p_1^{-1}(x)$ to $p_2^{-1}(x)$, from which a suitable f_E is easily fabricated.

(3) $\text{Hom}(E_1, E_2)$ may be topologized by either by (2) above, in which case we define an open neighborhood of a section of $\text{Lin}(E_1, E_2)$ in terms of a metric on E_2, or else, in a more elementary fashion, by choosing a positive continuous function $\varepsilon(x)$ on X and then saying that $f, g \in \text{Hom}(E_1, E_2)$ belongs to an $\varepsilon(x)$ open neighborhood, provided that for each $y \in p_1^{-1}(x) =$

$x \times \mathbb{R}^n$ whose distance from $x \times 0$ is less than 1, $f(y)$ and $g(y)$ are closer than $\varepsilon(x)$.

(4) For $m = n$, $\mathrm{Iso}(E_1, E_1)$ is nonempty, for it contains the identity. For any E_1 and E_2, $\mathrm{Hom}(E_1, E_2)$ is nonempty, for it contains the zero map.

$\mathrm{Iso}(E_1, E_2) \subseteq \mathrm{Hom}(E_1, E_2)$ is clearly open, because any map sufficiently near a linear transformation whose determinant is nonzero, is a linear transformation whose determinant is nonzero.

(5) If $f:(E_1, X, p_1) \to (E_2, X, p_2)$ is an isomorphism, then one may easily define f^{-1} by setting

$$f_E^{-1} | p_2^{-1}(x) = (f | p_1^{-1}(x))^{-1}.$$

The expression on the right is merely the inverse of the linear transformation of vector spaces $f | p_1^{-1}(x)$. It is clear that f_E^{-1} is continuous, and to check that it is a map of vector bundles, note simply that if $\gamma: O \to \mathrm{Gl}(n; \mathbb{R})$ is continuous, then $\bar{\gamma}(x) = (\gamma(x))^{-1}$ is also continuous. (This may be checked using a standard formula for expressing the inverse of an invertible matrix in terms of determinants.)

(6) Given a fixed $f \in \mathrm{Hom}(E_1, E_2)$

$$\{x \mid f | p_1^{-1}(x) \quad \text{is an isomorphism of vector spaces}\}$$

is visibly an open subset of X.

There are many other important constructions on vector bundles, such as the dual and the tensor product (consult [4, 51]). For our immediate purposes the next notion, the *induced bundle*, is basic. This will give a way to build many new vector bundles out of a fixed bundle. It is a special case of the abstract notion, from category theory, of a pull-back (see [79, 97]).

Definition 4.12 Let (E, X, p) be an n-dimensional vector bundle over a space X, and let $f: Y \to X$ be a continuous map.

Define

$$f^{-1}E = \{(y, e) \in Y \times E \mid f(y) = p(e)\}.$$

This set is given the relative topology of a subset of $Y \times E$.

We define $p': f^{-1}E \to Y$ by $p'(y, e) = y$.

Now, notice that if we have $O_\alpha \subseteq X$ and a homeomorphism $\psi_\alpha: p^{-1}(O_\alpha) \to O_\alpha \times \mathbb{R}^n$, then we may define another homeomorphism

$$\psi_\alpha': (p')^{-1}(f^{-1}(O_\alpha)) \to f^{-1}(O_\alpha) \times \mathbb{R}^n$$

by setting

$$\psi_\alpha'(x, e) = (x, \pi_2 \psi_\alpha(e)),$$

where $\pi_2: O_\alpha \times \mathbb{R}^n \to \mathbb{R}^n$ is projection onto the second factor. The map is evidently continuous and onto. Because x is the first factor in the definition of ψ'_α, the only way ψ'_α could fail to be 1–1 would be if $\psi'_\alpha(x, e_1) = \psi'_\alpha(x, e_2)$ for different e_1 and e_2. But $p(e_1) = p(e_2)$, since (x, e_1) and (x, e_2) both belong to $f^{-1}E$, so that $e_1, e_2 \in p^{-1}(x)$. Of course, $\pi_2 \circ \psi_\alpha$ is 1–1 on each vector space $p^{-1}(x)$. Finally, the reader may easily check that ψ'_α takes a small basis open set in $(p')^{-1}(f^{-1}(O_\alpha))$ into an open set, showing that ψ'_α is a homeomorphism.

The fact that the "interchange maps" $\psi'_\beta \circ \psi'^{-1}_\alpha$ come from continuous maps into the general linear group follows in a direct and formal way from the similar condition on $\psi_\beta \circ \psi^{-1}_\alpha$.

Hence $f^{-1}E$ is an n-dimensional vector bundle over X, which we write $(f^{-1}E, X, p')$ and call the *induced bundle*. Generally speaking, this terminology poses no danger of confusion with the inverse image of a set under a mapping.

Remarks (1) There are cases of the induced bundle, which we have already considered. For example, given (E, X, p) and a subspace $A \subseteq X$ with inclusion map $i: A \to X$, then the restriction (E_A, A, p_A), from Definition 4.8, is identical to the induced bundle $(i^{-1}E, A, p')$ (check this!).

(2) If $f: Y \to X$ is a constant map (that is, $f(y) = x_0$ for all $y \in Y$) then $(f^{-1}E, Y, p)$ is always a trivial bundle. To see this, define

$$h: Y \times p^{-1}(x_0) \to f^{-1}E$$

by $h(y, e) = (y, e)$. (Check details here.)

(3) Clearly, if one uses the identity map for f, the induced bundle is precisely the original bundle. In addition, one checks that if we are given two maps

$$Z \xrightarrow{h} Y \xrightarrow{g} X,$$

and a bundle (E, X, p), then there is a natural isomorphism of bundles

$$(g \circ h)^{-1}E \approx h^{-1}(g^{-1}E).$$

(4) Given two bundles over X, one may construct a product of these bundles over $X \times X$ in an obvious fashion. The Whitney sum of the two bundles then appears as the induced bundle from the product bundle by the diagonal map $\Delta: X \to X \times X$, defined by $\Delta(x) = (x, x)$.

(5) The basic classification theorem (in the compact case, our Theorem 4.2) shows how to recover any bundle in terms of bundles that are induced from a "universal" bundle. But before we can get to that stage, we need some further preparatory material. Just like with vector spaces, there are natural notions of sub and quotient bundles and 1–1 and onto maps.

Definition 4.13   (a)   A map or homomorphism of vector bundles over X (Definition 4.11), say $f:(E_1, X, p_1) \to (E_2, X, p_2)$, is called a *monomorphism* if each linear transformation $f_E|p_1^{-1}(x)$ is a 1–1 map (often called a monomorphism) of vector spaces for every $x \in X$. It is called an *epimorphism* if each $f_E|p_1^{-1}(x)$ is onto. It is called an *isomorphism* if it is both a monomorphism and a epimorphism (compare with Definition 4.11b).

(b)   A bundle (E_1, X, p_1) is a *subbundle* of (E_2, X, p_2) if each $p_1^{-1}(x)$ is a vector subspace of $p_2^{-1}(x)$ and there are maps (for suitable O_α) $\psi_\alpha: p_2^{-1}(O_\alpha) \xrightarrow{\equiv} O_\alpha \times \mathbb{R}^{\dim E_2}$ defining the bundle E_2 that, when restricted to $p_1^{-1}(O_\alpha)$, map $p_1^{-1}(O_\alpha)$ homeomorphically to $O_\alpha \times \mathbb{R}^{\dim E_1}$, where $\mathbb{R}^{\dim E_1}$ is a *fixed*, chosen subspace of $\mathbb{R}^{\dim E_2}$, thus defining the structure of the bundle E_1.

For example, both E_1 and E_2 would be subbundles of the vector bundle $E_1 \oplus E_2$, in a natural way. Now, in the category of vector spaces, a vector subspace of a vector space $F \subseteq E$ is the image of a 1–1 map, the inclusion, and conversely, the image of a 1–1 map is a vector subspace. This sort of characterization is also valid for vector bundles, but as we now see it is a bit more difficult to prove. We shall now assume that our bundles are over X, and for maps f, f_X is the identity.

Proposition 4.2 (E_1, X, p_1) is a subbundle of (E_2, X, p_2) if and only if (E_1, X, p_1) is the image of a 1–1 map or monomorphism of a vector bundle to (E_2, X, p_2) (that is, there is a 1–1 map of bundles, f, such that the image of f is precisely E_1).

Proof   The necessity is easy. A subbundle of a vector bundle is clearly a vector bundle, and it is the image of the inclusion map $i: E_1 \xrightarrow{\subseteq} E_2$.

On the other hand, suppose we are given a monomorphism of vector bundles

$$f:(E_1, X, p_1) \to (E_2, X, p_2)$$

with our convention that $f_X = 1_X$. It follows that $p_2 \circ f_E = p_1$, or in other words, the following diagram is commutative:

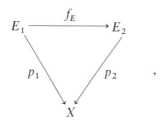

in the sense that if one follows the maps from E_1 to X around all possible routes in the diagram, then the maps (two in this case) with common domain and range are equal.

Now, for $x \in O_\alpha \subseteq X$, suitable O_α, we have homeomorphisms

$$\psi_\alpha : p_2^{-1}(O_\alpha) \overset{\equiv}{\to} O_\alpha \times \mathbb{R}^n, \qquad \tilde{\psi}_\alpha : p_1^{-1}(O_\alpha) \overset{\equiv}{\to} O_\alpha \times \mathbb{R}^m,$$

with $n = \dim E_2$ and $m = \dim E_1$. The map ψ_α sends $f(p_1^{-1}(x))$ onto an m-dimensional linear subspace of $x \times \mathbb{R}^n$, which we may write $x \times \mathbb{R}^m$, for some $\mathbb{R}^m \subseteq \mathbb{R}^n$. It follows that in some small open set about x, ψ_α maps every subset $p_1^{-1}(y)$ (often called the fiber over y) to some m-dimensional subset of $y \times \mathbb{R}^n$ whose projection to $y \times \mathbb{R}^m$, chosen above, is onto. (This is the usual remark that if one has a continuous family of linear maps and one is onto, then all others sufficiently near to it are onto.)

It follows that near x one may construct a continuous family of linear transformations $l(x) \in \mathrm{Gl}(n; \mathbb{R})$ that send $f(p_1^{-1}(y))$ by projection onto $y \times \mathbb{R}^m$ and map the orthogonal complement of $\mathbb{R}^m$ identically to itself. Define

$$\psi'_\alpha(z) = 1 \times l(p_1(z)) \circ \psi_\alpha.$$

This is visibly a homeomorphism near any point in $p_1^{-1}(x)$, and because l is a continuous homomorphism in $\mathrm{Gl}(n; \mathbb{R})$, this new coordinate neighborhood is trivially checked to be compatible with the structure of the bundle (E_2, X, p_2). But our construction is precisely what is needed to exhibit (E_1, X, p_1) as a subbundle of the vector bundle.

Remark   The reader should compare this material with the definition of a submanifold (Definition 1.5). There in order to describe a submanifold one might need to take more charts for the larger manifold than in the original presentation; an extreme case would be submanifolds of $\mathbb{R}^n$, since this latter needs only one chart.

In the case of a vector subbundle constructed from the image of a monomorphism of vector bundles, in the Proposition 4.2, we see that we may need to pass to smaller neighborhoods, and hence more neighborhoods, to exhibit the subbundle structure. This is why in Definition 4.13b we speak of suitable neighborhoods defining the bundle structure and in Proposition 4.2 we pass to a smaller neighborhood and take new coordinate homeomorphisms in terms of $l(x)$, which are thus compatible with the original bundle structure. Naturally, with either the original or the new neighborhoods the identity map from E_2 to itself is an isomorphism of vector bundles (Definition 4.3a). One often hears this situation described in terms of the different neighborhoods yielding the same vector bundle structure on E_2.

Finally, we observe that in the extreme case where (E_1, X, p_1) is a non-trivial bundle that is a subbundle of the trivial bundle (E_2, X, p_2), one needs only one neighborhood, all of X, to describe the bundle structure on E_2, but more than one neighborhood if one wishes to describe the subbundle structure of (E_1, X, p_1). A specific example of this interesting phenomenon is TS^2 as a subbundle of $T\mathbb{R}^3 | S^2$, where S^2 is, of course, a submanifold of $\mathbb{R}^3$.

We are now ready to define a quotient bundle.

Definition 4.14 Let (E_1, X, p_1) be a subbundle of the bundle (E_2, X, p_2). The *quotient bundle* $(E_2/E_1, X, p_3)$ is then defined as follows:

(a) For each $x \in X$, set $p_3^{-1}(x) = p_2^{-1}(x)/p_1^{-1}(x)$, simply the quotient of one vector space by a vector subspace. Let

$$E_2/E_1 = \bigcup_{x \in X} p_3^{-1}(x).$$

(b) Topologize E_2/E_1 by choosing for each $x \in X$ an open set O_α, $x \in O_\alpha$, with a homeomorphism

$$\psi_\alpha : p_2^{-1}(O_\alpha) \xrightarrow{\equiv} O_\alpha \times \mathbb{R}^n,$$

$n = \dim E_2$, such that $\psi_\alpha | p_1^{-1}(O_\alpha)$ is a suitable homeomorphism for E_1, from $p_1^{-1}(O_\alpha)$ to $O_\alpha \times \mathbb{R}^m$. Then we set

$$\tilde{\psi}_\alpha : p_3^{-1}(O_\alpha) \to O_\alpha \times \mathbb{R}^{n-m},$$

$m = \dim E_1$, by selecting a copy of $\mathbb{R}^{n-m}$ that is the orthogonal complement to $\mathbb{R}^m \subseteq \mathbb{R}^n$ with projection map $\pi : \mathbb{R}^n \to \mathbb{R}^{n-m}$ and putting

$$\tilde{\psi}_\alpha(\{y\}) = (1 \times \pi) \circ \psi_\alpha(y).$$

This map is clearly well defined, because the addition of any element in $p_1^{-1}(O_\alpha)$ is not detected after one applies the map $1 \times \pi$. It is visibly also $1-1$ and onto.

The topology on E_2/E_1 is then defined by giving $p_3^{-1}(O_\alpha)$ the open sets that result from applying $\tilde{\psi}_\alpha^{-1}$ to open sets in $O_\alpha \times \mathbb{R}^{n-m}$, and extending this to topologize E_2/E_1.

We now note

(c) If $\{y\} \in p_3^{-1}(O_\alpha) \cap p_3^{-1}(O_\beta)$, $\tilde{\psi}_\beta \circ \tilde{\psi}_\alpha^{-1}$ is clearly evaluated on a pair (x, z) as

$$(1 \times \pi) \circ \psi_\beta \circ \psi_\alpha^{-1}(1 \times \pi)^{-1}(x, z) = (1 \times \pi) \circ \psi_\beta \circ \psi_\alpha^{-1}(x, z')$$
$$= (1 \times \pi)(x, g_{\alpha\beta}(x)z') = (x, \pi g_{\alpha\beta}(x)z'),$$

where $\pi(z') = z$. But because the action of $g_{\alpha\beta}$ sends $p_1^{-1}(x)$ to itself, this is well defined and therefore creates a bundle structure on E_2/E_1.

Finally,

(d)   Define $p_3(\{y\}) = p_2(y)$. Then the final verifications that we have a bundle are immediate.

Of course, there is always a bundle map

$$(E_2, X, p_2) \rightarrow (E_2/E_1, X, p_3),$$

that is, for every x, a linear transformation of $p_2^{-1}(x)$ onto $p_3^{-1}(x)$. It is defined by sending each point to its equivalence class.

Remarks   (1)   For our purposes the most important examples will be furnished by manifolds and their tangent bundles. Let $M^n \subseteq \mathbb{R}^k$ be a submanifold. Write TM^n for the tangent bundle, thought of as the collection of tangent spaces to M^n in $\mathbb{R}^k$; write NM^n for the $(k - n)$-dimensional normal bundle and $T\mathbb{R}^k|M^n$ for the restriction of the (trivial) tangent bundle to M^n. Then we have the following isomorphisms:

$$T\mathbb{R}^k|M^n = TM^n \oplus NM^n, \; TM^n = T\mathbb{R}^k|M^n/NM^n, \; NM^n = T\mathbb{R}^k|M^n/TM^n.$$

(2)   Let $s: M^n \rightarrow TM^n$ be a nonzero section. Then s defines a 1-dimensional vector bundle, which over each point x consists of the 1-dimensional subspace of $(TM^n)_x$ containing the vector $s(x)$. As before, this line bundle is trivial, the map to $M^n \times \mathbb{R}$ being that map which associates to such $z \in (TM^n)_x$ the pair (x, α), where $z = \alpha s(x)$.

If for a tangent bundle there is a continuous inner product defined on the tangent space at each point, then there will be a well-defined notion of orthogonal complement. If we have a r-dimensional subbundle of TM^n, say E, then there will be an $(n - r)$-dimensional subbundle that consists at each point of the orthogonal complement. The orthogonal complement to a line bundle will therefore be, for example, an $(n - 1)$-dimensional subbundle. The continuous inner product will exist under very general circumstances; see Chapter V or [103]. In the compact case (X being compact), one may also infer the existence of an orthogonal complement from Lemma 4.6.

In other terms, for a given subbundle (E_1, X, p_1) of the bundle (E_2, X, p_2), one seeks a bundle $E_1^\perp$, where $E_2 = E_1 \oplus E_1^\perp$. One may then easily find suitable $E_1^\perp$ locally, and the full bundle $E_1^\perp$ can also be fabricated from the local pieces, using a partition of unity. The question of how to find an orthogonal complement for a subbundle is often called the problem of splitting, and it is known to be solvable for any vector bundle over a paracompact base space.

THE CLASSIFICATION OF VECTOR BUNDLES

Our next goal is to classify the n-dimensional vector bundles over a fixed space X. Naturally, we really want the *isomorphism classes* of such bundles, for which we write $v^n(X)$. We first observe that the correspondence that passes from X to $v^n(X)$ gives us something more than a mere set, for if $f: Y \rightarrow X$ is a continuous map, the induced bundle construction (Definition 4.12) gives a function, called $v(f)$,

$$v(f): v^n(X) \rightarrow v^n(Y).$$

This construction has the following already verified properties:

$$v(1_X) = 1_{v^n(X)}, \tag{1}$$

that is, it transforms identity maps to identity maps;

$$v(g \circ f) = v(f) \circ v(g) \tag{2}$$

(this was Remark (3) after Definition 4.12).

In the language of category theory, our construction is a contravariant functor from the category of topological spaces and continuous maps to the category of sets and maps. All this not withstanding, it is of great importance to gain an understanding of $v^n(X)$ in terms of the space X. Such an understanding should subsume in one result the basic facts alluded to above, such as that every bundle over $\mathbb{R}^k$ is trivial, i.e., $v^n(\mathbb{R}^k)$ consists of a single element, whereas $v^1(S^1)$ has two elements, the ordinary and Möbius bands, etc. We should also be able to build up $v^n(X)$ in terms of v^n applied to special, simple subspaces of X.

The answer that we shall get for the structure of $v^n(X)$ (see Theorem 4.1) will not always be easy to calculate, but it will have a simple and beautiful format. In rough terms, we shall construct a space B_n so that, for each X, $v^n(X)$ will be a sort of natural quotient of the set of continuous maps from X to B_n. One will be able to infer, in particular, that for many X, $v^n(X)$ is finitely computable.

For reasons of space and convenience, we shall present our main theorem when X is a compact Hausdorff space. An interested reader might pursue the most general theorem in the beautiful paper of Dold [26]. Our basic theorem will pass from maps to bundles, using the induced bundle construction. For that reason the initial part of our presentation will treat properties of maps. Then, we shall need to take a close look at the way in which two bundles defined over two subspaces of X may be glued together to form one bundle over their union.

Definition 4.15   Let $f, g: X \to Y$ be two continuous maps from X to Y. We say that f and g are *homotopic*, written $f \simeq g$, if there is a continuous map

$$F: X \times I \to Y, \qquad \text{where} \quad I = [0, 1],$$

such that for any $x \in X$

$$F(x, 0) = f(x), \qquad F(x, 1) = g(x).$$

F is often referred to as the homotopy between f and g.

Remarks   (1)   It is easy to check that this is an equivalence relation. Use $F(x, 1 - t)$ to check symmetry, and change scale from I to $[0, 1/2]$ and glue together two maps of $X \times [0, 1/2]$ to show that the relation is transitive.

(2)   One denotes by $[X, Y]$ the set of homotopy classes of maps, that is, equivalence classes of maps under the relation "homotopic," from X to Y.

(3)   If we consider maps from X to $\mathbb{R}^n$, then any two are homotopic. Simply define, for each $x \in X$, the map $F(x, t)$ by the formula

$$F(x, t) = tf(x) + (1 - t)g(x).$$

This argument immediately generalizes to maps from X to any convex subset of $\mathbb{R}^n$.

(4)   If $f \simeq g$ and Y has several connected components, then for any x, $f(x)$ and $g(x)$ belong to the same connected component.

(5)   If $n > 0$, then the identity map $1: S^n \to S^n$ is never homotopic to the constant map g, defined by $g(x) = a$, for some fixed a and $x \in S^n$. For if $f \simeq g$, we have

$$F: S^n \times I \to S^n$$

with $F(x, 1) = a$ for all $x \in S^n$. But then F defines a map on the quotient space of $S^n \times I$, with the subset $\{(x, 1) \,|\, x \in S^n\}$ shrunk to a point, that is,

$$\tilde{F}: D^{n+1} \to S^n,$$

because the quotient space is identified with either the upper hemisphere of S^{n+1} or, equivalently, D^{n+1} (check!). This map is still the identity on all pairs $\{(x, 0)\}$, that is, the boundary of D^{n+1}. But we proved [Corollary 3.3 (lemma to the Brouwer fixed point theorem)] that there is no such map.

The interest, at this time, in homotopic maps, is the following remarkable proposition.

Proposition 4.3   Let X be a compact Hausdorff space and $f, g: X \to Y$ be two homotopic maps, or in symbols, $f \simeq g$. Let (E, Y, p) be an n-dimensional vector bundle over Y.

Then the two induced bundles $f^{-1}E$ and $g^{-1}E$ (Definition 4.12) are isomorphic (Definition 4.11b).

Proof We need two preliminary steps.

(a) Given a vector bundle (F, X, q) and a closed subset $B \subseteq X$. Let $s: B \to F|B$ be a section over B, i.e., $s: B \to F$ is a continuous map such that whenever $b \in B$, $q(s(b)) = b$. Then there is a section $s_0: X \to F$ of the bundle such that $s_0|B = s$; i.e., if $b \in B$, $s(b) = s_0(b)$.

Of course, locally F is a product $O_\alpha \times \mathbb{R}^n$ for suitable $O_\alpha \subseteq X$. Then, the usual Tietze extension theorem, applied to maps to $\mathbb{R}^n$, shows that there is a section

$$s_{O_\alpha}: O_\alpha \to F$$

such that

$$s_{O_\alpha}|B \cap O_\alpha = s|B \cap O_\alpha.$$

This resolves the question in the small. We remark that, by compactness, we need only consider a finite number of sets O_α, say $O_{\alpha_1}, \ldots, O_{\alpha_k}$.

Let $\phi_1, \ldots, \phi_k$ be a partition of unity (Theorem 1.4) subordinate to the open cover $O_{\alpha_1}, \ldots, O_{\alpha_n}$. That is,

(i) $0 \le \phi_i \le 1$,

(ii) $\phi_i(x) \ne 0$ implies $x \in O_{\alpha_i}$, and

(iii) $\sum_i \phi_i(x) = 1$.

We define

$$s_i: X \to F$$

by setting

$$s_i(y) = \begin{cases} \phi_i(y)s_{O_{\alpha_i}}(y) & \text{if } y \in O_{\alpha_i} \\ 0 & \text{otherwise.} \end{cases}$$

s_i is trivially checked to be a section, with continuity coming from the fact that ϕ_i must vanish at the boundary of O_{α_i}.

Finally, we define

$$s_0(y) = \sum_i s_i(y).$$

The addition makes sense, because each $s_i(y)$ lies in the vector space $q^{-1}(y)$, and continuity is clear.

Now, if $y \in B$, we may calculate

$$s_0(y) = \sum_i s_i(y) = \sum_{i, y \in O_{\alpha_i}} \phi_i(y)s_{O_{\alpha_i}}(y) = \sum_{i, y \in O_{\alpha_i}} \phi_i(y)s(y),$$

because when $y \in B \cap O_{\alpha_i}$, $s_{O_{\alpha_i}}$ and s agree on y. But this last sum is just

$$\left(\sum_{i, y \in O_{\alpha_i}} \phi_i(y) \right) s(y) = \left(\sum_i \phi_i(y) \right) s(y) = s(y),$$

because the ϕ_i's are a partition of unity.

Our second preliminary step is now the following:

(b)   Given (E_1, X, p_1) and (E_2, X, p_2) as two n-dimensional vector bundles over a compact Hausdorff space X, let $A \subseteq X$ be a nonempty closed subset. Let $E_i | A$ denote the restriction of E_i to A (Definition 4.8). Suppose we have an isomorphism

$$f : E_1 | A \overset{\approx}{\to} E_2 | A.$$

Then there is an open set U, $A \subseteq U \subseteq X$, and an isomorphism

$$f_1 : E_1 | U \overset{\approx}{\to} E_2 | U$$

such that f_1 and f agree on all points of $p_1^{-1}(A)$. In other words, the iso-morphism over A extends to an isomorphism over U.

The easy way to see this is to use the remark following Definition 4.11. f is merely a section of the bundle $\mathrm{Lin}(E_1 | A, E_2 | A)$, which is an isomorphism on each fiber $p_1^{-1}(x)$, $x \in A$. By step (a) above the section extends, so that we have $f_1 : E_1 \to E_2$, which is a map of bundles so that over A, $f_1(x) = f(x)$. But f_1 is a map of bundles, and points y near enough points where $f_1 | p_1^{-1}(x)$ has nonzero determinant are points for which $f_1 | p_1^{-1}(y)$ has nonzero deter-minant [see Remark (6) following Definition 4.11]. Thus, f_1 is an isomor-phism over the desired open set. If one wishes to avoid the bundle $\mathrm{Lin}(E_1 | A, E_2 | A)$ and extensions to $\mathrm{Lin}(E_1, E_2)$, a local argument, much as in (a) above, may also be given.

To complete the proof, let

$$F : X \times I \to Y$$

describe the homotopy between our given maps f and g, with $F(x, 0) = f(x)$. Let $\pi : X \times I \to X$ be projection on the first factor. Define then, for any fixed $t \in I$,

$$F_t(x) = F(x, t).$$

We consider the two bundles $F^{-1}E$ and $\pi^{-1} F_t^{-1} E$. One may check at once from the definitions that they are isomorphic on the subset $X \times \{t\} \subseteq X \times I$. Note that bundles induced by π are essentially constant in the second coordinate.

We may use (b) above to conclude that $F^{-1}E$ and $\pi^{-1}F_t^{-1}E$ are isomorphic on an open set U containing $X \times \{t\}$. Because X is compact we see at once that there is some $\varepsilon > 0$, depending on t, such that

$$X \times (t - \varepsilon, t + \varepsilon) \subseteq U,$$

and consequently $F^{-1}E$ and $\pi^{-1}F_t^{-1}E$ are isomorphic on $X \times (t - \varepsilon, t + \varepsilon)$.

On the set $X \times \{t_1\}$, $F^{-1}E$ is the same as $F_{t_1}^{-1}E$, identifying $X \times \{t_1\}$ with X. On the set $X \times \{t_1\}$, $\pi^{-1}F_t^{-1}E$ is the same as $\pi^{-1}F_t^{-1}E$ on $X \times \{t\}$, because of the definitions of π and of an induced bundle. Thus, for all t_1, $|t - t_1| < \varepsilon$, the bundles $F_{t_1}^{-1}E$ are isomorphic.

Let $v^n(X)$ be the set of classes of isomorphic, n-dimensional vector bundles over X, given the discrete topology. Define

$$\tau: I \to v^n(X)$$

by $\tau(t_1) = \{F_{t_1}^{-1}E\}$. We have just proved that τ is locally constant. Therefore, τ is continuous. Since I is connected and $v^n(X)$ is discrete, τ must be constant. Therefore $F_{t_1}^{-1}E$ are isomorphic for all choices of $t_1 \in [0, 1]$. Select t_1 to be 0 and 1, and the proof of the proposition is complete.

Remarks (1) Proposition 4.3 can be used to show that in various cases $v^n(X)$ contains one element, or since there is always the trivial bundle $X \times \mathbb{R}^n$, every bundle is isomorphic to the trivial bundle; i.e., every bundle is trivial.

For example, let $X \subseteq \mathbb{R}^k$ be a convex subset. Suppose (E, X, p) is an n-dimensional vector bundle, which we write E for short. If $1: X \to X$ is the identity map, then by Remark (3) after Definition 4.12, $1^{-1}E$ is isomorphic to E. But by Remark (2) after Definition 4.15, 1 is homotopic to a constant map $g: X \to X$, where $g(x) = b$ for some fixed $b \in X$. According to Remark (2) after Definition 4.12, $g^{-1}E$ is a trivial bundle, and thus so is E.

More generally, if A is a subspace of $\mathbb{R}^m$, form the cone on A, denoted $C(x, A)$, which consists of all straight line segments from x, where $x \notin A$ is fixed, to a point in A. Then it is easily shown that the identity map from $C(x, A)$ to itself is homotopic to the constant map that sends everything to x. Consequently $v^n(C(x, A))$ consists of a single element in this case as well.

(2) If $n > 0$ and X is homeomorphic to Y, then there is a 1–1 correspondence (isomorphism of sets) between the sets $v^n(X)$ and $v^n(Y)$, for if the spaces are homeomorphic, there are continuous maps

$$f: X \to Y \quad \text{and} \quad g: Y \to X$$

such that $g \circ f = 1_X$, the identity on X, and $f \circ g = 1_Y$. We define maps

$$\phi: v^n(Y) \to v^n(X) \quad \text{and} \quad \psi: v^n(X) \to v^n(Y)$$

by the induced bundle construction, that is,

$$\phi(\{E\}) = \{f^{-1}E\} \qquad \text{and} \qquad \psi(\{F\}) = \{g^{-1}F\}.$$

The obvious fact that the induced bundles, via a given map, of two isomorphic bundles, yield isomorphic bundles, means that these maps are well defined. But then it is clear that the compositions $\psi \circ \phi$ and $\phi \circ \psi$ are the identity maps, so that ϕ (or ψ) is a 1–1 correspondence. (Compare Remark (3) following Definition 4.12.)

(3)   Our Proposition 4.3 now suggests a very important generalization of Remark (2) above. We say that the spaces X and Y have the same *homotopy type* if there are continuous maps $f: X \to Y$ and $g: Y \to X$ whose respective compositions are homotopic to (rather than merely equal to) the identity; specifically, we require $g \circ f \simeq 1_X$ and $f \circ g \simeq 1_Y$. For example, $\mathbb{R}$ and a single point have the same homotopy type because we set

$$f: \text{pt} \to \mathbb{R} \qquad \text{by} \quad f(\text{pt}) = 0$$

and

$$g: \mathbb{R} \to \text{pt}$$

as the unique such map. Then $g \circ f = 1_{\text{pt}}$, so trivially $g \circ f \simeq 1_{\text{pt}}$, and the homotopy between $f \circ g$ and $1_{\mathbb{R}}$ is given by

$$F(x, t) = xt.$$

If X has the homotopy type of a point, we shall say that X is *contractible*. Similarly, S^2 has the homotopy type of $\mathbb{R}^3 - (0, 0, 0)$. Both the Möbius and the ordinary band have the same homotopy type as the equatorial circle that runs around them. Clearly, we see from these examples that non-homeomorphic spaces may have the same homotopy type, while trivially homeomorphic spaces always have the same homotopy type.

Because (Proposition 4.3) the passage to induced bundle does not see a difference between homotopic maps, one may directly adopt the proof in (2) above to prove that, if X and Y have the same homotopy type, then $v^n(X)$ and $v^n(Y)$ are in 1–1 correspondence. If X is contractible, $v^n(X)$ is a single element.

For example, one may check without much difficulty that $v^1(S^1)$ has two elements, one represented by the ordinary band B and one represented by the Möbius band M; one thinks in terms of $\mathbb{R}$ identified with $(-1, 1)$, and $B = S^1 \times (-1, 1)$. Then one has $v^1(S^1) \equiv v^1(\mathbb{R}^2 - (0, 0)) \equiv v^1(B) \equiv v^1(M)$, etc.

In our drive towards the main theorem, we shall want to form quotient spaces of spaces over which we have bundles, and we shall want to glue together spaces over which we have bundles and get nice, well-defined bundles as an end product. The following set of lemmas systematically treats these basic technical questions and forms the foundation of our main results.

Lemma 4.1   Let (E, X, p) be an n-dimensional vector bundle over a compact Hausdroff space X. Let A be a closed subset over which the restricted bundle E_A or $E|A$ (Definition 4.8) is trivial, i.e., there is an isomorphism of vector bundles f from $p^{-1}(A)$ to $A \times \mathbb{R}^n$.

Then there is an n-dimensional vector bundle $E/p^{-1}(A)$, over the space X/A, defined as follows:

(a) $E/p^{-1}(A) = E/\sim$, where $\sim$ is the equivalence relation that identifies two points of $p^{-1}(A)$, if their images under the isomorphism map f have the same second coordinates.

(b) $\tilde{p}: E/p^{-1}(A) \to X/A$ is defined by

$$\tilde{p}(\{e\}) = \begin{cases} p(e) & \text{if} \quad p(e) \notin A \\ \text{the point } A/A \text{ in } X/A & \text{if} \quad p(e) \in A. \end{cases}$$

(Here X/A means the quotient space of X under the equivalence relation that identifies all of A with a single point, and A/A simply refers to that point.)

Proof   It is clear from the definitions that $E/p^{-1}(A)$ is a space, $\tilde{p}$ is a continuous map onto X/A, and $\tilde{p}^{-1}(\text{pt})$ is homeomorphic to $\mathbb{R}^n$ for any point, which again we simply abbreviate pt. It is also obvious that if O_α is an open set, from the definition of (E, X, p), that does not meet A, then $\tilde{p}^{-1}(O_\alpha) = p^{-1}(O_\alpha) \equiv O_\alpha \times \mathbb{R}^n$. We must check that $\tilde{p}$ is locally a Cartesian product in a neighborhood of A/A; i.e., we must find an open $O \subseteq X/A$, $A/A \in O$, with a suitable homeomorphism $\tilde{p}^{-1}(O) \xrightarrow{\cong} O \times \mathbb{R}^n$. Now, according to step (b) in the proof of Proposition 4.3, the isomorphism $f: p^{-1}(A) \to A \times \mathbb{R}^n$ extends to an isomorphism f_1 of bundles over some open set U, with $A \subseteq U$. Let O be the image of U in X/A; it follows immediately, from the definition of the quotient topology, that O is open. We then see at once that f_1 yields the desired homeomorphism

$$\tilde{p}^{-1}(O) \to O \times \mathbb{R}^n$$

It is routine to check that the coordinate transformation maps $g_{\alpha\beta}$ for this new bundle are linear and have the properties required by Definition 4.5.

Definition 4.16   Let (E, X, p) be a bundle and $A \subseteq X$ a closed subset.
(a)   An isomorphism of bundles $f: E_A \to A \times \mathbb{R}^n$ (just as in the previous lemma) is called a *trivialization* of E over A.

(b)   Two trivializations of E over A,

$$f, g : E_A \to A \times \mathbb{R}^n$$

are called *homotopic* if there is a trivialization of the n-dimensional vector bundle $(E \times I, X \times I, p \times 1_I)$, $I = [0, 1]$, over the subspace $A \times I$, written $F : (E \times I)_{A \times I} \to (A \times I) \times \mathbb{R}^n$, such that if $e \in p^{-1}(A)$,

$$F(e, 0) = f(e) \qquad \text{and} \qquad F(e, 1) = g(e).$$

Clearly, the concept is inspired by the notion of homotopic maps and offers a relevant way to compare trivializations. Note also that for any continuous map $l : X \to \mathrm{Gl}(n; \mathbb{R})$, the map

$$f_l : X \times \mathbb{R}^n \to X \times \mathbb{R}^n$$

defined by $f_l(x, u) = (x, l(x)u)$ is a trivialization of the vector bundle that is its domain; consequently, there is a great wealth of possible trivializations of a vector bundle, and hence a need to look at homotopic ones.

The next lemma shows that if we replace one trivilazation by a homotopic one then the construction of Lemma 4.1 does not suffer.

Lemma 4.2   If f and g are homotopic trivializations of (E, X, p) over the closed subset A of the compact Hausdorff space X, then the bundles $E/p^{-1}(A)$ constructed by Lemma 4.1 from either f or g are isomorphic.

Proof   Define

$$\rho : (X/A) \times I \to X \times I/A \times I$$

by the formula $\rho(\{x\}, t) = \{x, t\}$. It is the obvious map induced from the projection $X \times I \to X \times I/A \times I$.

If we use Lemma 4.1 on the trivialization of $E \times I$ over $A \times I$, guaranteed by the assumption that f and g are homotopic (Definition 4.16b), then we may define a bundle $(E \times I)/(p \times 1_I)^{-1}(A \times I)$ over $X \times I/A \times I$. We may then form the induced bundle

$$\rho^{-1}((E \times I)/(p \times 1_I)^{-1}(A \times I))$$

over the space $(X/A) \times I$. We note that the restriction of this bundle to $(X/A) \times \{0\}$ is precisely the construction of Lemma 4.1 according to the map f, while the restriction to $(X/A) \times \{1\}$ is that construction according to g (direct verification). We might also recall [Remark (1) after Definition 4.12] that the restriction of a bundle to a subspace is precisely the same bundle as the induced bundle formed by the inclusion map from the subspace to the whole space.

Therefore, because of Proposition 4.3, we shall only need to check that the inclusion maps

$$i_0 : X/A \to (X/A) \times I \qquad \text{and} \qquad i_1 : X/A \to (X/A) \times I$$

defined by $i_j(\{e\}) = (\{e\}, j)$ are homotopic. But that is completely trivial, for if we define

$$G : (X/A) \times I \to (X/A) \times I$$

to be the identity, $G(\{e\}, t) = (\{e\}, t)$, then $G(\{e\}, 0) = i_0(\{e\})$ and $G(\{e\}, 1) = i_1(\{e\})$, completing the proof.

As an application, we may now prove the following result:

Lemma 4.3 Let X be a compact Hausdorff space and A a closed subspace. We suppose that A is contractible; that is, A has the homotopy type of a single point. Let $f : X \to X/A$ be the projection map.
Then the map

$$\bar{f} : v^n(X/A) \to v^n(X)$$

defined by $\bar{f}(\{E\}) = \{f^{-1}E\}$, the induced bundle construction, is a 1–1 correspondence of sets.

Proof We need to construct an inverse for $\bar{f}$. Let (F, X, q) be an n-dimensional vector bundle over X. Then the restriction, F_A is automatically trivial by Remark (3) following Proposition 4.3. Let two trivializations

$$h : F_A \to A \times \mathbb{R}^n, \qquad g : F_A \to A \times \mathbb{R}^n$$

be given. Then $g \circ h^{-1}$ is an isomorphism of trival bundles, so that (by Definition 4.10) there is a (necessarily continuous) map

$$\gamma : A \to \mathrm{Gl}(n; \mathbb{R})$$

such that

$$g \circ h^{-1}(a, y) = (a, \gamma(a)y).$$

Because A is contractible, 1_A is homotopic to a constant map. But then $\gamma = \gamma \circ 1_A$ must also be homotopic to a constant map. Now $\mathrm{Gl}(n; \mathbb{R})$ has two pathwise connected components. If we require h and g to preserve orientation on $q^{-1}(\mathrm{pt})$ for any fixed point of A, $g \circ h^{-1}$ will lie in the component of the identity matrix. (This is no loss of generality because $h : F_A \to A \times \mathbb{R}^n$ can be followed by $1_A \times r$, when $r(x_1, \ldots, x_n) = (-x_1, x_2, \ldots, x_n)$, in case of reversed orientation.) But then γ is homotopic to a constant map in the path component of the identity matrix. Any two such maps into a path connected space are homotopic (reparametrize the path to have length 1 and move down the path), so γ is homotopic to the constant map that takes all A to the identity matrix. We conclude at once that $g \circ h^{-1}$ is homotopic to the identity 1.

But if $g \circ h^{-1} \simeq 1, g \simeq 1 \circ h = h$, or $g \simeq h$. Thus, we may use Lemma 4.2 to conclude that—up to isomorphism—there is a unique bundle $F/q^{-1}(A)$ over X/A. We must verify that this construction is the inverse of the map $\bar{f}$ so as to get our 1–1 correspondence or isomorphism of sets.

If we are given (F, X, q), then $f^{-1}(F/q^{-1}(A))$ is precisely

$$\{(x, \{e\}) | x \in X, \quad \{e\} \in F/q^{-1}(A), \quad f(x) = \tilde{q}(\{e\})\}.$$

But since $\tilde{q}(\{e\}) = q(e)$, the condition so expressed simply means $q(e) = f(x)$. It is clear that

$$\phi : (F, X, q) \to f^{-1}(F/q^{-1}(A))$$

defined by the formula $\phi(e) = (q(e), \{e\})$ is a map of bundles. Now, if $x \notin A$, ϕ restricted to $q^{-1}(x)$ is trivially an isomorphism of vector spaces. But when $x \in A$, it is also an isomorphism, because $F/q^{-1}(A)$ is defined by an equivalence relation that keeps the second coordinates in $A \times \mathbb{R}^n$ fixed. Therefore, ϕ is an isomorphism of vector bundles. This shows that $\bar{f}$ applied to the construction $F/q^{-1}(A)$ yields the identical bundle, up to isomorphism, to that which we started with.

To complete the proof that we have an inverse map to $\bar{f}$, we must show that $\bar{f}(E)$ followed by our construction from Lemma 4.1, that is, $\bar{f}(E)/p^{-1}(A)$, is isomorphic to E for any bundle E over X/A. If we take $(E, X/A, p)$ and form $f^{-1}E$, this then consists of

$$\{(x, e) | x \in X, \quad e \in E, \quad p(e) = f(x)\}.$$

If we pass to $f^{-1}E/p^{-1}(A)$, we may form a map

$$\psi : f^{-1}E/p^{-1}(A) \to E$$

by $\psi(\{x, e\}) = e$. This is well defined because different but equivalent pairs (x, e), with $x \in A$, have the same second coordinate after applying the trivialization. Once again, one may check that ψ is an isomorphism on every fiber; for example, if we apply ψ to say $\tilde{p}^{-1}(A/A)$, our formula for ψ is the identity map. Hence $f^{-1}E/p^{-1}(A)$ is isomorphic to E, as desired.

Remark   For a large class of spaces, A contractible implies that X and X/A have the same homotopy type. In such cases, Lemma 4.3 would follow at once from Remark (2) after Proposition 4.3. (This is a well known, but infrequently published fact in algebraic topology. The reader may consult the general material in [113].)

As an easy example, take the plane $\mathbb{R}^2$, let the closed interval $A = [-2,2]$ live in the x-axis, and let B be the union of two closed line segments, from $(-1,0)$ to $(0,1)$ and from $(+1,0)$ to $(0,1)$. Let $X = A \cup B$. Then X/A is homeomorphic to S^1, so that $v^n(X) = v^n(S^1)$ for all $n \geq 0$.

The construction of Lemma 4.1 is often referred to as *shrinking* or *collapsing*. We must now study in a similar fashion another basic construction, often called *gluing* or *clutching*.

Lemma 4.4 Let X be a compact Hausdorff space with X_1 and X_2 being closed subsets. Suppose $X_1 \cap X_2$ is not empty; i.e., $X_1 \cap X_2 \neq \phi$.

We shall assume that we are given two n-dimensional vector bundles (E_1, X_1, p_1) and (E_2, X_2, p_2) and an isomorphism of the bundles restricted to $X_1 \cap X_2$, say

$$f : (E_1)_{X_1 \cap X_2} \xrightarrow{\approx} (E_2)_{X_1 \cap X_2}.$$

Then there is an n-dimensional vector bundle

$$E_1 \cup_f E_2$$

over $X_1 \cup X_2$ whose restriction to X_1 (resp. X_2) is isomorphic to E_1 (resp. E_2). We shall write the projection map as

$$p_{1,2} : E_1 \cup_f E_2 \to X_1 \cup X_2.$$

Proof This is quite straightforward. We introduce an equivalence relation in the disjoint union $E_1 \cup E_2$ by asking that $e_1 \sim e_2$, if either (a) $e_1 = e_2$, or (b) $e_1 \in (E_1)_{X_1 \cap X_2}$, $e_2 \in (E_2)_{X_1 \cap X_2}$, and $f(e_1) = e_2$, or (c) precisely the same as (b) except that the subscripts are interchanged and f is replaced with f^{-1}. Set $E_1 \cup_F E_2$ to be the quotient space of the disjoint union by this equivalence relation. Because f is a homomorphism (indeed isomorphism) of vector bundles, one may define the obvious projection map $p_{1,2} : E_1 \cup_f E_2 \to X_1 \cup X_2$, by $p_{1,2}(e) = p_1(e)$ if $e \in E_1$, otherwise $p_2(e)$.

One should verify local triviality, i.e., the existence of suitable open sets O_α, with $p^{-1}(O_\alpha)$ homeomorphic to $O_\alpha \times \mathbb{R}^n$. First, if $x \in (X_1 \cup X_2) - (X_1 \cap X_2)$, we choose an open neighborhood O'_α of x that lies in one of the sets X_1 or X_2, but not both. If necessary, choose a smaller open set O_α over which the relevant bundle, E_1 or E_2, is trivial; i.e., say $p_1^{-1}(O_\alpha) = O_\alpha \times \mathbb{R}^n$. Then, it is clear from our definition that $E_1 \cup_f E_2$ is also trivial over O_α.

On the other hand, if $x \in X_1 \cap X_2$, one begins with a closed set $\bar{O}_\alpha$ whose interior contains x over which, say, E_1 is trivial. Using the isomorphism f and part (b) of the proof of Proposition 4.3, one easily sees that this trivialization may be extended from a trivialization of E_1 over $\bar{O}_\alpha \cap X_1 \cap X_2$ to a trivialization of E_2 over $(O_\alpha \cap X_1 \cap X_2) \cup (O_\beta \cap X_1 \cap X_2)$, for some

$O_\beta \subseteq X_2$, $x \in O_\beta$, over which E_2 is trivial, i.e., isomorphic to the bundle $O_\beta \times \mathbb{R}^n$. But then it is clear that we have a trivialization of $E_1 \cup_f E_2$ over $O_\alpha \cup O_\beta$.

The final assertion, about the restrictions to X_1 and X_2, is immediate.

Remark If $g_1: E_1 \to E_1'$ is an isomorphism of vector bundles over X_1, $g_2: E_2 \to E_2'$ is an isomorphism of vector bundles over X_2, and we have two isomorphisms of the restrictions

$$f: (E_1)_{X_1 \cap X_2} \to (E_2)_{X_1 \cap X_2}, \qquad f': (E_1')_{X_1 \cap X_2} \to (E_2')_{X_1 \cap X_2}$$

that satisfy $g_2 \circ f = f' \circ g_1 | X_1 \cap X_2$, then g_1 and g_2 define an isomorphism

$$E_1 \cup_f E_2 \xrightarrow{\approx} E_1' \cup_{f'} E_2'.$$

Definition 4.17 With the above data, suppose that we are given two isomorphisms

$$f_0, f_1: (E_1)_{X_1 \cap X_2} \to (E_2)_{X_1 \cap X_2}.$$

We say that these isomorphisms are *homotopic* if, when we write $\pi: X \times I \to X$ for the projection, we have an isomorphism

$$H: (\pi^{-1} E_1)_{(X_1 \cap X_2) \times I} \to (\pi^{-1} E_2)_{(X_1 \cap X_2) \times I}$$

that restricts to f_0 and f_1 on the endpoints, that is, the values 0 and 1 in $I = [0, 1]$.

We need one more basic lemma.

Lemma 4.5 Two homotopic isomorphisms

$$f_0, f_1: (E_1)_{X_1 \cap X_2} \to (E_2)_{X_1 \cap X_2}$$

determine *glued bundles* that are isomorphic, that is,

$$E_1 \cup_{f_0} E_2 \approx E_1 \cup_{f_1} E_2.$$

Proof Define $i_t: X \to X \times I$ by setting $i_t(x) = (x, t)$, so that i_0 and i_1 are the inclusions of the faces of the cylinder.

Because f_0 and f_1 are homotopic, we have an isomorphism

$$H: (\pi^{-1} E_1)_{(X_1 \cap X_2) \times I} \to (\pi^{-1} E_2)_{(X_1 \cap X_2) \times I}.$$

At any specific value of $t \in I$, H describes an isomorphism

$$H_t: (E_1)_{X_1 \cap X_2} \to (E_2)_{X_1 \cap X_2},$$

which is the same as the isomorphism obtained after applying i_t^{-1} to each bundle (because the restriction of a bundle to a subspace is the same as the induced bundle obtained from the inclusion map of the subspace).

I claim that for any $t \in I$

$$E_1 \cup_{H_t} E_2 \approx i_t^{-1}((\pi^{-1}E_1) \cup_H (\pi^{-1}E_2)).$$

For given $e \in E_1 \cup_{H_t} E_2$, if $p_{1,2}(e) \notin X_1 \cap X_2$, we may map e to the same point, in the restriction of $\pi^{-1}E_1 \cup_H \pi^{-1}E_2$ to $\pi^{-1}E_1$ or $\pi^{-1}E_2$ at time t. But if $p_{1,2}(e) \in X_1 \cap X_2$, since the identification map H in $\pi^{-1}E_1 \cup_H \pi^{-1}E_2$ is precisely H_t for this value t, the desired map from $E_1 \cup_{H_t} E_2$ to $i_t^{-1}((\pi^{-1}E_1) \cup_H (\pi^{-1}E_2))$ may therefore be chosen to be the identity. The resulting map is visibly an isomorphism of each fiber, as desired.

Returning to $i_t : X \to X \times I$, clearly all these maps i_t are homotopic (by simple reparametrization). By Proposition 4.3, all the bundles $i_t^{-1}((\pi^{-1}E_1) \cup_H (\pi^{-1}E_2))$ are isomorphic for any choice of t. Taking $t = 0$ or 1 and using the isomorphism from the previous paragraph gives our desired result.

We now have completed our analysis of how to build bundles over unions, and we have a good understanding on how to cope with contractible subspaces, or subspaces over which a given bundle is trivial. The most logical problem to tackle next, indeed an important case of the classification theorem, is the problem of classifying bundles over a space that is the union of two contractible subspaces. The simplest example is the sphere S^n, which is the union of the closed upper and lower hemispheres D_+^n and D_-^n, each of which is homeomorphic to the disk

$$D^n = \{X \in \mathbb{R}^n \,|\, \|X\| \leq 1\}.$$

Of course D^n is a convex subset of $\mathbb{R}^n$ and hence is contractible.

The classification of bundles over spheres was achieved in the 1930s, and with little more difficulty we may extend it to bundles over suspensions, defined as follows:

Definition 4.18 If X is a topological space, we define the suspension of X, written SX, to be the quotient of $X \times I$ by the equivalence relation

$$(x, 1) \sim (y, 1) \qquad \text{for all} \quad x, y \in X,$$
$$(u, 0) \sim (v, 0) \qquad \text{for all} \quad u, v \in X.$$

In other words, we collapse (separately) the upper and lower faces of $X \times I$ to single points. We observe that the subspaces of SX defined by

$$C_+ X = \{\{(x, t)\} \in SX \,|\, t \geq \tfrac{1}{2}\} \qquad \text{and} \qquad C_- X = \{\{(x, t)\} \in SX \,|\, t \leq \tfrac{1}{2}\}$$

are both cones over the set of all $\{(x, \frac{1}{2})\}$, which is obviously homeomorphic to X because no distinct pairs (x, t) with $t \neq 0$, $t \neq 1$ are equivalent. In terms of symbols, we then may write

$$SX = C_+X \cup C_-X \qquad \text{and} \qquad X = C_+X \cap C_-X.$$

Remark   If X is nonempty, SX is always pathwise connected. (Check!) If X is a compact, Hausdorff space, then so is SX. Naturally, $S(S^n) \equiv S^{n+1}$.

Proposition 4.4   If X is a compact Hausdorff space and $x_0 \in X$, then there is a natural isomorphism of sets

$$v^n(SX) \approx [X; \mathrm{Gl}(n; \mathbb{R})]_0 / \sim,$$

where $[X, \mathrm{Gl}(n; \mathbb{R})]_0$ is the set of those homotopy classes of maps $X \to \mathrm{Gl}(n; \mathbb{R})$ that send x_0 into the identity matrix, and $\sim$ identifies $\{g\}$ with the class of $x \to ag(x)a^{-1}$ for any $a \in \mathrm{Gl}(n; \mathbb{R})$ not in the path component of the identity. Since any two such a may be joined by a path, yielding a homotopy, the choice of a is immaterial.

Proof   We shall construct two maps between these sets that are inverses of one another. We begin with the easier.

Given a homotopy class $\{g\}$ in $[X; \mathrm{Gl}(N; \mathbb{R})]_0$, we consider the trivial bundles $C_+X \times \mathbb{R}^n$ and $C_-X \times \mathbb{R}^n$ over C_+X and C_-X. We define an isomorphism f over the restrictions to $C_+X \cap C_-X = X$, by

$$(x, u) \to (x, g(x)u).$$

Using Lemma 4.4, we see that we have an n-dimensional vector bundle over SX, namely

$$(C_+X \times \mathbb{R}^n) \cup_f (C_-X \times \mathbb{R}^n).$$

Observe that Lemma 4.5 assures us that the construction does not depend on the choice of g, in its homotopy class (check). If we replace g by an equivalent map, in the sense of $\sim$, then we change the orientation on each piece of the bundle, $C_+X \times \mathbb{R}^n$ and $C_-X \times \mathbb{R}^n$, but the resulting bundle is of course isomorphic to the original. We write this class of bundles, coming from $\{g\}$, as $\tau(\{g\})$.

To define a second map, going in the other direction, we let (E, SX, p) be a bundle over SX. The subspaces C_+X and C_-X being contractible, it follows at once that the restricted bundles $(E)_{C_+X}$ and $(E)_{C_-X}$ are trivial; we choose two trivializations

$$f^+ : (E)_{C_+X} \to C_+X \times \mathbb{R}^n \qquad \text{and}$$

$$f^- : (E)_{C_-X} \to C_-X \times \mathbb{R}^n.$$

We assume that f^- is chosen such that f^+ and f^- are the same map of $p^{-1}(\{(x_0, \frac{1}{2})\})$ to $\mathbb{R}^n$. Write $f_0^+ = f^+|p^{-1}(C_+X \cap C_-X)$ and $f_0^- = f^-|p^{-1}(C_+X \cap C_-X)$.

Now, the map

$$(f_0^+) \circ (f_0^-)^{-1}: X \times \mathbb{R}^n \to X \times \mathbb{R}^n$$

is clearly a bundle map of trivial bundles. By Definition 4.10, we must have a (continuous) map $g: X \to \text{Gl}(n; \mathbb{R})$, so that

$$(f_0^+) \circ (f_0^-)^{-1}(x, u) = (x, g(x)u).$$

Because of the way in which we chose f^+ and f^-, it follows that $g(x_0)$ is the identity matrix.

It is now necessary to see that this construction is independant of our choices of trivialization. Suppose therefore that f^+ was replaced by another trivialization $g^+: (E)_{C_+X} \to C_+X \times \mathbb{R}^n$ preserving the original map on the inverse image of $(x_0, \frac{1}{2})$. Then as before

$$f^+ \circ (g^+)^{-1}: C_+X \times \mathbb{R}^n \to C_+X \times \mathbb{R}^n$$

defines a map $\gamma: C_1 X \to \text{Gl}(n; \mathbb{R})$, which (because C_+X is contractible) must be homotopic to a constant map. Our conventions to the effect that f^+ and g^+ agree on one fiber (inverse image of $(x_0, \frac{1}{2})$) show that this map may be assumed constant at the identity matrix. Restricting attention to $C_+X \cap C_-X = X$, the map $f_0^+(g_0^+)^{-1}$, which is precisely the map

$$(x, u) \to (x, \gamma(x)u),$$

must be homotopic as an isomorphism of bundles (Definition 4.17), to the identity map. We may also easily do this in such a way that $\gamma(x_0)$ remains at the identity matrix during the homotopy.

It follows at once that the two bundle isomorphisms

$$g_0^+ \circ (f_0^-)^{-1} \qquad \text{and} \qquad (f_0^+) \circ (g_0^+)^{-1} \circ (g_0^+) \circ (f_0^-)^{-1} = (f_0^+) \circ (f_0^-)^{-1}$$

are homotopic isomorphisms, and therefore the homotopy class of g is unchanged if f^+ is replaced by g^+. A similar argument may be given for f^-. It is easy to see that choosing the reverse orientations on $p^{-1}(C_+X)$ and $p^{-1}(C_-X)$ produces a map equivalent to g, in that it has the form $x \to a \circ g(x) \circ a^{-1}$ (for we can let a be a fixed orientation reversing the linear isomorphism of $\mathbb{R}^n$ to itself and proceed as above).

Therefore, the equivalence class of $g: X \to \text{Gl}(n; \mathbb{R})$ is well defined, starting from a bundle (E, SX, p). Using the remark after Lemma 4.4, one checks without difficulty that an isomorphic bundle yields the same element of $[X, \text{Gl}(n; \mathbb{R})]_0/\sim$. Thus, we have our map

$$\sigma: v^n(SX) \to [X, \text{Gl}(n; \mathbb{R})]_0/\sim.$$

We must check that the compositions of τ and σ are the respective identities. Given $g : X \to \mathrm{Gl}(n; \mathbb{R})$, construct $f : X \times \mathbb{R}^n \to X \times \mathbb{R}^n$ by $f(x, u) = (x, g(x)u)$, and a bundle over $C_+ X \cup C_- X = SX$ built from the isomorphism f, i.e.,

$$E = (C_+ X \times \mathbb{R}^n) \cup_f (C_- X \times \mathbb{R}^n).$$

If we choose f^+ and f^- to be the identity maps in each case, we clearly get $\sigma(\{E\}) = \{g\}$, proving that $\sigma \circ \tau$ is the identity. Equivalent g's yield isomorphic bundles by reversing orientation, again getting equivalent g's.

On the other hand, given a bundle (E, SX, p), choose suitable trivializations f^+ and f^- and form g by $(f_0^+) \circ (f_0^-)^{-1}(x, u) = (x, g(x)u)$. To pass back to a bundle from g, one forms

$$(C_+ X \times \mathbb{R}^n) \cup_f (C_- X \times \mathbb{R}^n),$$

where $f = (f_0^+) \circ (f_0^-)^{-1}$. An isomorphism from (E, SX, p) to this latter bundle is formed by taking f^+ on $(E)_{C_+ X}$ and f^- on $(E)_{C_- X}$, compatibility on the overlap being immediate. This shows that $\tau \circ \sigma$ is also the identity, completing our proof.

Remark　The statement and proof of this proposition are complicated by the equivalence $\sim$, and automorphisms $a \circ g(x) \circ a^{-1}$. A simpler result may be had if one restricts attention to oriented vector bundles, whereby an orientation is chosen over some point and all coordinate transformations $g_{\alpha\beta}$ are required to lie in the connected component of the identity matrix (compare [103], as well as our problems and projects at the end of this chapter).

While Proposition 4.4 is a satisfactory classification theorem for understanding vector bundles over suspensions, we note that few spaces are suspensions, and therefore one needs a modified approach to get the general result. The next two lemmas deal with the splitting of bundles into separate pieces; they are important technical tools.

The following lemma is actually valid for paracompact X, although we shall need it only when X is a compact Hausdorff space.

Lemma 4.6　Assume we are given two vector bundles over X, say (E_2, X, p_2) and (E_3, X, p_3), and a map of bundles

$$f : E_2 \to E_3$$

(naturally, we assume $f_X = 1_X$) that maps every $p_2^{-1}(x)$ onto $p_3^{-1}(x)$, $x \in X$.
Then there is a subbundle E_1 of E_2 and an isomorphism

$$E_1 \oplus E_3 \approx E_2.$$

E_1 is the subset of elements in each $p_2^{-1}(x)$ that are mapped to $0 \in p_3^{-1}(x)$ (clearly a subbundle).

Proof Suppose that we are able to find a map of vector bundles

$$\rho : E_3 \to E_2$$

such that $f \circ \rho = 1_{E_3}$, where 1_{E_3} means the identity map on E_3. Let then $i : E_1 \to E_2$ be the inclusion of the subbundle E_1, defined in the statement of the lemma, into E_2. We form the map

$$i + \rho : E_1 \oplus E_3 \to E_2,$$

where $(i + \rho)(a, b) = i(a) + \rho(b)$ and addition takes place in the fiber over each point $p_2^{-1}(x)$. There is no difficulty in checking that $i + \rho$ is a map of bundles. If we temporarily restrict attention to points lying over a given $x_0 \in X$, we note the following:

(a) $i + \rho$ is 1–1, because if $(i + \rho)(e_1, e_3) = 0$, $i(e_1) + \rho(e_3) = 0$, and hence $f(i(e_1) + \rho(e_3)) = 0$. But since $f \circ i(e_1) = 0$, by definition of E_1, and $f \circ \rho(e_3) = e_3$, we must have $e_3 = 0$. But then, because $i(e_1) + \rho(e_3) = 0$, we have $i(e_1) = 0$, or $e_1 = 0$. Thus, $(i + \rho)(e_1, e_3) = 0$ implies $e_1 = 0$, $e_3 = 0$, proving the claim.

(b) Given $e_2 \in p_2^{-1}(x_0) \subseteq E_2$, we observe that $f(\rho(e_3)) = e_3$, so that $f(e_2 - \rho \circ f(e_2)) = f(e_2) - f(e_2) = 0$. Then $e_2 - \rho \circ f(e_2)$ belongs to E_1, so we have

$$(i + \rho)(e_2 - \rho \circ f(e_2), f(e_2)) = e_2 - \rho \circ f(e_2) + \rho \circ f(e_2) = e_2.$$

Thus, $i + \rho$ is onto.

We conclude that for any $x_0 \in X$ $i + \rho$ is an isomorphism of fibers over x_0, and hence $i + \rho$ is an isomorphism of vector bundles. Thus, we have to construct our map ρ, often called a right inverse for f. We shall use a partition of unity argument, passing from the local to the global.

For every $x \in X$, select an open neighborhood O_x such that both bundles E_2 and E_3 are trivial over O_x, i.e.,

$$\phi_2 : p_2^{-1}(O_x) \overset{\equiv}{\to} O_x \times \mathbb{R}^n, \qquad \phi_3 : p_3^{-1}(O_x) \overset{\equiv}{\to} O_x \times \mathbb{R}^k.$$

Naturally, $k \leq n$. The map

$$\phi_3 \circ f \circ \phi_2^{-1} : O_x \times \mathbb{R}^n \to O_x \times \mathbb{R}^k$$

is onto.

Let $\mathbb{R}^{n-k}$ be the subspace of $\mathbb{R}^n$, annihilated by $\phi_3 \circ f \circ \phi_2^{-1}$, for first coordinate x; let π be the projection of $\mathbb{R}^n$ onto $\mathbb{R}^{n-k}$. Form the sum map

$$h = (\phi_3 \circ f \circ \phi_2^{-1} + 1 \times \pi) : O_x \times \mathbb{R}^n \to (O_x \times \mathbb{R}^k) \oplus (O_x \times \mathbb{R}^{n-k}).$$

Of course at x the map h is visibly onto, and for this fixed x the domain and range are vector spaces of the same dimension. Thus, h is an isomorphism of vector spaces at x. By step (b) in the proof of Proposition 4.3, h is an isomorphism of vector bundles in a neighborhood of x.

Therefore, in this possibly smaller neighborhood a right inverse to $\phi_3 \circ f \circ \phi_2^{-1}$ may be found by taking the composition of the inclusion of $O_x \times \mathbb{R}^k$ into the Whitney sum $(O_x \times \mathbb{R}^k) \oplus (O_x \times \mathbb{R}^{n-k})$ and then applying h^{-1}. $(1 \times \pi$ annihilates such an element, and h applied to it is the same as $\phi_3 \circ f \circ \phi_2^{-1}$ applied to it.)

Using the homeomorphisms ϕ_2 and ϕ_3, we easily convert this to a right inverse for

$$f : p_2^{-1}(O) \to p_3^{-1}(O)$$

for some suitably small neighborhood of x, say O.

Finally, choose a finite cover by such O's, a partition of unity, and add up the products of the functions in the partition of unity and the right inverses for f in the respective sets.

Lemma 4.7   Let (E, X, p) be a vector bundle over a compact Hausdorff space. Then there is an integer $N > 0$ and an onto map of bundles

$$f : X \times \mathbb{R}^N \to E.$$

That is, f is a map of bundles, which is a linear transformation that maps the points over x in the domain onto those over x in the range, for all $x \in X$. We continue to assume $f_X = 1_X$.

Proof   This, too, is a rather standard localization argument. Let $S(E)$ be the vector space of section of E, i.e., continuous maps $s : X \to E$ with $p \circ s = 1_X$. Addition and multiplication are carried out in each set $\psi_\beta(p^{-1}(x))$ just as they would be in a vector space and then carried back to E by ψ_β^{-1}. Of course $S(E)$ is almost always an infinite-dimensional vector space over the real numbers.

But note that there is a natural map

$$f_1 : X \times S(E) \to E$$

defined by $f_1(x, s) = s(x)$; such a map is often called an evaluation map.

If we can find a finite-dimensional subspace of $S(E)$, say F, such that the restriction of f_1 to $X \times F$ maps $X \times F$ onto E, and if we can show that f_1 is linear on each fiber, i.e., linear over each point $x \in X$, we will be done. But the linearity on each fiber is immediate, because if $\psi_\beta : p^{-1}(O_\beta) \xrightarrow{\cong} O_\beta \times \mathbb{R}^n$

is a homeomorphism coming from the definition of E and $x \in O_\beta$, then the composition

$$X \times S(E) \xrightarrow{f_1} E \xrightarrow{\psi_\beta} O_\beta \times \mathbb{R}^n$$

is defined for given $x \in O_\beta$ and $s \in S(E)$, and is given by

$$(\psi_\beta \circ f_1)(x, s) = \psi_\beta(s(x));$$

therefore,

$$(\psi_\beta \circ f_1)(x, s_1 + s_2) = \psi_\beta(s_1(x) + s_2(x)) = \psi_\beta(s_1(x)) + \psi_\beta(s_2(x))$$

because of the definition of addition. Similarly, $(\psi_\beta \circ f_1)(x, as) = a(\psi_\beta \circ f_1)(x, s)$, for any real number a. (Compare Definition 4.10 with the identity map on X.)

Thus, we are reduced to locating our finite-dimensional subspace F. For each point $x \in X$, choose a suitable O_{β_j} and define sections $s_1^j, \ldots, s_n^j$ as follows:

$$s_i^j(x) = \psi_{\beta_j}^{-1}(x, (0, \ldots, 1, \ldots, 0)),$$

where the 1 occurs in the ith place. Notice that the $s_1^j, \ldots, s_n^j$ form a basis for $p^{-1}(x)$ for any point $x \in O_{\beta_j}$, and that if α is any positive number, $\alpha s_1^j, \ldots, \alpha s_n^j$ would also be a basis for $p^{-1}(x)$. Select a finite cover by $O_{\beta_1}, \ldots, O_{\beta_m}$ and a partition of unity $\{\phi_j\}$ subordinate to this cover. Then the collection of sections obtained by multiplying all the sections obtained for the $O_{\beta_1}, \ldots, O_{\beta_m}$ by the respective $\phi_1, \ldots, \phi_m$ gives a finite family of functions, so that for any $y \in X$ the family contains a positive multiple of a basis for $p^{-1}(y)$. If we choose F to be the subspace spanned by these sections, $\phi_j(x)s_1^j(x), \ldots, \phi_j(x)s_n^j(x)$, for $j = 1, \ldots, m$, then f_1 clearly maps $X \times F$ onto E, completing our proof.

Corollary 4.1 If E is a vector bundle over a compact Hausdorff space X, then there is another vector bundle over X, say E_1, such that $E \oplus E_1$ is trivial.

Proof Apply Lemma 4.6 to the conclusion to Lemma 4.7.

Remarks Corollary 4.1 suggests an interesting algebraic interpretation for vector bundles. We think of all trivial vector bundles as a sort of zero element, and the Whitney sum of bundles as an addition. Then the previous corollary shows that we have an additive inverse operation. With proper definitions this approach gives rise to the important K-theory (see [4, 88]), which asociates an abelian group with every X. Unfortunately, this is largely beyond the scope of this book.

We are now ready to examine the *universal bundles* and to tackle the basic classification theorem.

Definition 4.19   We recall (from Definition 2.4) that $G_{n,m}$ is the Grassman manifold of n-dimensional vector subspaces of $\mathbb{R}^{n+m}$.

(a)   If $f: X \times \mathbb{R}^k \to E$ is an onto mapping of vector bundles over X (compact Hausdorff here) and dim $E = n$, we define a map

$$f': X \to G_{k-n,n}$$

that assigns to each point $x \in X$ the null space or kernel of the linear transformation f as a map of the set $x \times \mathbb{R}^k$ to $p^{-1}(x)$.

(b)   Set $F \subseteq G_{k-n,n} \times \mathbb{R}^k$ to be subspace consisting of all (u, x) with u containing the point x; this makes sense because u is a $(k - n)$-dimensional subspace of $\mathbb{R}^k$. F is trivially checked to be a subbundle of the trivial vector bundle $(G_{k-n,n} \times \mathbb{R}^k, G_{k-n,n}, p)$ with p having projection on the first factor.

We form the quotient bundle

$$E_{k-n,n} = G_{k-n,n} \times \mathbb{R}^k/F.$$

It is an n-dimensional vector bundle over $G_{k-n,n}$, which we write

$$(E_{k-n,n}, G_{k-n,n}, p_n).$$

It is called the *universal* or *classifying* bundle for n-dimensional vector bundles.

Remarks   (1)   The map in Definition 4.19a has some remarkable properties. For example, if F is any n-dimensional bundle over X and the bundle map

$$f: X \times \mathbb{R}^k \to F$$

is onto, then

$$f': X \to G_{k-n,n}$$

has the property that

$$f'^{-1}(E_{k-n,n}) = F,$$

where $E_{k-n,n}$ is the universal bundle defined in Definition 4.19b. This is immediate because the left-hand side is by definition

$$\{y, \{u, t\} \mid y \in X, (u, t) \in G_{k-n,k} \times \mathbb{R}^k, \qquad f'(y) = u\},$$

that is, the quotient in which are identified to a point those pairs $(y, t) \in X \times \mathbb{R}^k$ such that all $t \in f'(y)$, i.e., $f(t) = 0$. But these pairs, with this identification, yield precisely the quotient of $X \times \mathbb{R}^k$ with the subbundle, which is the null space of f, identified to a point; since f is onto, this is isomorphic to F (check).

(2) Let $M^n \subseteq \mathbb{R}^k$ be a smooth submanifold, TM^n and NM^n be the tangent and normal bundles. As usual, let $T\mathbb{R}^k | M^n$ be the restriction of the tangent bundle to $\mathbb{R}^k$ (which is trivial) to the submanifold M^n. Then there is a natural projection

$$f: T\mathbb{R}^k | M^n \to TM^n$$

equivalent to taking the quotient bundle of the domain by NM^n, to which we may apply Definition 4.19a. The resulting map f' assigns to each $x \in M^n$ the translation of the normal space at x down to the origin. The induced bundle from the universal bundle must then be, according to Remark (1) above, isomorphic to the tangent bundle to M^n.

There remains one small hurdle between us and the statement of the main classification theorem. That is, we must consider the limit of a family of Grassman manifolds. For any $k > 0$, let

$$p: \mathbb{R}^{n+k} \to \mathbb{R}^{n+k-1}$$

be the projection map that ignores the last coordinate. If l is an $(n-1)$-dimensional subspace of $\mathbb{R}^{n+k-1}$, its inverse image $p^{-1}(l)$ is an n-dimensional subspace of $\mathbb{R}^{n+k}$. This procedure clearly defines a continuous inclusion map

$$j_{n-1}: G_{n-1,k} \to G_{n,k}.$$

It is easy to check that if $(E_{k-n,n}, G_{k-n,n}, p_n)$ is the universal or classifying bundle (Definition 4.19b) then the bundle induced from this by j_{k-n-1} is precisely $(E_{k-n-1,n}, G_{k-n-1,n}, p_n)$.

Now j_{n-1} defines a map of sets of homotopy classes by $(j_{n-1})_*(\{f\}) = \{j_{n-1} \circ f\}$, i.e.,

$$(j_{n-1})_*: [X, G_{n-1,k}] \to [X, G_{n,k}]$$

is defined by composition with j_{n-1}.

Definition 4.20   We define the *limit* (direct limit or injective limit)

$$\varinjlim [X, G_{i,m}]$$

to be the quotient of the disjoint union

$$\bigcup_i [X, G_{i,m}]$$

by the equivalence relation that identifies

$$\{f_{i_1}\} \in [X, G_{i_1,m}] \qquad \text{with} \qquad \{f_{i_2}\} \in [X, G_{i_2,m}]$$

when there is an $i_3 \geq \max(i_1, i_2)$ such that

$$(j_{i_3})_* \cdots (j_{i_1+1})_*(j_{i_1})_*(\{f_1\}) = (j_{i_3})_* \cdots (j_{i_2+1})_*(j_{i_2})_*(\{f_2\}).$$

In other words, two elements represent the same element, in the limit, if after a suitable number of $(j_s)_*$'s they have a common image in some $[X, G_{s,m}]$.

We note that there is a natural map

$$[X; G_{i,m}] \to \varinjlim[X, G_{i,m}]$$

that associates to every class of maps its equivalence class in the limit.

We may now state our main theorem.

Theorem 4.1   If X is a compact Hausdorff space, there is an isomorphism of sets (i.e., a 1–1 correspondence)

$$\varinjlim[X; G_{k-n,n}] \approx v^n(X).$$

(Here k is the variable for the purpose of taking the limit. We write $k - n$ because that form occurs naturally in Definition 4.19.)

Proof   An element in the limit is represented as a homotopy class

$$\{f\} \in [X, G_{k-n,n}]$$

for some k, perhaps very large. The induced bundle $f^{-1}(E_{k-n,n})$ is an n-dimensional vector bundle over X. By Proposition 4.3 and the fact that the maps j_{k-n-1} carry, by the induced bundle construction, universal bundles into universal bundles, we see that this construction yields a well-defined map

$$\Phi: \varinjlim[X, G_{k-n,n}] \to v^n(X).$$

We shall define a map

$$\Psi: v^n(X) \to \varinjlim[X, G_{k-n,n}]$$

that will be the inverse for Φ. Given an n-dimensional vector bundle E, we may use Lemma 4.7 to find a map

$$g: X \times \mathbb{R}^k \to E$$

that is onto. Using Definition 4.19a, we obtain a map

$$g': X \to G_{k-n,n}.$$

We shall show that for sufficiently large i the homotopy class of

$$(j_i)_* \cdots (j_{k-n})_*(\{g'\})$$

depends only on the original bundle, and not on the choice of g or k. For brevity write

$$(j_{i,k-n})_* = (j_i)_* \cdots (j_{k-n})_*.$$

Suppose that we are given two onto mappings of vector bundles

$$g_1 : X \times \mathbb{R}^{k_1} \to E \qquad \text{and} \qquad g_2 : X \times \mathbb{R}^{k_2} \to E.$$

For each t, $0 \le t \le 1$, form

$$G_t : X \times \mathbb{R}^{k_1} \times \mathbb{R}^{k_2} \to E,$$

defined by $G_t(x, y_1, y_2) = (1 - t)g_1(x, y_1) + tg_2(x, y_2)$. This makes sense because the operations take place in $p^{-1}(x) \subseteq E$, which is a vector space, for each x. It is completely trivial to check that this a bundle map, or homomorphism from $X \times \mathbb{R}^{k_1 + k_2}$ onto E. By definition 4.19, we arrive at a homotopy of maps

$$G'_t : X \to G_{k_1 + k_2 - n, n}$$

or

$$G' : X \times I \to G_{k_1 + k_2 - n, n}$$

defined by $G'(x, t) = G'_t(x)$.

It is clear that G'_0 is the same as $(j_{k_1 + k_2 - n, k_1 - n})_*$ applied to g'_1. But unfortunately G'_1 differs from the map $(j_{k_1 + k_2 - n, k_2 - n})_*$ applied to g'_2 in that G_1 constitutes projection onto the last, rather than the first, coordinates, acting on g'_2.

The proof will be complete if we can find a homotopy of the identity map of $G_{k_1 + k_2 - n, n}$ to a map that permutes the coordinates in such a way that the last k_2 coordinates become the first k_2 coordinates, etc. To do this, observe that for a map of $\mathbb{R}^{k_1 + k_2}$ to itself to induce a map of $G_{k_1 + k_2 - n, n}$ to itself the map must take linear subspaces, such as the $(k_1 + k_2 - n)$-dimensional subspace, to themselves. Any orthogonal transformation will have this property. Select an orthogonal transformation of $\mathbb{R}^{k_1 + k_2}$ that takes the last k_2 coordinates to the first k_2 coordinates. Assuming that $k_1 > 1$, which is no loss of generality here, we may further arrange the matter so that the determinant of the orthogonal transformation is $+1$. Because the orthogonal matrices of determinant $+1$ are pathwise connected, we may connect this orthogonal transformation (viewed as a matrix) to the identity by a path of orthogonal matrices of determinant $+1$. This gives a homotopy of our original orthogonal transformation, which takes the last k_2 coordinates to the first, to the identity transformation, and correspondingly a suitable homotopy of maps of $G_{k_1 + k_2 - n, n}$ to itself.

We conclude that

$$(j_{k_1 + k_2 - n, k_1 - n})_*(g'_1) \simeq (j_{k_1 + k_2 - n, k_2 - n})_*(g'_2),$$

showing that Ψ is well defined.

We must finally check that Φ and Ψ are inverses of one another. Consider a map $f: X \to G_{k-n,n}$; then $\Phi(\{f\})$ is represented by the induced bundle $f^{-1}(E_{k-n,n})$, where we recall that $E_{k-n,n}$ is the quotient of the trivial bundle $G_{k-n,n} \times \mathbb{R}^k$ by the subbundle made up of pairs (x, y) with $x \in G_{k-n,n}$ and $y \in x$. Of course, $f^{-1}(E_{k-n,n})$ is then a quotient bundle of the bundle induced by the map f from the trivial bundle $G_{k-n,n} \times \mathbb{R}^k$. Consider the projection map from $f^{-1}(G_{k-n,n} \times \mathbb{R}^k)$ to $f^{-1}(E_{k-n,n})$ defined by $(x, e) \to (x, \{e\})$, and call it g. Because

$$f^{-1}(E_{k-n,n}) = \{(x, \{e\}) \mid x \in X, \quad \{e\} \in E_{k-n,n}, \quad f(x) = p_n(e)\}$$

and

$$f^{-1}(G_{k-n,n} \times \mathbb{R}^k) = \{(x, e) \mid x \in X, \quad e \in \mathbb{R}^k, \quad f(x) = \pi(e)\},$$

π being projection on first factor, we see that for each $x \in X$ the null space or kernel of g consists of all $\omega \in \mathbb{R}^k$ such that $\omega \in f(x)$. But this means (by Definition 4.19a) that $g'(x) = f(x)$, and thus, by checking the definitions,

$$\Psi \circ \Phi(\{f\}) = \{f\}.$$

On the other hand, given the bundle (E, X, p), we select a map

$$g: X \times \mathbb{R}^k \to E$$

that is onto (lemma 4.7). Form $g': X \to G_{k-n,n}$ and consider

$$(g')^{-1}(E_{k-n,n}),$$

with $E_{k-n,n}$ the universal bundle as defined above. We then define

$$h: (g')^{-1}(E_{k-n,n}) \to E$$

by $h(x, (g'(x), \{y\})) = g(x, y)$, where $x \in X$, $(g'(x), \{y\}) \in E_{k-n,n}$, with $g'(x) = p_n(y)$. Because the map g annihilates its own kernel or nullspace, this is then trivially checked to be a well-defined bundle map. Now, given $z \in E$, we select $(x, \omega) \in X \times \mathbb{R}^k$ such that $g(x, \omega) = z$. Then immediately $h(x, g(x), \{\omega\}) = z$, proving that h is onto. But the domain and range of h are both n-dimensional vector bundles over X, so that, over each point x, h is an onto map from a finite-dimensional vector space to one of the same dimension. Since such a linear map is always an isomorphism, by basic linear algebra, we see that h is an isomorphism. This means that

$$\Phi \circ \Psi(\{E\}) = \{E\},$$

completing our proof.

Note that the last part of the proof points to the need to use the limit in the proof, for we cannot control how large a k is needed in the map $g: X \times \mathbb{R}^k \to E$ without some knowledge of X, nor can we say for what sort of k is a given class in the limit represented by a specific map.

EXAMPLES OF CLASSIFICATION

To achieve applications of the main theorem, one must naturally know facts about $[X, G_{k-n,n}]$, or more exactly

$$\varinjlim[X, G_{k-n,n}].$$

If X is in some sense a finite-dimensional space, then there will be a large $\bar{k}$ such that

$$[X, G_{k-n,n}] = \varinjlim[X, G_{k-n,n}].$$

Furthermore, $G_{k-n,n}$ is well known in many respects, so that one may often say a great deal about $v''(X)$.

I propose to list here, with precise references, many of the basic facts about $G_{k-n,n}$.

(i) $G_{k-n,n}$ is a homogeneous space (see [47, 103]). The orthogonal group $O(k)$, as in Remark (5) following Definition 4.2, acts on $\mathbb{R}^k$ as a transformation group in such a way that a linear space of a given dimension is mapped into a linear space of the same dimension. Hence a transformation from $O(k)$ maps $G_{k-n,n}$ into itself.

I claim that the action is transitive, for given two $(k - n)$-dimensional subspaces of $\mathbb{R}^k$, let us say y_1 and y_2, we choose bases for y_1 and y_2, say $v_1^1, \ldots, v_1^{k-n}$ and $v_2^1, \ldots, v_2^{k-n}$, as well as bases for the orthogonal complements, say $w_1^1, \ldots, w_1^n$ and $w_2^1, \ldots, w_2^n$. Without loss of generality, these bases may be assumed to be orthonormal (i.e., mutually perpendicular and of length 1). Define an orthogonal transformation

$$\phi: \mathbb{R}^k \to \mathbb{R}^k$$

by setting $\phi(v_1^i) = v_2^i$ and $\phi(w_1^j) = w_2^j$ and extending by linearity. It is then clear that when this orthogonal transformation acts on $G_{k-n,n}$ it takes y_1 to y_2, as we wished. We write this action as $\mu: O(k) \times G_{k-n,n} \to G_{k-n,n}$.

Since $O(k)$ is compact, $G_{k-n,n}$ is homeomorphic to

$$O(k)/G_0,$$

where $G_0 = \{g \in O(k) \mid \mu(g, x_0) = x_0\}$ for some fixed $x_0 \in G_{k-n,n}$ (see Proposition 4.1). We suppose, as a matter of convenience, that

$$x_0 = \{t_0, \ldots, t_k) \mid t_{k-n+1} = t_{k-n+2} = \cdots = t_k = 0\}.$$

Naturally, an orthogonal transformation will take the $(k - n)$-dimensional subspace to itself if and only if it takes the orthogonal complement $x_0^\perp$ to itself. The restriction of an orthogonal transformation to a subspace that it carries into itself is still an orthogonal transformation. It follows that G_0 in our basis must be the group of orthogonal matrices of the form

$$\left(\begin{array}{c|c} A_1 & 0 \\ \hline 0 & A_2 \end{array} \right),$$

where $A_1 \in O(k - n)$ and $A_2 \in O(n)$. In other words

$$G_0 \equiv O(k - n) \times O(n),$$

and identifying these groups as subgroups of $O(k)$ as we have done, we may write

$$G_{k-n,n} = O(k)/O(k - n) \times O(n).$$

(ii)   For certain nice spaces X, one may say a great deal about $[X, G_{k-n,n}]$. These X are built up out of closed disks $D^m = \{X \in \mathbb{R}^m \mid \|X\| \le 1\}$. The details are as follows (see $[25, 113]$).

A finite cell complex is a finite union of spaces $X^0, X^1, \ldots, X^p$. X^0 is a finite disjoint union of points.

Given X^k, X^{k+1} may be formed as follows. Take a finite collection of maps

$$\alpha_1, \ldots, \alpha_r \colon S^k \to X^k,$$

where S^k is the k-sphere. Form the disjoint union

$$Z = X^k \cup e_1^{k+1} \cup \cdots \cup e_r^{k+1},$$

where the e_j^{k+1} are disjoint copies of D^{k+1}. X^{k+1} is the quotient space of Z by the equivalence relation that identifies any point y lying in the boundary of some e_i^{k+1} (which is precisely S^k) with the point $\alpha_i(y) \in X^k$. In other words, points interior to any e_i^{k+1} are identified only to themselves, while points on the boundary of e_i^{k+1} are attached to their images, under the maps α_i, in X^k.

For example, we may write the circle S^1 as a cell complex in several ways: $S^1 = e^0 \cup e^1$, as in Fig. 4.2a, or $S^1 = e_a^0 \cup e_b^0 \cup e_a^1 \cup e_b^1$, as in Fig. 4.2b. There are naturally many other possibilities using more cells.

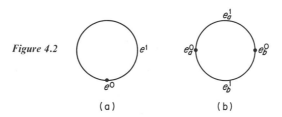

Figure 4.2

(a)   (b)

For the torus, $S^1 \times S^1$, we may describe the structure of a cell complex as shown in Fig. 4.3. Here the torus is X^2; X^0 consists of two points. X^1 is three circles, and may be displayed as a cell complex as shown in Fig. 4.4. There are two cells of dimension 2 in this description of $S^1 \times S^1$. Of course, there are many other ways to write $S^1 \times S^1$ as a cell complex.

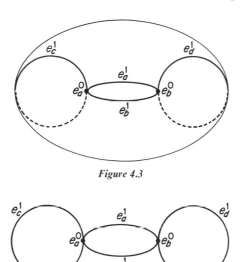

Figure 4.3

Figure 4.4

Observe that if X is a cell complex, then each subspace X^i (called the i-skeleton) is also a cell complex. Also, any quotient space X/X^i becomes a cell complex when the single point, which is the equivalence class of X_i, is treated as a single zero cell.

The Grassman manifolds $G_{n,k}$ may be given the structure of a cell complex in such a way that the maps

$$j_{n-1} : G_{n-1,k} \to G_{n,k}$$

take individual cells into individual cells of no greater dimension. This procedure, due to Ehresmann ([31, 78]), is as follows.

Fix as filtration

$$\mathbb{R}^0 \subseteq \mathbb{R}^1 \subseteq \cdots \subseteq \mathbb{R}^k.$$

Select a sequence of positive integers σ_i, $1 \leq i \leq k - n$, so that

$$1 \leq \sigma_1 \leq \cdots \leq \sigma_{k-n} \leq k.$$

Let $e(\sigma_1, \ldots, \sigma_{k-n})$ denote the set of all $(k - n)$-dimensional subspaces of X in $\mathbb{R}^k$, so that X meets $\mathbb{R}^{\sigma_i}$ in an i-dimensional subspace and meets $\mathbb{R}^{\sigma_i - 1}$ in $(i - 1)$-dimensional subspace for all i. The sets $e(\sigma_1, \ldots, \sigma_{k-n})$ may be shown to give a cell decomposition of $G_{k-n,n}$. The number of j-cells in $G_{k-n,n}$ is easily expressed in terms of partitions.

(iii)   Note that if X and Y are cell complexes and $f : X \to Y$ is a continuous map, then f is homotopic to a map that takes i cells into a union of cells of no higher dimension (see [102, 113] for this cellular approximation theorem). It is not difficult to see from (ii) above that the cells of $G_{k-n,n}$ of dimension $\leq p$, for $p < k - n$, are independent of k or n. We may then infer

(iv)   If X is a finite complex all of whose cells have dimension less that $k - n$, then

$$(j_n)_* : [X, G_{k-n,n}] \to [X, G_{k-n+1,n}]$$

is an isomorphism of sets.

Referring to the definition of limit, we see that in this case

$$[X, G_{k-n,n}] \equiv \lim[X, G_{i-n,n}].$$

This fact may now be exploited to make a full analysis of the situation in low-dimensional cases.

(v)   (a)   If X is a cell complex of dimension less than or equal to 2 (i.e., no cells of dimension 3 or greater), then line bundles over X are given by

$$\varinjlim[X, G_{k-1,1}].$$

By (iv) above this is isomorphic as a set to

$$[X, G_{3,1}] \equiv [X, G_{1,3}] \equiv [X, \mathbb{R}P^3]$$

(see the remarks after Definition 2.4).

In particular, we can check that there are two isomorphism classes of line bundles over the circle by noting that there are two homotopy classes of maps $S^1 \to \mathbb{R}P^3$ (in fact, $\pi_1(\mathbb{R}P^3)$ has two elements (see [59, 102]).

(b)   To get at $v^2(X)$, consider $[X, G_{3,2}] = [X, G_{2,3}]$. Some easy calculations are now possible. Suppose, for example, that X is a suspension such as $S^2 = S(S^1)$.

$$v^2(S^2) = [S^2, G_{3,2}] = [S^1, \text{Gl}(2; \mathbb{R})]_0 / \sim$$

by Proposition 4.4. $\text{Gl}(2; \mathbb{R})$ is homeomorphic to $O(2) \times \mathbb{R}^3$, so it has the homotopy type of $O(2)$, which is homeomorphic to two disjoint copies of S^1 (see [20]).

Then

$$v^2(S^2) = [S^1, S^1]_0 / \sim.$$

$[S^1, S^1]_0$ is the integers, and $\sim$ identifies at most two elements together, so $v^2(S^2)$ is in 1–1 correspondence with the integers.

In this correspondence, the tangent bundle to S^2 corresponds to ± 2 (this may be checked by examining the identification of two disks along the common boundary and examining the overlap map, as in Proposition 4.4).

For further calculations and applications, consult [103, 53].

Lastly, a few remarks about orientable bundles and the case of complex vector bundles are in order here. By our basic Definition 4.5., a vector bundle (E, X, p) has a local structure given by homeomorphisms

$$\psi_\alpha : p^{-1}(U_\alpha) \to U_\alpha \times \mathbb{R}^n.$$

The coordinate overlap maps $\psi_\beta \circ \psi_\alpha^{-1}$ are completely determined by functions

$$g_{\alpha\beta} : U_\alpha \cap U_\beta \to \mathrm{Gl}(n; \mathbb{R}).$$

If the structure of the vector bundle may be determined in such a way that every $g_{\alpha\beta}(x)$ has positive determinant, then a choice of an orientation in any $p^{-1}(x)$ may be carried over to a choice of an orientation in $p^{-1}(y)$ for any y in the same connected component of X as the point x. Hence, if X is connected and every $g_{\alpha\beta}(x)$ is of positive determinant, then every $p^{-1}(x)$ may be given a specific, well-defined orientation. In this case the bundle is called *orientable*, and if in addition an orientation is specified on some $p^{-1}(x)$, say by giving a definite homeomorphism $h : p^{-1}(x) \xrightarrow{\equiv} \mathbb{R}^n$, then we call the bundle *oriented*.

For example, every bundle over S^2 is orientable. We simply cover S^2 with two overlapping open sets, each slightly larger than a hemisphere, and observe the overlap may be assumed to be homeomorphic to $S^1 \times (-\varepsilon, \varepsilon)$ for some small $\varepsilon > 0$. In particular, the overlap of the two open sets is connected. Therefore, it we choose the homeomorphisms ψ_α and ψ_β so that at one point $\psi_\beta \circ \psi_\alpha^{-1}$ preserves orientation, it will preserve orientation at every other point in $S^1 \times (-\varepsilon, \varepsilon)$. Equivalently, the determinant of $g_{\alpha\beta}(x)$ will be positive for all x in the overlap. (Since any homeomorphism ψ_α may may be composed with

$$1_{U_\alpha} \times h : U_\alpha \times \mathbb{R}^n \to U_\alpha \times \mathbb{R}^n$$

for any linear homeomorphism $h : \mathbb{R}^n \to \mathbb{R}^n$, consistent with the definition of the bundle, we may choose any orientation we wish for ψ_α or ψ_β restricted to a given point.)

On the other hand, the two orientations on TS^2 give equivalent vector bundles (see problem 2), although inequivalent *oriented* bundles.

For oriented bundles, one may develop a theory of classification similar to Theorem 4.1. One works with oriented Grassman manifolds, defined as homogeneous spaces [see Remark (1), after Theorem 4.1] in terms of the special orthogonal groups SO(n), consisting of orthogonal matrices of determinant $+1$. The resultant classifying spaces are two-sheeted covering spaces of our Grassman manifolds, because for each $(k - n)$-dimensional subspace of $\mathbb{R}^k$ there are two oriented $(k - n)$-dimensional subspaces (consult [103]).

The theory extends to vector bundles defined over any topological field or skew field, but the case of complex vector bundles defined over the field of complex numbers is especially important and interesting. All the basic definitions carry over immediately. The classification is, in fact, slightly simpler, because there are no difficulties of orientation, as we had in Propositions 4.4, in the case of complex vector bundles. This results from the fact that the general linear group over the complex numbers, Gl($n; C$), is connected (see problem 3). In this case, the classifying space is

$$U(k)/U(k - n) \times U(n),$$

where the unitary group U(k) is the group of $k \times k$ complex matrices whose inverses equal the conjugates of their transposes. U(k) is connected, so that all bundles and homomorphisms are automatically oriented. For further developments of this beautiful and interesting subject, consult [4, 103].

The classification of bundles is an important and interesting topic, and it has been studied in depth by many authors. Generalizations of Theorem 4.1 include the classification of a general principal bundle (the group being any topological group), as well as extending the results to cover more general X than compact Hausdorff spaces. For the early history, see Steenrod's book [103]. The most beautiful and general result in the area of classification is due to Dold [26].

PROBLEMS AND PROJECTS

1. (a)   Use our results to show that there is no nontrivial line bundle over the sphere S^2. One may show that $[S^1, O(1)]_0$ is trivial or that $[S^2, \mathbb{R}P^3]$ is trivial (the latter because $[S^2, \mathbb{R}P^3] = [S^2, S^3]$, since S^3 is a covering space of $\mathbb{R}P^3$ and, from "cellular approximation," any map $f: S^2 \to S^3$ is homotopic to one that misses a point and thus factors through $S^3 - \text{pt} \equiv D^3$, which is contractible).

(b)   Describe the line bundles over $S^1 \cup S^1$, consisting of two copies of a circle attached at a single point.

(c)   Let X and Y be spaces, and $x_0 \in X$, $y_0 \in Y$ be single points. Define $i_1 : X \to X \times Y$ and $i_2 : Y \to X \times Y$ by $i_1(x) = (x, y_0)$ and $i_2(y) = (x_0, y)$. Let π_1 and π_2 be the projections such that $\pi_1 \circ i_1 = 1_X$ and $\pi_2 \circ i_2 = 1_Y$.

Deduce from Theorem 4.1 that if E is a nontrivial vector bundle over X (resp. Y), then $\pi_1^{-1}(E)$ (resp. $\pi_2^{-1}E$) is a nontrivial vector bundle over $X \times Y$.

As an example, exhibit nontrivial line bundles over the torus $S^1 \times S^1$.

(d)   In remark (1b) following Theorem 4.1, we describe a cell complex structure on $S^1 \times S^1$. If we identify to a single point all cells except one 2-dimensional cell, we have a map $g : S^1 \times S^1 \to S^2$. Examine the induced bundle $g^{-1}(TS^2)$. Does it have a nonzero section? (Is it trivial or nontrivial?) What is the bundle $i_1^{-1}(g^{-1}TS^2)$, with $i_1 : S^1 \to S^1 \times S^1$ as in (c) above?

2.   (a)   The maps

$$g_1 : S^1 \to O(2) \qquad \text{by} \quad g_1(\theta) = \begin{pmatrix} \cos\theta & -\sin\theta \\ \sin\theta & \cos\theta \end{pmatrix}$$

and

$$g_2 : S^1 \to O(2) \qquad \text{by} \quad g_2(\theta) = \begin{pmatrix} \cos\theta & \sin\theta \\ -\sin\theta & \cos\theta \end{pmatrix}$$

are not homotopic. Following each map by the inclusion of $O(2) \subseteq Gl(2; \mathbb{R})$ we get two bundles in $v^2(S^2)$, which are equivalent because g_1 and g_2 are related by

$$g_1(\theta) = A g_2(\theta) A^{-1}, \qquad \text{where} \qquad A = \begin{pmatrix} 0 & 1 \\ 1 & 0 \end{pmatrix}.$$

Conclude as a consequence of this sort of analysis that the bundles TS^2, with different orientations, are equivalent vector bundles (identical bundles) but inequivalent oriented bundles, and that the maps $S^1 \to Gl(2; \mathbb{R})$, from Proposition 4.4, for the two orientations are not homotopic but are equivalent under $\sim$.

(b)   If n is odd and $I \in Gl(n; \mathbb{R})$ is the identity matrix, then $\det(-I) = -1$. Nevertheless, $-I$, commutes with every matrix in $Gl(n; \mathbb{R})$.

Conclude that for n odd the equivalence relation in Proposition 4.4 is trivial, and that

$$v^n(SX) = [X, Gl(n; \mathbb{R})]_0.$$

(c)   If $X = S^m$ and n is odd, $[X, Gl(n; \mathbb{R})]_0 = \pi_m(Gl(n; \mathbb{R}))$ the mth homotopy group. $Gl(3; \mathbb{R})$ has the homotopy type of $O(3)$, which is two disjoint copies of $SO(3)$. $SO(3)$ is homeomorphic to $\mathbb{R}P^3$, and S^3 is a twofold covering space of $\mathbb{R}P^3$. Thus conclude

$$v^3(S^3) = \pi_2(S^3),$$

which by "cellular approximation" must be trivial, since any cellular map misses a point (and thus lies in the contractible $D^3 = S^3 - \text{pt}$). Hence $v^3(S^3)$ has only one (trivial) element. Consult [103] for details here.

(d)   Show that every map $S^2 \rightarrow S^1$ is homotopic to a constant (because $\mathbb{R} \rightarrow S^1$ is a covering space) and, thus that $v^2(S^3)$ and $v^1(S^3)$ must also consist of a single (trivial) element.

3.   The unitary group $U(n)$ is the group of isomorphisms of the n-dimensional complex vector space $\mathbb{C}^n$ that preserve the inner product

$$(Y, Z) = y_1\bar{z}_1 + \cdots + y_n\bar{z}_n,$$

where $Y = (y_1, \ldots, y_n)$, etc, and $\bar{z}_i$ means the complex conjugate of z_i. But the set

$$\{Y \in \mathbb{C}^n | \sqrt{(Y, Y)} = 1\}$$

is clearly the sphere S^{2n-1}, because $\mathbb{C}^n$ is homeomorphic to $\mathbb{R}^{2n}$. Fixing $Y \in S^{2n-1}$, verify that the action of $U(n)$ on S^{2n-1} defined by

$$A \rightarrow AY$$

is transitive and effective. Conclude that

$$S^{2n-1} = U(n)/U(n-1)$$

(choose suitable Y using Proposition 4.1).

Verify that if G_{x_0} and G/G_{x_0} are path connected, so is G, in this case.

Calculate $U(1)$ and conclude that $U(n)$ is path connected for all n.

State the correct version of Proposition 4.4 for complex vector bundles over a suspension (compare [4]).

4.   Show that an n-dimensional vector bundle (E, X, p) is trivial if and only if there are n sections $s_1, \ldots, s_n$ and for each x the vectors $s_1(x), \ldots, s_n(x)$ are linearly independent.

Show that if the bundle is orientable, then it is trivial if and only if there are $n - 1$ such sections, linearly independent at a point. (The last section may be determined from orientability.)

Find 2-dimensional vector bundles with 0, 1, or 2 sections that are linearly independent at every point (0 such sections means simply that there is no everywhere nonvanishing section of the bundle). (These examples can be built with tangent bundles.)

5

Differentiation and Integration

on Manifolds

In order to understand analysis on manifolds, one must surely understand differentiation and integration over a manifold. In its most general form, this subject combines the difficulties of Lebesgue theory with some of the more subtle geometric properties (see [117]). For our immediate purposes, it will suffice to consider the smooth case; here there is still much interplay between the analysis and geometry, yet one can get to important results with reasonable speed.

In many respects, the theory of integration is easier and more basic than differentiation, and we shall treat it first for the case of open sets in Euclidean space. The interplay between differentiation and geometry is only clarified with the introduction of differential forms, and for this reason we shall then need a disgression into the area of exterior algebra and the construction of various bundles starting from the tangent bundle to a manifold. Then we can consider exterior differentiation and the integration over a manifold. With all this complete, we may prove a very simple and elegant generalization of Stokes' theorem, which clarifies and includes all the basic integration theorems and formulas from the traditional vector analysis. We also include the famous Poincaré lemma, which relates to exact forms and generalizes the basic theory of exact, first-order, ordinary differential equations.

The material in this chapter points to possible generalizations in many different directions. One may, via deRham's theorem [23], use this analysis to study the topology of smooth manifolds. We describe this briefly (no proofs). On the other hand, this material leads to basic examples for the study of differential operators on manifolds, which will be the topic of our next chapter.

The reader who wants a more complete treatment of some of these topics might look at the following:

(i)   for integration theory [84, 117].
(ii)   for exterior algebra and differential forms [20, 47, 104].
(iii)   for Stokes' theorem and deRham theory [23, 53, 18].

Integration theory for continuous functions over open sets in Euclidean space is just a part of the usual advanced calculus. We review the essentials here.

INTEGRATION IN SEVERAL VARIABLES

Let $O \subseteq \mathbb{R}^n$ be open, and let $C(O)$ denote the continuous functions from O to $\mathbb{R}$. If $f \in C(O)$, define the support of f to be

$$\operatorname{supp} f = \overline{\{x \mid f(x) \neq 0\}},$$

that is, the closure of the set where f is nonzero. We denote the subset of $C(O)$, consisting of those functions whose support is compact, by $C_c(O)$. Note that with the usual notions of pointwise addition and multiplication, $C(O)$ is a ring (or even an algebra over the real numbers); $C_c(O)$ is an ideal in this ring.

In general, we can only expect to be able to integrate a function with compact support. Over any entire open set, integrals may exist in the sense of the usual improper integral, for example

$$\int_{-\infty}^{\infty} e^{-x^2}\, dx,$$

but these integrals are defined as limits of integrals of functions with compact support. Therefore, we shall consider the case of $C_c(O)$ first.

We recall the definition of the integral of $f \in C_c(O)$. Choose a box (Cartesian product of closed intervals) containing supp f. We consider partitions of this box by looking at arbitrary finite partitions of each interval that is a factor in our box. Define f to vanish outside the set O. Number the boxes in the partition P of our original box, say, B_i.

We then define the lower and upper sums for f with respect to the partition P as

$$L(f, P) = \sum_{\substack{B_i \in P \\ B_i \cap \text{supp } f \neq \varnothing}} \left(\left(\underset{x \in B_i}{\text{g.l.b.}(f(x))} \right)(\text{volume of } B_i) \right),$$

$$U(f, P) = \sum_{\substack{B_i \in P \\ B_i \cap \text{supp } f \neq \varnothing}} \left(\left(\underset{x \in B_i}{\text{l.u.b.}(f(x))} \right)(\text{volume of } B_i) \right).$$

Of course g.l.b. and l.u.b. abbreviate greatest lower and least upper bound.

We define the *lower integral* of f to be the least upper bound of all the lower sums $L(f, P)$, for any partition P, and the *upper integral* to be the greatest lower bound of all the possible upper sums $U(f, P)$.

Note that if f is merely bounded and has compact support, then the lower and upper integrals must necessarily exist, and the former is always less than or equal to the latter. If, in addition, f is continuous, then a standard argument shows that the two integrals, the lower and upper, are actually equal. In the case where the lower and upper integrals agree, one speaks of f as being *integrable*, and one writes the common value of the upper and lower integrals as

$$\int_O f \, dx_1 \cdots dx_n.$$

The integral as we have defined it is a real number, as in definite integration. (Of course, it is not necessary for f to be continuous in order that it be integrable. For example, functions that are continuous except at a finite (or even countable) number of points are integrable.) When we introduce the language of forms, our $dx_1 \cdots dx_n$ will be written $dx_1 \wedge \cdots \wedge dx_n$, but the basic idea of the integral, as developed in advanced calculus courses, will carry over.

We should now examine some basic properties of the integral of functions in $C_c(O)$.

(i) *Linearity.*   If $f, g \in C_c(O)$ and $\alpha, \beta \in \mathbb{R}$,

$$\int_O (\alpha f(x) + \beta g(x)) \, dx_1 \cdots dx_n$$

$$= \alpha \int_O f(x) \, dx_1 \cdots dx_n + \beta \int_O g(x) \, dx_1 \cdots dx_n.$$

Note that we shall write f or $f(x)$, $x = (x_1, \ldots, x_n)$, interchangeably under the integral sign (the variable x occasionally adding clarity).

(ii) *Preservation of order.*   If $f \in C_c(O)$, and $f(x) \geq 0$, for all $x \in O$, then

$$\int_O f(x) \, dx_1 \cdots dx_n \geq 0.$$

It follows at once that if f, $g \in C_c(O)$ and $f(x) \geq g(x)$, for all $x \in O$, then

$$\int_O f(x)\,dx_1 \cdots dx_n \geq \int_O g(x)\,dx_1 \cdots dx_n.$$

We also note that it is an easy consequence of this that when $f \in C_c(O)$,

$$\int_O |f(x)|\,dx_1 \cdots dx_n \geq \left| \int_O f(x)\,dx_1 \cdots dx_n \right|.$$

(iii)   Suppose $O_1 \subseteq O$ and $O_2 \subseteq O$ are two open subsets, and suppose further that $\bar{O}_1 \cap \bar{O}_2$ has measure zero, in the sense of Definition 3.2. Then if $f \in C_c(O)$,

$$\int_{O_1} (f\,|\,O_1)\,dx_1 \cdots dx_n + \int_{O_2} (f\,|\,O_2)\,dx_1 \cdots dx_n$$

$$= \int_{O_1 \cup O_2} (f\,|\,O_1 \cup O_2)\,dx_1 \cdots dx_n.$$

(This property is usually applied when $\bar{O}_1 \cap \bar{O}_2$ is a submanifold of $\mathbb{R}^n$ of lower dimension than n.)

(iv)   *Change of variables in multiple integration.*   Suppose O_1 and O_2 are open in $\mathbb{R}^n$, and assume that

$$g : O_1 \to O_2$$

is a diffeomorphism of at least class C^1 (i.e., the function and the first partials exists and are continuous). Let $f \in C_c(O_2)$. Finally, we use $x = (x_1, \ldots, x_n)$ for the coordinates in O_2 and $y = (y_1, \ldots, y_n)$ for the coordinates in O_1.

Then we have the following formula:

$$\int_{O_2} f\,dx_1 \cdots dx_n = \int_{O_1} f(g(y)) \,|\det(J_g(y))|\,dy_1 \cdots dy_n.$$

Here, $J_g(y)$ is the Jacobian, det means determinant, and the bars refer to absolute value.

For the details of these properties, consult any basic advanced calculus book, or texts such as [93, 24].

The key to integration on manifolds is to be able to piece together the value of the integral in different coordinate neighborhoods. Unfortunately, the coordinate charts are merely homeomorphisms

$$\phi_\alpha : O_\alpha \to \mathbb{R}^n,$$

where the domains do not have, in any intrinsic sense, notions of distance, volume, etc. The key is then to carry out the integration in the range of such coordinate homeomorphisms and then piece together the results for the different, often overlapping, coordinate neighborhoods. This is, on the one hand, a problem in the careful use of (iv) above. But it can be also be viewed as a problem in giving some sense to an expression like

$$f\,dx_1 \cdots dx_n$$

over an entire manifold, or at least in an open neighborhood of supp f. The need to create such a theory of those expressions that occur under integral signs, over an entire manifold, is the basic motivation for the theory of differential forms.

But before we define differential forms, we need to examine the basic exterior algebra. This is the algebraic foundation of the theory of forms, which arises out of the need to keep the orientations of overlapping sets correct. In integration in a single variable we distinguish between $\int_a^b f\,dx$ and $\int_b^a f\,dx$; that is, we distinguish between the different orientations of the x-axis. In several variables, the problems are more complex, and a more elaborate theory is necessary.

EXTERIOR ALGEBRA AND FORMS

Definition 5.1   Let V be a finite dimensional vector space over $\mathbb{R}$. For each positive integer i, let S^i be the vector space over $\mathbb{R}$ which is spanned by the i-tuples of elements in V, that is,

$$\{(y_1, \ldots, y_i)\}$$

with each $y_j \in V$. S^i is infinite dimensional unless V has dimension 0.

In S^i we introduce an equivalence relation, which is most easily expressed in terms of three elementary equivalence relations. That is to say, two elements of S^i are called equivalent if there are a finite number of elements in S^i, beginning with the first original element and ending with the second original element, and each successive pair of elements is equivalent in one of the following three senses:

$$(y_1, \ldots, y_r, \ldots, y_s, \ldots, y_i) \sim -(y_1, \ldots, y_{r-1}, y_s, \ldots, y_r, y_{s+1}, \ldots, y_i). \quad (1)$$

That is, if one i-tuple is obtained from another by a simple permutation of two elements, then the original is equivalent to the negative of the permuted i-tuple. For example $(y_1, y_2, y_3) \sim -(y_2, y_1, y_3)$.

$$\alpha(y_1, \ldots, y_r, \ldots, y_i) \sim (y_1, \ldots, \alpha y_r, \ldots, y_i) \quad (2)$$

for any real α and r, $1 \le r \le i$.

$$(y_1, \ldots, y_r', \ldots, y_i) + (y_1, \ldots, y_r'', \ldots, y_i)$$
$$\sim (y_1, \ldots, y_{r-1}, y_r' + y_r'', y_{r+1}, \ldots, y_i). \quad (3)$$

We write the equivalence class of $(y_1, \ldots, y_i)$ as $y_1 \wedge \cdots \wedge y_i$, and define the quotient space of S^i by this equivalence relation to be $\wedge^i V$, the i-fold exterior power of V.

We set $\wedge^0 V = \mathbb{R}$ (as a convention), and we write

$$\wedge V = \sum_{i \geq 0} \oplus \wedge^i V$$

as the direct sum of these vector spaces. We shall soon define a multiplication that will make $\wedge V$ an algebra, the exterior algebra on V.

Remarks (1) Each $\wedge^i V$ is clearly a vector space, and clearly $\wedge^1 V = V$. But note that by (1) in Definition 5.1 if any two of the elements $y_1, \ldots, y_i$ are the same, then $y_1 \wedge \cdots \wedge y_i = 0$ in $\wedge^i V$. If $\{y_1, \ldots, y_n\}$ is a basis for V, then it follows at once that a basis for $\wedge^i V$ is given by the elements

$$y_{n_1} \wedge \cdots \wedge y_{n_i},$$

where $n_1 < n_2 < \cdots < n_i$. But then, by the usual elementary combinatorics, the number of such elements is the binomial coefficient $\binom{n}{i}$, so that

$$\dim \wedge^i V = \binom{n}{i}$$

for $0 < i \leq n$.

In particular $\dim \wedge^n V = 1$ whenever $\dim V = n$, and a basis for $\wedge^n V$ is given by the single vector $y_1 \wedge \cdots \wedge y_n$ for any basis $\{y_i\}$ of V.

(2) $\wedge^i V$ may be described in terms of universal properties, which frequently add insight into what is happening in these constructions. Consider maps from the i-fold Cartesian product of V with itself to the vector space W,

$$f : V \times \cdots \times V \to W.$$

The domain is regarded here as a set, not as a vector space. Such a map is called *multilinear* if it is linear in each individual factor of its domain. Specifically, we require that for any r, $1 \leq r \leq i$, we have

$$f(x_1, \ldots, x'_r, \ldots, x_i) + f(x_1, \ldots, x''_r, \ldots, x_i) = f(x_1, \ldots, x'_r + x''_r, \ldots, x_i)$$

and

$$f(x_1, \ldots, \alpha x_r, \ldots, x_i) = \alpha f(x_1, \ldots, x_i)$$

for any $\alpha \in \mathbb{R}$.

Such a map is called *alternating* if it changes sign when any two entries are permuted, i.e.,

$$f(x_1, \ldots, x_r, \ldots, x_s, \ldots, x_i) = -f(x_1, \ldots, x_s, \ldots, x_r, \ldots, x_i).$$

Suppose f is a multilinear and alternating map, and suppose

$$\phi: V \times \cdots \times V \to \wedge^i V$$

is the obvious multilinear and alternating map defined by

$$\phi(x_1, \ldots, x_i) = x_1 \wedge \cdots \wedge x_i,$$

where of course the domain of ϕ is assumed to have i factors.

Then there is a unique linear map (i.e., homomorphism of vector spaces) $\tilde{f}: \wedge^i V \to W$ that makes the following diagram commutative

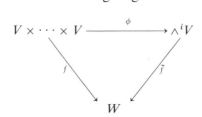

(that the diagram commutes means simply that $f = \tilde{f} \circ \phi$). In fact, we may define $\tilde{f}$ by setting

$$\tilde{f}(x_1 \wedge \cdots \wedge x_i) = f(x_1, \ldots, x_i).$$

(Check that this is well defined and works!) This gives a characterization of $\wedge^i V$ as a vehicle by which multilinear and alternating maps are converted to linear maps with $\wedge^i V$ as domain.

It is easy to work out low-dimensional examples of these phenomena. Suppose $V = \mathbb{R}^2$, $W = \mathbb{R}$. Then

$$f((x_1, x_2), (x_3, x_4)) = x_1 x_4 - x_2 x_3$$

is multilinear and alternating, while

$$g((x_1, x_2), (x_3, x_4)) = x_1 x_4 + 5 x_2 x_3$$

is only multilinear.

If $\dim V = n$, then $\wedge^n V$ is one dimensional, and the map

$$\phi: V \times \cdots \times V \to \wedge^n V$$

is easily calculated, in terms of a basis $\{y_1, \ldots, y_n\}$ of V, to be

$$\phi(\alpha_{11} y_1 + \cdots + \alpha_{1n} y_n, \ldots, \alpha_{n1} y_1 + \cdots + \alpha_{nn} y_n) = (\det \alpha_{ij}) y_1 \wedge \cdots \wedge y_n.$$

An arbitrary multilinear, alternating map

$$f: V \times \cdots \times V \to W$$

with n factors in the domain and dim $V = n$ is therefore given by sending $(\alpha_{11}y_1 + \cdots + \alpha_{1n}y_n, \ldots, \alpha_{n1}y_1 + \cdots + \alpha_{nn}y_n)$ to $\det(\alpha_{ij})$ times some fixed element in W (namely, $\tilde{f}(y_1 \wedge \cdots \wedge y_n)$).

(3)   We now must consider *products* in $\wedge V$. With the same general given data as above, take $\omega \in \wedge^i V$, $\sigma \in \wedge^j V$. We wish to define $\omega \wedge \sigma \in \wedge^{i+j} V$, so as to make $\wedge V$ an algebra over the real numbers. Take a basis $\{y_1, \ldots, y_n\}$ for V, so that we write (check)

$$\omega = \sum_{(r_1, \ldots, r_i)} a_{r_1, \ldots, r_i} \, y_{r_1} \wedge \cdots \wedge y_{r_i}$$

and

$$\sigma = \sum_{(s_1, \ldots, s_j)} b_{s_1, \ldots, s_j} \, y_{s_1} \wedge \cdots \wedge y_{s_j}.$$

Here $(r_1, \ldots, r_i)$ is a i-tuple of distinct integers from 1 to n (inclusive), $a_{r_1, \ldots, r_i}$ is a real number depending on $(r_1, \ldots, r_i)$, etc. If we wish, these presentations may be made unique by requiring $r_1 < \cdots < r_i$, but this is not advantageous here.

It will thus suffice to define the product on one term from ω and one term from σ, and then extend the definition to ω and σ entirely by imposing the distributive law for the algebra. With this in mind, we set

$$(a_{r_1, \ldots, r_i} y_{r_1} \wedge \cdots \wedge y_{r_i})(b_{s_1, \ldots, s_j} y_{s_1} \wedge \cdots \wedge y_{s_j})$$
$$= (a_{r_1, \ldots, r_i})(b_{s_1, \ldots, s_j}) y_{r_1} \wedge \cdots \wedge y_{r_i} \wedge y_{s_1} \wedge \cdots \wedge y_{s_j}.$$

Here, the numbers $a_{r_1, \ldots, r_i}$ and $b_{s_1, \ldots, s_j}$ are to be multiplied like ordinary real numbers. Note that if one r_p is the same as one s_q, then the product is automatically zero, for $y \wedge y$ is always zero in $\wedge V$.

Of course, one should verify (routinely) that the definition is independent of the representation of ω and σ, which can vary according to the definition of $\wedge V$ by writing $-x \wedge y$ for $y \wedge x$, etc. It then follows at once that $\omega \wedge \sigma = -\sigma \wedge \omega$, or that the algebra $\wedge V$ is anticommutative. The product $\omega \wedge \sigma$ is on occasion referred to as the *exterior product*.

(4)   Finally, we observe that our definitions do not require that V be a finite-dimensional vector space. In the infinite-dimensional case, we consider only *finite* sums of multiples of terms $y_{r_1} \wedge \cdots \wedge y_{r_i}$, in our definition of $\wedge^i V$, as well as the definition of the exterior product. Of course, if V is infinite dimensional, so will be $\wedge^i V$, $i > 0$.

Now, to get at the notion of a differential form on a manifold, observe that a tangent vector is effectively a directional derivative, or a linear combination of partial derivatives, while a differential form [as inferred from property (iv) of integration] should be a function of a derivative, multiplied by expressions such as dx. In other words, forms are like linear functions of derivatives. This necessitates one more technical prerequisite, namely the cotangent bundle to a manifold, where the vector space of tangent vectors at a point is replaced by its dual, the space of linear maps from it to $\mathbb{R}$. The relationship of tangent space to cotangent space is the same sort of relationship as that of a matrix to its transpose or adjoint.

Definition 5.2 (a) If V is a vector space, V^* is the dual vector space whose elements are the linear maps $V \to \mathbb{R}$.
   (b) If M^n is a differentiable manifold ($C^r, r \geq 1$), with $(TM^n)_x$ the tangent space at x, or equivalently the elements in the tangent bundle over x, we write

$$(T^*M^n)_x = ((TM^n)_x)^*.$$

   (c) Define $T^*M^n = \bigcup_{x \in M^n} (T^*M^n)_x$.

Just as in Definition 2.5, we topologize T^*M^n by choosing open coordinate neighborhoods U_α with homeomorphisms $\phi_\alpha: U_\alpha \to \mathbb{R}^n$ and expressing T^*M^n as the quotient space of

$$\bigcup_\alpha U_\alpha \times (\mathbb{R}^n)^*$$

under the equivalence relation that identifies $(x_1, f_1) \in U_{\alpha_1} \times (\mathbb{R}^n)^*$ with $(x_2, f_2) \in U_{\alpha_2} \times (\mathbb{R}^n)^*$ whenever (1) $x_1 = x_2$ and (2) $f_2 \circ (J_{\phi_{\alpha_2} \circ \phi_{\alpha_1}^{-1}}) = f_1$ (the left side being the composition of the maps). Note that if we use the usual dual basis in $(\mathbb{R}^n)^*$, we may write

$$(J_{\phi_{\alpha_2} \circ \phi_{\alpha_1}^{-1}})^t f_2 = f_1,$$

where t means transpose. Finally, we define the projection $\pi(\{(x, f)\}) = x$.

Remarks (1) As before, if M^n is a C^r-manifold, $r \geq 1$, then T^*M^n is a C^{r-1}-manifold.
   (2) The only real difference between T^*M^n and TM^n is that in the former we make identifications by composing with the Jacobian on the right and in the latter by composing with the Jacobian on the left. Or, in other words, in the latter we operate with the Jacobian on the left, and in the former we operate with the transpose of the Jacobian on the left on dual spaces.
   (3) It can be shown (see [103, 53]) that any vector bundle is isomorphic to one for which the overlap maps $g_{\alpha\beta}: U_\alpha \cap U_\beta \to \mathrm{Gl}(n; \mathbb{R})$ actually lie in the subgroup of orthogonal matrices (in fact, $\mathrm{Gl}(n; \mathbb{R})$ and $O(n)$ have the same homotopy type, because $\mathrm{Gl}(n; \mathbb{R})$ is homeomorphic, by a map which is not a homomorphism, to the Cartesian product of $O(n)$ and some Euclidean space). But for such a bundle the identifications of TM^n as a map of $\mathbb{R}^n$ and

T^*M^n as a map of the dual of $\mathbb{R}^n$ are precisely the same (check!) Therefore, the bundles TM^n and T^*M^n are equivalent.

(4) If $f: M^n \to N^n$ is a diffeomorphism, we may define, for $f(x) = y$,

$$d^*f: T^*N^k \to T^*M^n$$

as follows: if we are given a coordinate chart $\psi_\beta: V_\beta \xrightarrow{\cong} \mathbb{R}^n$, for N^n and $(y, g) \in V_\beta \times (\mathbb{R}^n)^*$, and similarly $x \in U_\alpha$, $\phi_\alpha: U_\alpha \xrightarrow{\cong} (\mathbb{R}^n)^*$, in M^n, then we set

$$d^*f\{(y, g)\} = \{(x, g \circ J_{\psi_\beta \circ f \circ \phi_\alpha^{-1}})\}.$$

Note that such a definition is not possible if f is not $1-1$, for then there is no specific x that we may associate with $f^{-1}y$, and thus no specific map $J_{\psi_\beta \circ f \circ \phi_\alpha^{-1}}$. The situation is most clearly brought out by setting M^n equal to a disjoint union of two copies of N^n and letting f be a map which is different on each copy.

Since T^*M^n is a dual to the tangent bundle, we build differential forms by taking exterior products and then looking at sections. The following two definitions give the details.

Definition 5.3 (a)   Given a linear map of vector spaces $f: V \to W$, we define a linear map of exterior powers

$$\wedge^i f: \wedge^i V \to \wedge^i W$$

by setting $(\wedge^i f)(y_1 \wedge \cdots \wedge y_i) = f(y_1) \wedge \cdots \wedge f(y_i)$.

(b)   Let M^n be a differentiable manifold. The bundle of i-covectors, or i-forms, is

$$\wedge^i(T^*M^n) = \bigcup_{x \in M^n} \wedge^i((T^*M^n)_x).$$

It is topologized just as in Definition 5.2c, using the equivalence relation obtained by applying $\wedge^i$ to the Jacobian. Specifically, $(x, u) \in U_\alpha \times \wedge^i(\mathbb{R}^n)^*$ is equivalent to $(x, v) \in U_\beta \times \wedge^i(\mathbb{R}^n)^*$ when $\wedge^i[(J_{\phi_\beta \circ \phi_\alpha^{-1}}(x))^t](v) = u$.

The projection $\pi: \wedge^i(T^*M^n) \to M^n$ takes z into the unique x for which $z \in \wedge^i((T^*M^n)_x)$.

Naturally, $\wedge^i(T^*M^n)$ is a vector bundle of dimension $\binom{n}{i}$.

Definition 5.4 A *differential i-form*, or a *differential form of degree i*, is a section of the bundle defined in Definition 5.3b. Specifically, it is a continuous map

$$s: M^n \to \wedge^i(T^*M^n)$$

such that $\pi \circ s = 1_{M^n}$ i.e., $\pi(s(x)) = x$, for all $x \in M^n$ (compare Definition 4.6).

We shall usually assume that M^n is smooth, so that $\wedge^i(T^*M^n)$ is smooth, in which case we shall assume that forms s are smooth also.

Examples   *1.* A zero-form associates to each $x \in M^n$, an element in $\wedge^0((T^*M^n)_x) = \mathbb{R}$; thus, a zero-form is simply a smooth function.

2. If $M^n = \mathbb{R}^2$, there is only one coordinate chart or neighborhood to consider. Forms are then expressions like

$$f(x, y)\, dx + g(x, y)\, dy \qquad \text{or} \qquad h(x, y)\, dx \wedge dy,$$

where the usual conventions at a point P take dx to be the linear function from $(T\mathbb{R}^2)_p$ to $\mathbb{R}$ that is 1 on $(1, 0)$ and 0 on $(0, 1)$, and the reverse for dy.

3. Cover the circle S^1 by the sets U_1 and U_2 which are respectively the complements of the south and north "poles." Select coordinate homeomorphisms so that the overlap map has a simple positive derivative. Because $U_1 \cap U_2$ constitutes all of U_1 or U_2 except a single point, a constant form on U_1 completely determines one on U_2, and visa versa, yielding a form on S^1.

4. In general, if M^n is orientable, (Definition 2.2), there is a nowhere vanishing n-form defined on M^n. To see this, select a locally finite cover of of M^n by coordinate charts U_α such that each U_α contains a compact K_α and the K_α's also cover M^n. Select sections of $\pi : \wedge^n M^n \to M^n$ that are non zero constants over K_α and vanish outside U_α. This is possible because the vector bundle in question is trivial over each U_α. Call the sections s_α, and assume s_α to be a positive multiple of $dx_1 \wedge \cdots \wedge dx_n$, where $x_1, \ldots, x_n$ are coordinate axes that specify some given orientation on the manifold restricted to U_α. In a region of overlap $U_\alpha \cap U_\beta$, this condition (of being a positive multiple of such a n-form) is well defined because the determinants of the Jacobians of the overlap maps are positive. In fact, if V is an n-dimensional vector space and $f : V \to V$ is a linear transformation, then one easily calculates the effect of $\wedge^n f$ on a basis vector, say $v_1 \wedge \cdots \wedge v_n$ (for the 1-dimensional vector space $\wedge^n V$) to be multiplication by the determinant of f (compare Remark (2) after Definition 5.1).

Since the s_α vanish outside U_α, which are elements of a locally finite cover, and the K_α cover M^n, the form

$$s(x) = \sum_\alpha s_\alpha(x)$$

is well defined and nowhere vanishing on M^n.

5. It follows immediately from Definition 5.3b that every form on M^n has, locally, an expression like it would have on Euclidean space. Specifically, if $x \in M^n$, U_α is a coordinate neighborhood at x, and s is an i-form on M^n, then there is a basis of T^*M^n at u, say $dx_1, \ldots, dx_n$, and functions

$$f_{m_1, \ldots, m_i}(x)$$

for every i-tuple $(m_1, \ldots, m_i)$, $m_1 < \cdots < m_i$, such that in this trivialization of $\pi^{-1}(U_\alpha) \subseteq \wedge^i(T^*M^n)$ we have

$$s(x) = \sum_{m_1 < \cdots < m_i} f_{m_1, \ldots, m_i}(x)\, dx_{m_1} \wedge \cdots \wedge dx_{m_i}.$$

Of course, here $dx_{m_j} \in (T^*M^n)_x$, and is a linear map $(TM^n)_x \to \mathbb{R}$. With our restrictions on $m_1, \ldots, m_i$, the representation is unique (for our choice of basis $dx_1, \ldots, dx_n$).

   6. It is clear that differential i-forms may be added, or multiplied by a scalar, just as any sections of a given vector bundle. We shall routinely require that forms be as differentiable as possible; i.e., if M^n is C^r, then $\wedge^i(T^*M^n)$ is C^{r-1}, and we assume that forms are C^{r-1}. In the case of the local representation of Example 5, the functions $f_{m_1, \ldots, m_i}(x)$ would automatically be C^{r-1}. In most cases, we take everything smooth.

   Finally, following Remark (3) after Definition 5.1, we may clearly multiply an i-form (differential form) with a j-form at every point, getting an $(i+j)$-form, defined by

$$(\sigma \wedge \omega)(x) = (\sigma(x)) \wedge (\omega(x)).$$

This multiplication is trivially checked to be anticommutative, in the sense that $\sigma \wedge \omega = -\omega \wedge \sigma$. With these definitions, we set the following definition:

Definition 5.5   Let $D^i(M^n)$ be the vector space of differential forms of degree i on M^n (i-forms). We set

$$D^*(M^n) = \sum_{i \geq 0} \oplus D^i(M^n),$$

the direct sum of all the $D^i(M)$. It is a graded, anticommutative algebra over the real numbers, called the *algebra of differential forms* on M^n.

   Remarks   (1) $D^0(M^n)$ is the real valued, differentiable functions on M^n.
   (2) $D^i(M^n) = 0$, for $i > n$, since at any point such a form would lie in a vector space $\wedge^i V$, where $i > \dim V$, and this is automatically zero. [See Remark (1) after Definition 5.1.]
   (3)   If M^n is orientable, Example 4 above implies that $D^n(M^n)$ contains a nowhere vanishing form. Of course, a simple local argument shows that for any $M^n, n > 0$, there are forms in $D^i(M^n), i \leq n$, that are nonvanishing at some points.

$D^*(M^n)$ associates to every differentiable manifold a graded algebra over the real numbers, a direct sum of vector spaces. It is only natural to expect that for a differentiable map

$$f: M^n \to N^k,$$

we should get an algebra homomorphism of the algebras of differential forms. If we look at 0-forms, $\omega \in D^0(N^k)$, there is a natural construction of a 0-form or function on M^n, namely, $\omega \circ f$, the composition of functions. Therefore, we should expect the homomorphism to be

$$D(f): D^*(N^k) \to D^*(M^n).$$

We give an elementary—if rather crude—definition of this now. After the next proposition, we shall be able to give a more perceptive definition.

Definition 5.6   Let $f: M^n \to N^k$ be a differentiable map, and let $s \in D^i(N^k)$. Then s is a section of the bundle

$$\pi: \wedge^i(T^*N^k) \to N^k.$$

One checks at once that $s \circ f$ defines a section of the induced bundle

$$f^{-1}(\wedge^i(T^*N^k)).$$

(See Definition 4.12 and consider the map $x \to (x, s(f(x)))$.)
    At any point x, the adjoint or dual of df, the differential of f, defines a map (linear)

$$(df)_x^*: f^{-1}(T^*N^k)_x \to (T^*M^n)_x$$

by $(df)_x^*(x, \alpha) = \alpha \circ (df)$. Taking the i-fold exterior power yields a map

$$\wedge^i((df)^*)_x: f^{-1}(\wedge^i(T^*N^k))_x \to \wedge^i(T^*M^n)_x.$$

If $x \in M^n$, we set

$$D(f)(s) = \wedge^i((df)^*)(s \circ f)$$

by the immediately above map at x.

    This is trivially checked to be a section of $\wedge^i(T^*M^n) \to M^n$, and hence an i-form on M^n as desired.
    It is a routine verification to check that $D(f)$ is an algebra homomorphism, preserving degree. Note that we do *not* claim that $(df)^*$ is defined for all x as a map of cotangent bundles.
    This definition is cumbersome for practical calculations. The alternate description, after the next proposition, is much more workable.

The situation of Definition 5.6, where the constructed map goes in the opposite direction from f, is called *contravariance*. The differential of a map going the same way, is an example of *covariance*. These terms, borrowed from classical differential geometry, are now part of the standard language of categories and functors (see [79, 97]).

One easily checks that if $f: M^n \to N^k$ is the inclusion of a submanifold, $D(f)$ is a restriction map, which in suitable local coordinates amounts to setting certain variables equal to zero.

It is desirable to have an alternative view of differential forms that places the emphasis on vector fields (see Definition 2.6). This adds clarity and gives an alternative to Definition 5.6. We sketch this briefly now.

Recall that a vector field on M^n is a cross section of the tangent bundle. The collection of all (smooth) vector fields on a smooth manifold forms a vector space under pointwise addition and scalar multiplication, say $S(M^n)$. This vector space $S(M^n)$ over the real numbers is also a module over the ring of smooth functions defined on M^n. If $S: M^n \to TM^n$ is a vector field and $f: M^n \to \mathbb{R}$ is a smooth function, then we set

$$(f \cdot s)(x) = f(x) \cdot s(x).$$

Recall also that an alternating multilinear i-form, with values in $C^\infty(M^n)(= D^0(M^n))$, the smooth functions on M^n, means a map

$$A: S(M^n) \times \cdots \times S(M^n) \to C^\infty(M^n),$$

where the domain has i factors [refer to Remark (2) after Definition 5.1]. We shall also require that A be linear, in any variable, with respect to multiplication by a function in $C^\infty(M^n)$. That is,

$$A(s_1, \ldots, s_{r-1}, f s_r, s_{r+1}, \ldots, s_i) = f A(s_1, \ldots, s_i).$$

Such A are called the exterior i-forms on vector fields, and we write them as $A^i(M^n)$ (and refer to them as exterior forms, for short). The relation of these with differential forms is the following:

Proposition 5.1   Let M^n be an n-manifold (smooth) and U_α a coordinate chart, in which $dx_1, \ldots, dx_n$ are differential 1-forms parallel to the coordinate axes. Let

$$\omega = f(x)\, dx_{m_1} \wedge \cdots \wedge dx_{m_i}$$

be an i-form in this chart. Pass to an exterior i-form on vector fields by taking i vector fields $s_{m_1}, \ldots, s_{m_i}$ and forming the exterior i-form

$$f(x) \sum_\rho \text{sign } \rho(dx_{m_1}(s_{\rho(m_1)}(x))) \cdots (ds_{m_i}(s_{\rho(m_i)}(x))),$$

where ρ are the permutations of the i symbols, and sign ρ is $+1$ is ρ is even, -1 if ρ is odd. Of course, $dx_{m_1}(s_{\rho(m_1)}(x))$ means dx_{m_1} on the vector.

This map extends to a linear map

$$\psi : D^i(M^n) \rightarrow A^i(M^n)$$

which is an isomorphism of vector spaces.

Proof The map is clearly well defined and multilinear, and extends obviously to a map on all $D^i(M^n)$. To see that it is alternating, note that if we permute two of the vector fields we have the same sum, except that sign p is changed for each term; hence the result is the negative of the original.

To see that this map ψ is 1–1, take as our form

$$\sum_{m_1 < \cdots < m_i} f_{m_1, \ldots, m_i} dx_{m_1} \wedge \cdots \wedge dx_{m_i}$$

locally, and select i vector fields parallel to $x_{m_1}, \ldots, x_{m_i}$ locally for some fixed $m_1, \ldots, m_i$, say $s_{m_1}, \ldots, s_{m_i}$. Then, for ψ evaluated on these vector fields, all terms but this $f_{m_1, \ldots, m_i} dx_{m_1} \wedge \cdots \wedge dx_{m_i}$, give zero, and this term gives $f_{m_1, \ldots, m_i}(x)$ times the product of the values of dx_{m_i} on $s_{m_i}(x)$, etc. Hence, our map is 1–1.

To check that it is onto, take an exterior i-form A locally, select elements of T^*M^n and vector fields, $dx_{m_1}, \ldots, dx_{m_i}$ and $s_{m_1}, \ldots, s_{m_i}$, respectively, such that

$$dx_{m_p}(s_{m_q}(x)) = \begin{cases} 1 & \text{if } p = q \\ 0 & \text{otherwise.} \end{cases}$$

Consider the differential form

$$\omega = A(s_{m_1}(x), \ldots, s_{m_i}(x)) \, dx_{m_1}, \ldots, dx_{m_i}.$$

Passing back to exterior forms,

$$\psi(\omega)(s_{m_1}(x), \ldots, s_{m_i}(x))$$
$$= A(s_{m_1}(x), \ldots, s_{m_i}(x)) \sum \text{sign } \rho (dx_{m_1}(s_{\rho(m_1)}(x)) \cdots dx_{m_i}(s_{\rho(m_i)}(x)))$$
$$= A(s_{m_1}(x), \ldots, s_{m_i}(x))(+1)(dx_{m_1})(s_{m_1}(x)) \cdots (dx_{m_i})(s_{m_i}(x))$$
$$= A(s_{m_1}(x), \ldots, s_{m_i}(x)).$$

Of course, we must check this on i arbitrary vector fields. But everything is linear over $C^\infty(M^n)$, and locally any vector field s may be written

$$s(x) = \alpha_1(x)s_1(x) + \cdots + \alpha_n(x)s_n(x)$$

for $\alpha_1, \ldots, \alpha_n \in C^\infty(M^n)$ (check this). Thus, $\psi(\omega) = A$, and ψ is onto.

The passage from local to global is routine, with partitions of unity.

To study the exterior product of forms from the point of view of exterior forms, consult [47].

The above proposition, interpreting a differential i-form as an exterior i-form on vector fields, may now be used to give an alternate description of Definition 5.6. Suppose $f: M^n \to N^k$ is a smooth map of smooth manifolds. Let s be a smooth vector field on M^n. For each point $x \in M^n$, $(df)_x s(x)$ is a vector in (TN^k). Now suppose $\sigma \in D^i(N^k)$ and $s_{m_1}, \ldots, s_{m_i}$ are i-vector fields on M^n. For each $x \in M^n$, $(df)_x s_{m_1}(x), \ldots, (df)_x s_{m_i}(x)$ are i tangent vectors to N^k at $f(x)$. Extend each $(df)_x s_{m_j}(x)$ to a smooth vector field $\tilde{s}_{m_j}$ on N^k that agrees with $(df)_x s_{m_j}(x)$ at $f(x)$, and set

$$D(f)(\sigma)(s_{m_1}, \ldots, s_{m_j})(x) = \sigma(\tilde{s}_{m_1}, \ldots, \tilde{s}_{m_j})(f(x)).$$

It is easy to verify that this is independent of the choice of the extensions of each vector to a vector field and gives an alternate description of $D(f)(\sigma)$.

There remains one further property of differential forms, which is basic yet a bit deeper than the previous properties. We suppose, as an example, that $M^n = \mathbb{R}^1$ and $f: \mathbb{R}^1 \to \mathbb{R}^1$ is a smooth function, i.e., a 0-form. The expression $f'(x)\, dx$ immediately defines a 1-form, the usual differential. The amazing thing is that this generalizes to an arbitrary i-form on any smooth manifold. It is a sort of super differentiation, called the exterior derivative, and it has many interesting and useful properties.

Rather than state a definition immediately, we shall prove that there exists one and only one linear map

$$d: D^i(M^n) \to D^{i+1}(M^n)$$

with the desired properties.

Proposition 5.2   Let M^n be a smooth manifold, with $D^i(M^n)$ being the smooth differential i-forms on M^n.

Then for each $i \geq 0$, there is a unique linear map

$$d: D^i(M^n) \to D^{i+1}(M^n)$$

satisfying the following:

(i)   If $\omega \in D^i(M^n)$, $d(d\omega) = 0 \in D^{i+2}(M^n)$.

(ii) $d(\omega_1 \wedge \omega_2) = d\omega_1 \wedge \omega_2 + (-1)^i \omega_1 \wedge d\omega_2$ for all $\omega_1 \in D^i(M^n)$, $\omega_2 \in D^j(M^n)$.

(iii)   In a coordinate chart U_α, choose a basis of tangent vectors $X_1, \ldots, X_n$ parallel to the coordinate axes, and choose a dual basis dx_i by $(dx_i)(X_j) = 1$ if $i = j$, 0 otherwise. Given $f \in C^\infty(M^n) = D^0(M^n)$. Then in this chart U_α, with respect to our bases, we have

$$df = \sum_{i=1}^n \frac{\partial f}{\partial x_i}\, dx_i.$$

Proof We prove uniqueness first. Note that if $\omega = 0$ in an open set U $d\omega = 0$ in U, because we can select a C^∞ function f that is positive at any given, fixed $x \in U$ and 0 outside U, and then calculate [with (ii)] at $x \in U$,

$$0 = d(0) = d(f \circ \omega) = d(f \wedge \omega) = df \wedge \omega + f \wedge d\omega;$$

since f is positive at x, and ω is 0 on U, $d\omega = 0$ at x.

Now, given two maps d_1 and d_2 meeting the conditions of the proposition, but for which $d_1 \neq d_2$, there must be some form ω for which $d_1\omega \neq d_2\omega$.

Our first remarks, which use only (ii), may be immediately applied to $d_1 - d_2$, which satisfies (ii) to get that $d_1 - d_2 = 0$, provided $d_1 - d_2$ is 0 in some open neighborhood of any point (because these first remarks permit the use of a partition of unity). In other words, we need only show that in a coordinate neighborhood of any point $d_1 = d_2$. For forms of degree 1, this is immediate from property (iii). For general forms ω, in a coordinate chart, which are simply sums of terms like

$$f(x)\, dx_{r_1} \wedge \cdots \wedge dx_{r_i},$$

it follows by an obvious induction using (i) and (ii).

We prove existence by first working locally, where we may assume that our form is

$$f(x)\, dx_{r_1} \wedge \cdots \wedge dx_{r_i}.$$

Using (iii), we make the definition

$$d(f\, dx_{r_1} \wedge \cdots \wedge dx_{r_i}) = (df) \wedge dx_{r_1} \wedge \cdots \wedge dx_{r_i}.$$

We must now prove (ii) and (i). First we set

$$\omega_1 = f_1\, dx_{r_1} \wedge \cdots \wedge dx_{r_i}, \qquad \omega_2 = f_2\, dx_{s_1} \wedge \cdots \wedge dx_{s_j}.$$

We calculate

$$\begin{aligned}
d(\omega_1 \wedge \omega_2) &= d((f_1 \cdot f_2)\, dx_{r_1} \wedge \cdots \wedge dx_{r_i} \wedge dx_{s_1} \wedge \cdots \wedge dx_{s_j}) \\
&= d(f_1 \cdot f_2) \wedge dx_{r_1} \wedge \cdots \wedge dx_{r_i} \wedge dx_{s_1} \wedge \cdots \wedge dx_{s_j} \\
&= [(df_1)f_2 + f_1(df_2)]\, dx_{r_1} \wedge \cdots \wedge dx_{r_i} \wedge dx_{s_1} \wedge \cdots \wedge dx_{s_j} \\
&= df_1 \wedge dx_{r_1} \wedge \cdots \wedge dx_{r_i} \wedge f_2 \wedge dx_{s_1} \wedge \cdots \wedge dx_{s_j} \\
&\quad + (-1)^i f_1\, dx_{r_1} \wedge \cdots \wedge dx_{r_i} \wedge df_2 \wedge dx_{s_1} \wedge \cdots \wedge dx_{s_j}.
\end{aligned}$$

We have used the fact that a function is a 0-form, whose product (exterior product) with any form is commutative, whereas to get df_2 past the forms $dx_{r_1} \wedge \cdots \wedge dx_{r_i}$ we need to use anticommuiativity i times. But the end product of this calculation is

$$d\omega_1 \wedge \omega_2 + (-1)^i \omega_1 \wedge d\omega_2,$$

as desired.

Finally, to prove (i), apply d to

$$(df)_x = \sum_{i=1}^{n} \frac{\partial f(x)}{\partial x_i} dx_i,$$

obtaining

$$d(df)_x = \sum_{i=1}^{n} \sum_{j=1}^{n} \frac{\partial^2 f}{\partial x_j \partial x_i} dx_j \wedge dx_i = \sum_{i,j=1}^{n} \frac{\partial^2 f}{\partial x_j \partial x_i} dx_j \wedge dx_i.$$

Since $\partial^2 f/\partial x_j \partial x_i = \partial^2 f/\partial x_i \partial x_j$ in our case and $dx_i \wedge dx_j = -dx_j \wedge dx_i$, we get term-by-term cancellation, and hence the desired result. In general, to check that $d(d\omega) = 0$, apply this to each term in a local expression for ω, observing that if

$$\omega = f\, dx_{r_1} \wedge \cdots \wedge dx_{r_i},$$

$$d\omega = \left(\sum_{j=1}^{n} \frac{\partial f}{\partial x_j} dx_j \right) \wedge dx_{r_1} \wedge \cdots \wedge dx_{r_i};$$

we know that d annihilates every term in this exterior product, completing the proof of the proposition. We then make the following definition:

Definition 5.7   The unique linear map $d: D^i(M^n) \to D^{i+1}(M^n)$ is called the *exterior derivative*.

In order to compare the exterior derivatives on different manifolds, it is important now to understand the relation between d and maps $D(f)$, as defined in Definition 5.6.

Proposition 5.3   Let $f: M^n \to N^k$ be smooth. Then for any $\omega \in D^i(N^k)$,

$$d(D(f))(\omega) = D(f)(d\omega).$$

Of course, the d on the left is the exterior derivative, in dimension i, on the manifold M^n, while that on the right refers to N^k.

Proof   As we have seen, it suffices to prove such a proposition locally, that is, for an expression of the form

$$f_{m_1,\ldots,m_i} dx_{m_1} \wedge \cdots \wedge dx_{m_i}.$$

But because $D(f)$ takes products to products and d on a product decomposes in terms of the different d's on each factor [Proposition 5.2, condition (ii)], it suffices to check this on a 0-form or function $\alpha \in D^0(N^k)$ (check!).

According to Definition 5.6, $D(f)(\alpha) = \alpha \circ f$. Using Proposition 5.2, we calculate as follows:

$$d(D(f)(\alpha)) = d(\alpha \circ f) = \sum_{i=1}^{n} \frac{\partial(\alpha \circ f)}{\partial x_i} dx_i.$$

Expressed locally, in terms of some given basis in N^k these partial derivatives must be calculated by the usual chain rule or, equivalently, as the product of the appropriate Jacobians (as in Proposition 1.2). But this means that the components $\partial(\alpha \circ f)/\partial x_i$ are the elements in the matrix product

$$(\text{grad } \alpha)J_f$$

(recall that $\alpha \circ f$ means apply f first and then α).

On the other hand, taking d first, we have

$$d\alpha = \sum_{j=1}^{k} \frac{\partial \alpha}{\partial y_i} dy_j$$

(fixing a basis $y_1, \ldots, y_k$ in N^k). To calculate $D(f)$ on this, by Definition 5.6, we apply f initially and compose the result with the dual of the Jacobian. But left multiplication by the transpose of the Jacobian is the same as right multiplication by the Jacobian, giving the same result, $(\text{grad } \alpha) \circ J_f$.

Of course, not only does the exterior derivative generalize the notion of differentiation on $D^0(M^n)$, but acting on $D^1(M^n)$ it relates to the theory of exact, first-order linear equations (such as $M + N \, dy/dx = 0$). We shall fully examine these ramifications in the Poincaré lemma (Proposition 5.4); at this point, before embarking on the integration of differential forms, we may give the basic definitions which relate to the exterior derivative.

Definition 5.8 A differential form $\omega \in D^i(M^n)$, $i \geq 0$, is *closed* if $d\omega = 0$. If $\omega = d\sigma$ for some $\sigma \in D^{i-1}(M^n)$, we call ω *exact*.

Note that a zero form is closed precisely when $\partial f/\partial x_i = 0$, for all i, in any coordinate chart, i.e., a zero form is closed if and only if it is constant. Now, Proposition 5.2(i), implies that an exact form is closed ($d\omega = d(d\sigma) = 0$). The subspace of exact forms in $D^i(M^n)$ is written $B^i(M^n)$. The subspace of closed forms is written $Z^i(M^n)$. As we have observed, $B^i(M^n) \subseteq Z^i(M^n)$. The quotient space

$$H^i(M^n) = Z^i(M^n)/B^i(M^n)$$

is called the i-dimensional deRham cohomology (compare [23, 53]). This is an important (advanced) invariant of a manifold. We shall discuss it a little further at the end of this chapter.

INTEGRATION ON MANIFOLDS

We now return to integration with the intent of developing the machinery to integrate an n-form over an n-dimensional manifold. One should be able to do this by piecing together integrals over little pieces of the manifold, the coordinate neighborhoods. But if this is to make sense globally, one must have a specific way of putting together the pieces, which means that the theory should be for orientable manifolds. (One can discuss the nonorientable case in terms of twofold covering spaces [23].) By Example 4 following Definition 5.4, we know that there is an everywhere nonvanishing n-form on an orientable n-manifold. This will serve as a frame of reference, or a measure of size, against which we may make comparisons. The details then are the following:

Let M^n be a smooth n-manifold (usually C^1 will suffice). Assuming M^n orientable, suppose we are given a nowhere vanishing n-form, $\mu \in D^n(M^n)$. For a given open set $U \subseteq M^n$, suppose we have two coordinate charts (with the same underlying set U)

$$\phi_1 : U \to \mathbb{R}^n \quad \text{and} \quad \phi_2 : U \to \mathbb{R}^n.$$

For convenience, we suppose that the coordinates in these charts are written $(x_1, \ldots, x_n)$ and $(y_1, \ldots, y_n)$. There are then nonvanishing functions f_1 and f_2 such that throughout U

$$D(\phi_1)(dx_1 \wedge \cdots \wedge dx_n) = f_1 \mu \quad \text{and} \quad D(\phi_2)(dy_1 \wedge \cdots \wedge dy_n) = f_2 \mu.$$

This is possible because the space of n-forms is one dimensional at a point and the forms in question are nowhere vanishing. In fact, we may also assume that the charts respect the orientation in the manifold, so that f_1 and f_2 are positive.

Now, let ω be an arbitrary n-form with compact support; ω will be the form that we wish to integrate (i.e., outside some compact K, ω is identically zero). In U, we write

$$\omega = g_1 \cdot D(\phi_1) \, dx_1 \wedge \cdots \wedge dx_n \quad \text{and} \quad \omega = g_2 \cdot D(\phi_2) \, dy_1 \wedge \cdots \wedge dy_n.$$

Naturally we may assume g_1 and g_2 are continuous and have compact support.

We claim (writing now $dx_1 \wedge \cdots \wedge dx_n$ for the usual $dx_1 \cdots dx_n$ in $\mathbb{R}^n$) that

$$\int_{\phi_1(U)} g_1 \circ \phi_1^{-1} \, dx_1 \wedge \cdots \wedge dx_n = \int_{\phi_2(U)} g_2 \circ \phi_2^{-1} \, dy_1 \wedge \cdots \wedge dy_n. \tag{*}$$

To see this, note first that by Example 4 after Definition 5.4, we have

$$D(\phi_1 \circ \phi_2^{-1}) \, dx_1 \wedge \cdots \wedge dx_n = \det(J_{\phi_1 \circ \phi_2^{-1}}) \, dy_1 \wedge \cdots \wedge dy_n,$$

where we may dispense with any absolute value symbols because both f_1 and f_2 are assumed positive. On the other hand, applying the *change of variable formula*, we get

$$\int_{\phi_1(U)} g_1 \circ \phi_1^{-1} \, dx_1 \wedge \cdots \wedge dx_n$$

$$= \int_{\phi_2(U)} g_1 \circ \phi_1^{-1} \circ \phi_1 \circ \phi_2^{-1} \big| (J_{\phi_1 \circ \phi_2^{-1}}) \big| \, dy_1 \wedge \cdots \wedge dy_n,$$

where the variable in the second integral at which one is to evaluate the function, as well as the Jacobian, is $\phi_2(\phi_1^{-1}(x))$.

Now, returning to our two formulas for ω, and recalling that on functions g(0-forms) $D(\phi)g = g \circ \phi$, we have

$$D(\phi_2^{-1})(\omega) = (g_2 \circ \phi_2^{-1}) \, dy_1 \wedge \cdots \wedge dy_n$$

because $D(\phi_2^{-1}) \circ D(\phi_2) = D(\phi_2 \circ \phi_2^{-1})$ is the identity, while using the other formula,

$$D(\phi_2^{-1})(\omega) = (g_1 \circ \phi_2^{-1}) D(\phi_2^{-1}) D(\phi_1) \, dx_1 \wedge \cdots \wedge dx_n.$$

Because $D(\phi_2^{-1}) D(\phi_1) = D(\phi_1 \circ \phi_2^{-1})$ and

$$D(\phi_1 \circ \phi_2^{-1}) \, dx_1 \wedge \cdots \wedge dx_n = \big| J_{\phi_1 \circ \phi_2^{-1}} \big| \, dy_1 \wedge \cdots \wedge dy_n,$$

we substitute $(g_2 \circ \phi_2^{-1}) \, dy_1 \wedge \cdots \wedge dy_n$ for

$$(g_1 \circ \phi_1^{-1} \circ \phi_1 \circ \phi_2^{-1}) \big| J_{\phi_1 \circ \phi_2^{-1}} \big| \, dy_1 \wedge \cdots \wedge dy_n,$$

that is, we substitute our first expression for $D(\phi_2^{-1})(\omega)$ for the second in the change of variable formula. Then our claim $(*)$ is immediate.

We are now in a position to make our definition.

Definition 5.9 Let $\{U_i, \phi_i\}$ be a locally finite coordinate cover of the orientable manifold M^n, by coordinate neighborhoods U_i, endowed with homeomorphisms ϕ_i. Suppose we have (with no loss of generality) a partition of unity $\{\zeta_i\}$, where each ζ_i has compact support in U_i.

Let μ be a nowhere vanishing form in $D^n(M^n)$. In each chart U_i, we suppose, by orientability, that

$$D(\phi_i) dx_1 \wedge \cdots \wedge dx_n = f_i \cdot \mu, \qquad f_i > 0.$$

Let ω be a continuous form, with compact support, and suppose that in the chart U_i,

$$\omega = g_i \cdot D(\phi_i) dx_1 \wedge \cdots \wedge dx_n.$$

We then define

$$\int_{M^n} \omega = \sum_i \int_{\phi(U_i)} (\zeta_i g_i) \circ \phi_i^{-1} \, dx_1 \wedge \cdots \wedge dx_n,$$

where $\zeta_i g_i$ means the product of these functions, and $(\zeta_i g_i) \circ \phi_i^{-1}$ means their composition with ϕ_i^{-1}.

Remarks   (1)   If ω is a differential form with compact support, there is no serious problem in showing that

$$\int_{M^n} \omega$$

exists. Because the cover in our definition is locally finite, only a finite number of terms in our sum are nonzero when the support is compact. Each term

$$\int_{\phi_i(U_i)} (\zeta_i g_i) \circ \phi_i^{-1} \, dx_1 \wedge \cdots \wedge dx_n$$

will exist if we can show that the integral of the absolute value of the integrand exists. Since we suppose that each ζ_i has compact support, and since the g_i may be assumed bounded (because ω) has compact support), the desired integral will clearly exist.

We must obviously also address the question of independence of the choice of cover and partition of unity.

(2)   Our earlier claim (∗) established the independence of our definition from the choice of homeomorphism $\phi_i: U_i \to \mathbb{R}^n$ for a given cover $\{U_i\}$.

Because any two coordinate covers have a common refinement, it will suffice to check the independence of our definition when our cover is replaced by a refinement $\{V_j\}$; i.e., if $I = \{i\}$, $J = \{j\}$, there is a cover $\{V_j\}$ and a map $r: J \to I$ such that for each j, $V_j \subseteq U_{r(j)}$. With no loss of generality assume that $\{V_j\}$ is locally finite, with homeomorphisms ψ_j, and $\{\eta_j\}$ is a suitable partition of unity subordinate to the cover $\{V_j\}$. Set P_i to be the set of j such that $V_j \subseteq U_i$. We calculate

$$\int_{M^n} \omega = \sum_j \int_{\psi_j(V_j)} (\eta_j g_j) \circ \psi_j^{-1} \, dy_1 \wedge \cdots \wedge dy_n$$

$$= \sum_i \sum_{j \in P_i} \int_{\psi_j(V_j)} (\eta_j g_j) \circ \psi_j^{-1} \, dy_1 \wedge \cdots \wedge dy_n.$$

Our formula (∗) above shows that this is precisely

$$\sum_i \sum_{j \in P_i} \int_{\phi_i(V_j)} (\eta_j g_j) \circ \phi_i^{-1} \, dx_1 \wedge \cdots \wedge dx_n,$$

where we use g_j to also denote the appropriate function in this set $\phi_i(V_j)$. And because supp $n_j \subseteq V_j$, we may rewrite this

$$\sum_i \int_{\phi_i(U_i)} \left(\sum_{j \in P_i} n_j \right) g_i \circ \phi_i^{-1} \, dx_1 \wedge \cdots \wedge dx_n.$$

Putting $\zeta_i = \sum_{j \in P_i} \eta_i$ (which is obviously a partition of unity) results in our original definition for this partition of unity.

(3) To address the question of the independence of a partition of unity for a given cover, we may use a standard approximation technique. Our integral may be approximated with arbitrary fine accuracy by integrals of forms ω_α that are constant on compact sets in U_i, say, constant on sets $\phi_i^{-1}(\square)$, where $\square$ is a compact cube in $\mathbb{R}^n$, and zero elsewhere. The fact that ω is discontinuous on the boundary of $\phi^{-1}(\square)$ causes no difficulty for the integration at hand.

Now, if $\{\zeta_i'\}$ and $\{\zeta_i''\}$ are two partitions of unity subordinate to $\{U_i\}$ and ω_α is a form as in the preceding paragraph, where we may clearly assume that supp ω_α is a compact subset of all the (finite in number) U_i which meet it, then we have according to $\{\zeta_i'\}$

$$\int_{M^n} \omega_\alpha = \sum_i \int_{\phi_i(U_i)} (\zeta_i' g_i) \circ \phi_i^{-1} \, dx_1 \wedge \cdots \wedge dx_n,$$

while according to $\{\zeta_i''\}$ we get

$$\sum_i \int_{\phi_i(U_i)} (\zeta_i'' g_i) \circ \phi_i^{-1} \, dx_1 \wedge \cdots \wedge dx_n.$$

If we assume that g_i is the constant 1 (no loss of generality), these sums are then

$$\sum_i \int_{\phi_i(\text{supp }\omega_\alpha)} \zeta_i' \circ \phi_i^{-1} \, dx_1 \wedge \cdots \wedge dx_n,$$

and the same with ζ_i'' replacing ζ_i'. Letting i_0 be the one such index with supp $\omega_\alpha \subseteq U_{i_0}$, we use $(*)$ to rewrite this as

$$\sum_i \int_{\phi_i(\text{supp }\omega_\alpha)} \zeta_i' \circ \phi_i^{-1} \, dx_1 \wedge \cdots \wedge dx_n = \sum_i \int_{\phi_{i_0}(\text{supp }\omega_\alpha)} \zeta_i' \circ \phi_{i_0}^{-1} \, dx_1 \wedge \cdots \wedge dx_n$$

$$= \int_{\phi_{i_0}(\text{supp }\omega_\alpha)} \left(\sum_i \zeta_i' \right) \circ \phi_{i_0}^{-1} \, dx_1 \wedge \cdots \wedge dx_n$$

$$= \int_{\phi_{i_0}(\text{supp }\omega_\alpha)} \phi_{i_0}^{-1} \, dx_1 \wedge \cdots \wedge dx_n.$$

This last inequality follows, of course, because $\{\zeta_i'\}$ is a partition of unity. The same obviously holds for $\{\zeta_i''\}$, proving the desired independence.

(4) Our original choice of the form μ is a question of the choice of orientation in the coordinate charts. The reader may easily check (see the remarks before Definition 5.9) that it does not figure in any way in the numerical value of the integral.

(5)   Of course, the numerical value of our integral is determined by the usual measure (or customary integration) in Euclidean space $\mathbb{R}^n$. Various other measures on $\mathbb{R}^n$ would change the value of the integral.

(6)   Finally, we remark that our approach is totally extrinsic, in the sense that we define the integral on a manifold entirely in terms of integrals off the manifold. A reader who has some familiarity with measure theory might wish to explore the alternative approach to integration, which would proceed from an assignment of a measure to nice subsets of the manifold and then develop the integral. One straightforward way to tackle this would be to use coordinate charts and a partition of unity to define an "average" measure for suitable nice subsets.

(7)   We emphasize that the integral is basically defined for n-forms with compact support. The reader is probably more comfortable thinking in terms of integration of functions. Such an approach would be the following:

(i)   Select a nowhere vanishing form μ using the orientability of M^n.

(ii)   For any continuous function f with compact support, form the n-form

$$\omega = f\mu,$$

i.e., $f \wedge \mu$. Define the integral of f, with respect to this choice of μ, to be the integral of ω.

One often sees this written as

$$\int_{M^n} f\mu$$

and the form μ referred to as a *volume element* on the manifold. Naturally, the numerical value of the integral of a function f will depend strongly on the choice of μ.

Because the space of n-forms at a given point x is one-dimensional, any form ω may be written

$$\omega = f\mu,$$

so the theories are totally equivalent.

THE POINCARÉ LEMMA

We now are ready for the Poincaré lemma. In our earlier language, an exact form is closed. This famous lemma shows that when $M^n = \mathbb{R}^n$ the converse is also true. This amazing result is a broad generalization of traditional results on exact, first-order differential equations.

Proposition 5.4 (Poincaré's Lemma)   Let ω be a smooth p-form on the manifold $\mathbb{R}^n$, with $p > 0$, $n > 0$. Suppose ω is closed, i.e., $d\omega = 0$.
Then ω is exact; i.e., there is $v \in D^{p-1}(\mathbb{R}^n)$ such that $dv = \omega$.

Proof   We write

$$\omega = \sum_{i_1 < \cdots < i_p} A_{i_1, \ldots, i_p}\, dx_{i_1}, \ldots, dx_{i_p}.$$

We must first define a linear map with certain nice properties,

$$\Delta: D^p(\mathbb{R}^n) \to D^{p-1}(\mathbb{R}^n).$$

Then, after a short calculation, our result will follow. We set Δ on a single term and extend to the general definition linearly.

$$\Delta(a(x)\, dx_{i_1} \wedge \cdots \wedge dx_{i_p}) =$$

$$\left(\int_0^1 a(tx)t^{p-1}\, dt\right) \sum_{j=1}^{p} (-1)^{j-1} x_{i_j}\, dx_{i_1} \wedge \cdots \wedge \widehat{dx}_{i_j} \wedge \cdots \wedge dx_{i_p}.$$

Here, the caret over a term means omit that term. This map is vaguely some sort of inverse to exterior differentiation. More precisely, we calculate, for $\omega = a(x)\, dx_{i_1} \wedge \cdots \wedge dx_{i_p}$,

$$\Delta d\omega = \Delta\left(\sum_k \frac{\partial a(x)}{\partial x_{i_k}} dx_{i_k} \wedge dx_{i_1} \wedge \cdots \wedge dx_{i_p}\right)$$

$$= \sum_k \left(\int_0^1 \frac{\partial a(tx)}{\partial x_{i_k}} t^p\, dt\right)\left(x_{i_k}\, dx_{i_1} \wedge \cdots \wedge dx_{i_p}\right.$$

$$\left. + \sum_j (-1)^j x_{i_j}\, dx_{i_k} \wedge (dx_{i_1} \wedge \cdots \wedge \widehat{dx}_{i_j} \wedge \cdots \wedge dx_{i_p})\right).$$

On the other hand,

$$d\Delta\omega = d\left(\sum_j (-1)^{j-1}\left(\int_0^1 a(tx)t^{p-1}\, dt\right) x_{i_j}\, dx_{i_1} \wedge \cdots \wedge \widehat{dx}_{i_j} \wedge \cdots \wedge dx_{i_p}\right)$$

$$= \sum_j (-1)^{j-1}\left[\sum_k \left(\int_0^1 \frac{\partial a(tx)}{\partial x_{i_k}} t^p\, dt\right) x_{i_j}\, dx_{i_k}\right.$$

$$\left. + \left(\int_0^1 a(tx)t^{p-1}\, dt\right) dx_{i_j}\right] \wedge dx_{i_1} \wedge \cdots \wedge \widehat{dx}_{i_j} \wedge \cdots \wedge dx_{i_p}$$

$$= \left[\sum_{j,k} (-1)^{j-1}\left(\int_0^1 \frac{\partial a(tx)}{\partial x_{i_k}} t^p\, dt\right) x_{i_j}\, dx_{i_k} \wedge dx_{i_1} \wedge \cdots \wedge \widehat{dx}_{i_j} \wedge \cdots \wedge dx_{i_p}\right]$$

$$+ p\left(\int_0^1 a(tx)t^{p-1}\, dt\right) dx_{i_1} \wedge \cdots \wedge dx_{i_p},$$

the last term arising because there are $j - 1$ simple permutations, and hence sign changes, to bring dx_{i_j} back to its conventional place and there is a sign $(-1)^{j-1}$ in front of the earlier formula.

Notice that the first term of $d\Delta\omega$ is precisely the negative of the second term of $\Delta d\omega$. Note also that we may write the last term of $d\Delta\omega$ using

$$\int_0^1 a(tx)pt^{p-1}\, dt = a(tx)t^p\Big|_0^1 - \int_0^1 \sum_k \frac{\partial a(tx)}{\partial x_{i_k}} x_{i_k} t^p\, dt,$$

and therefore calculate as follows:

$$\Delta d\omega + d\Delta\omega = \sum_k \left(\int_0^1 \frac{\partial a(tx)}{\partial x_{i_k}} t^p\, dt \right) x_{i_k}\, dx_{i_1} \wedge \cdots \wedge dx_{i_p}$$
$$+ a(x)\, dx_{i_1} \wedge \cdots \wedge dx_{i_p}$$
$$- \sum_k \left(\int_0^1 \frac{\partial a(tx)}{\partial x_{i_k}} x_{i_k} t^p\, dt \right) dx_{i_1} \wedge \cdots \wedge dx_{i_p}$$
$$= a(x)\, dx_{i_1} \wedge \cdots \wedge dx_{i_p}.$$

In other words, we have the following equation in the compositions of these linear maps:

$$\Delta d + d\Delta = 1,$$

where 1 is the identity map on $D^p(\mathbb{R}^n)$.

The remainder of the proof is remarkably easy. Suppose ω is a closed form. Then

$$\Delta d\omega + d\Delta\omega = d\Delta\omega \qquad \text{but} \qquad \Delta d\omega + d\Delta\omega = 1(\omega) = \omega.$$

Thus $\omega = d(\Delta\omega)$, and ω is exact.

Remarks (1) Δ is not a strict inverse for d, but it functions as one when the form ω is closed. In algebraic topology, Δ bears the technical name *chain homotopy* between the identity map and the zero map. More generally, suppose we have two (linear) maps

$$\phi_1, \phi_2 : D^*(N^k) \to D^*(M^n)$$

such that $d\phi_i = \phi_i d$ for $i = 1, 2$, the first d referring to M^n and the second to N^k; such maps are sometimes called chain maps (and by Proposition 5.3, we have seen that any $D(f)$ is always such a map). Then ϕ_1 and ϕ_2 are chain homotopic if there is Δ with

$$\Delta d + d\Delta = \phi_1 - \phi_2.$$

In the case of the Poincaré lemma, $\phi_1 = 1$ and $\phi_2 = 0$.

(2) The reader should examine—in detail—the construction of Δ in the case $\omega \in D^0(M^n)$.

(3) It is easy to see that the proof may be made to work when $\mathbb{R}^n$ is replaced by an open, convex subset of $\mathbb{R}^n$.

(4) We may examine the case of $\mathbb{R}^2$ in detail. Suppose

$$\omega = M(x, y)\,dx + N(x, y)\,dy \in D^1(\mathbb{R}^2),$$

where dx and dy are dual to $i = (1, 0)$ and $j = (0, 1)$. Assume that

$$\partial M/\partial y = \partial N/\partial x.$$

Then it is a trivial calculation to check that $d\omega = 0$. We then recover the well-known fact (from ordinary differential equations) that

$$\omega = d\phi$$

for some function $\phi(x, y)$. In practice, to solve $M + N\,dy/dx = 0$, one must solve

$$\phi(x, y) = \text{const}$$

for y as a function of x.

(5) Examining the remarks after Definition 5.8, we conclude from the Poincaré lemma, that the deRham cohomology groups vanish, i.e.,

$$H^i(\mathbb{R}^n) = 0, \qquad i > 0.$$

$Z^0(\mathbb{R}^n)$ is given by constant functions, while $B^0(\mathbb{R}^n)$ is zero, by definition. It follows that

$$H^0(\mathbb{R}^n) = \mathbb{R}.$$

STOKES' THEOREM

Finally, we wish to turn to the general Stokes' theorem, which generalizes the usual integral formulas from vector analysis. The most reasonable general context in which one might wish to prove such a theorem is that of a manifold with boundary.

Definition 5.10   A Hausdorff space V^n is an n-dimensional *manifold with boundary* if V^n has a countable base and every $x \in V^n$ has an open neighborhood O_x with either one or the other of the following:

(i)   a homeomorphism $\phi_x: O_x \overset{\equiv}{\to} \mathbb{R}^n$, or

(ii)   a homeomorphism $\psi_x: O_x \overset{\equiv}{\to} \mathbb{R}^n_+$,

where $\mathbb{R}^n_+ = \{(x_1, \ldots, x_n) \in \mathbb{R}^n \,|\, x_1 \geqq 0\}$.

The points lying in open sets that have homeomorphisms of the second kind, i.e., ψ_x, and $x_1 = 0$ are called the points of the boundary, written ∂V^n. Clearly ∂V^n is a manifold of dimension $n - 1$.

One may define differentiability (i.e., a differentiable manifold with boundary) as before. One simply demands that the overlap maps be differentiable, up to a certain order, in the usual sense of advanced calculus, i.e., a map on the set such as $\mathbb{R}^n_+$ is differentiable at a point if it extends to an open neighborhood of that point, in all of $\mathbb{R}^n$, so as to be differentiable.

For example, $D^n = \{x \in \mathbb{R}^n \,|\, \|x\| \leq 1\}$ is clearly a manifold with boundary, $\partial D^n = S^{n-1}$. Given any manifold with boundary one may pass to an ordinary manifold by gluing together two identical copies along the boundary. This construction is called the *double*.

Most earlier concepts have generalizations to manifolds with boundary. For example, we call such a manifold orientable if one can choose the charts of the first kind in Definition 5.10 to have overlap functions whose Jacobians have positive determinant. Equivalently, one may define orientability by the existence of a nowhere vanishing n-form on $M^n - \partial M^n$. The choice of such a specific n-form is called an *orientation*; once such a choice is made, we call the manifold *oriented*.

Given an oriented manifold M^n with nonempty boundary ∂M^n, consider a coordinate chart of the second type in Definition 5.10, i.e., $\psi_\alpha: O_\alpha \to \mathbb{R}^n_+$. Suppose $dx_1, \ldots, dx_n$ are the duals of the vectors, which point in the positive direction along the axes of $\mathbb{R}^n_+$, for any $(x_1, \ldots, x_n)$ with $x_1 > 0$. Suppose

$$dx_1 \wedge \cdots \wedge dx_n$$

is a positive multiple of the orientation on M^n. Then we define an *orientation in* ∂M^n, as a matter of convention, by the form

$$-dx_2 \wedge \cdots \wedge dx_n.$$

Clearly, this choice prescribes an orientation on ∂M^n. In other words, take the orientation of M^n where the first coordinate is an inward pointing normal, throw away the first coordinate, and change the sign of the orientation form to arrive at the orientation on ∂M^n. This convention yields the simplest form of Stokes' theorem.

One may obviously extend the theory of integration to manifolds with boundary, requiring that the forms in question approach a continuous limit as one approaches the boundary ∂M^n and then using the same formula as in Definition 5.9. To insure that the necessary point-set topology works, as with the need for locally finite covers to do integration, etc., we shall always assume that these manifolds with boundary have countable base.

We may now state the basic Stokes' theorem.

Theorem 5.1 (Stokes' Theorem) Let V^n be a smooth, oriented, n-dimensional manifold with boundary ∂V^n. Let $\omega \in D^{n-1}(V^n)$ be a smooth $(n-1)$-form with compact support. We write ω also for $D(h)(\omega)$, where h is the inclusion map $h: \partial V^n \to V^n$. In other words, we write ω for the restriction ω to ∂V^n.

Then we have

$$\int_{\partial V^n} \omega = \int_{V^n} d\omega.$$

Proof Let $\{\zeta_i\}$ be a partition of unity of V^n with respect to the cover $\{O_i\}$ of V^n, and assume supp ζ_i is a compact subset of O_i for each i. Note that locally, that is, in each O_i, the form $\zeta_i \omega$ may be written

$$\zeta_i \omega = \sum_{j=1}^{n} f_j \, dx_1 \wedge \cdots \wedge \widehat{dx_j} \wedge \cdots \wedge dx_n.$$

Of course, the caret over a symbol means one omits that symbol.

By the additivity of our integrals, it suffices to take $\zeta_i \omega$ to be our form, and we write

$$d(\zeta_i \omega) = \sum_{j=1}^{n} \frac{\partial f_i}{\partial x_j} \, dx_j \wedge dx_1 \wedge \cdots \wedge \widehat{dx_j} \wedge \cdots \wedge dx_n,$$

because the other terms in the expression for df_j contain some dx_k that is duplicated in the rest of the expression and hence yields zero.

Therefore, we calculate

$$\int_{V^n} d(\zeta_i \omega) = \int_{V^n} \sum_{j=1}^{n} \frac{\partial f_i}{\partial x_j} \, dx_j \wedge dx_1 \wedge \cdots \wedge \widehat{dx_j} \wedge \cdots \wedge dx_n$$

$$= \sum_{j=1}^{n} \int_{\mathbb{R}^n} \frac{\partial f_j}{\partial x_j} (-1)^{j-1} \, dx_1 \wedge \cdots \wedge dx_n,$$

in the case where O_i is homeomorphic to $\mathbb{R}^n$.

But if O_i is homeomorphic to $\mathbb{R}^n$, we write this as an iterated integral. Since each f_j has compact support, we see that

$$\int_{-\infty}^{\infty} \frac{\partial f_j}{\partial x_j} \, dx_j$$

vanishes, for any fixed values of $x_1, \ldots, x_{j-1}, x_{j+1}, \ldots, x_k$. In this case ($O_i$ homeomorphic to $\mathbb{R}^n$) in addition

$$\int_{\partial V^n} \zeta_i \omega$$

is trivially zero, because $\operatorname{supp}(\zeta_i \omega) \cap \partial V^n = \phi$. Thus, the theorem is proved for $\zeta_i \omega$ in such an open set.

On the other hand, consider an open coordinate neighborhood of the second kind, that is, O_i homeomorphic to $\mathbb{R}^n_+$. Once again, we write

$$\zeta_i \omega = \sum f_j \, dx_1 \wedge \cdots \wedge \widehat{dx_j} \wedge \cdots \wedge dx_n$$

and

$$d(\zeta_i \omega) = \sum_j \frac{\partial f_j}{\partial x_j} dx_j \wedge dx_1 \wedge \cdots \wedge \widehat{dx_j} \wedge \cdots \wedge dx_n.$$

If $j > 1$,

$$\int_{\mathbb{R}^n_+} \frac{\partial f_j}{\partial x_j} dx_j \wedge dx_1 \wedge \cdots \wedge \widehat{dx_j} \wedge \cdots \wedge dx_n;$$

this may be rewritten as an iterated integral, involving

$$\int_{-\infty}^{\infty} \frac{\partial f_j}{\partial x_j} dx_j.$$

This clearly vanishes when f_j has compact support.

But then, if $j = 1$, we calculate

$$\int_{\mathbb{R}^n} \frac{\partial f}{\partial x_1} dx_1 \wedge \cdots \wedge dx_n = \int_{-\infty}^{\infty} dx_n \int_{-\infty}^{\infty} dx_{n-1} \cdots \int_0^{\infty} \frac{\partial f_1}{\partial x_1} dx_1$$

$$= -\int_{\mathbb{R}^{n-1}_0} f_1(0, x_2, \ldots, x_n) \, dx_2 \wedge \cdots \wedge dx_n,$$

where $\mathbb{R}^{n-1}_0 = \{x \in \mathbb{R}^n_+ \,|\, x_1 = 0\}$ is the boundary of $\mathbb{R}^n_+$.

Therefore,

$$\int_{V^n} d(\zeta_i \omega) = -\int_{\mathbb{R}^{n-1}_0} f_1(0, x_2, \ldots, x_n) \, dx_2 \wedge \cdots \wedge dx_n.$$

Now if $h: \partial V^n \to V^n$ is the inclusion, one checks at once that $D(h)$ annihilates dx_1. In other words

$$D(h)(f_j(dx_1 \wedge \cdots \wedge \widehat{dx_j} \wedge \cdots \wedge dx_n)) = 0 \qquad \text{if} \quad j \neq 1$$

(compare the remarks following Definition 5.6). This then means that

$$\int_{\partial V^n} \zeta_i \omega,$$

which is our shorthand for

$$\int_{\partial V^n} D(h)(\zeta_i \omega),$$

must be equal to

$$\int_{R_0^n} f_1 \, dx_2 \wedge \cdots \wedge dx_n.$$

Recalling our conventions on orientation immediately preceding Theorem 5.1, we have

$$\int_{V^n} d(\zeta_i \omega) = \int_{\partial V^n} \zeta_i \omega,$$

which completes the proof.

Examples   1. Let $f(x, y)$ and $g(x, y)$ be smooth functions on $D^2 = \{(x, y) \,|\, x^2 + y^2 \leqq 1\}$. Set

$$\omega = f(x, y) \, dx + g(x, y) \, dy$$

and notice that

$$d\omega = -\frac{\partial f}{\partial y} \, dx \wedge dy + \frac{\partial g}{\partial x} \, dx \wedge dy$$

(because $dy \wedge dx = -dx \wedge dy$). According to Theorem 5.1,

$$\int_{D^2} d\omega = \int_{S^1 = \partial D^2} \omega$$

or

$$\int_{D^2} \left(\frac{\partial g}{\partial x} - \frac{\partial f}{\partial y} \right) dx \, dy = \int_{S^1} f \, dx + g \, dy.$$

This is, of course, the classical Green's formula. In a similar vein, one may recover the classical Stokes' theorem for an orientable 2-dimensional manifold with boundary in $\mathbb{R}^3$.

On the other hand, the basic Gauss theorem (divergence theorem) is on easy corollary once one knows the star operator on differential forms (see the next chapter, the material following Definition 6.5).

2. Various extensions of Stokes' theorem (Theorem 5.1) are possible; one may consider a polyhedron (union of simplexes) in Euclidean space. This is a case where the boundary is not a smooth submanifold, but rather several pieces of Euclidean space sewed together continuously along their boundaries. With suitable care about orientations, one can then generalize Stokes' theorem.

A general version of Stokes' theorem is a basic tool in the famous deRham theorem, which establishes the agreement between deRham cohomology and the ordinary cohomology, with real coefficients of a manifold. The map, which gives the isomorphism, is an integration map, and the needed commutation property, for exterior differentiation, is a direct result of Stokes' theorem.

For all this, see DeRham's book [23], and [53, 18]. The latter two works discuss this result from the point of view of sheaf theory. This is a powerful machine, combining algebra and geometry, that is especially useful for these purposes.

One immediate corollary of Stokes' theorem is

Corollary 5.1 If M^n is a compact orientable n-manifold (smooth and without boundary) and $\omega \in D^{n-1}(M^n)$, then

$$\int_{M^n} d\omega = 0.$$

Proof Let B_ε be a small open ball within M^n and set $V^n = M^n - B_\varepsilon$, so that V^n is a manifold with boundary. Stokes' theorem then says that

$$\int_{V^n} d\omega = \int_{\partial V^n} \omega = \pm \int_{\partial B_\varepsilon} \omega = \pm \int_{B_\varepsilon} d\omega,$$

depending on orientation.

Because the right-hand side is bounded by the maximum of $d\omega$ on the compact B_ε multiplied by the volume of B_ε, i.e., the integral of $1 \cdot \mu$ over the small ball, it is clear that the right-hand integral goes to zero with ε. Since the left does not depend on ε, we conclude that it must be zero, proving the corollary.

In conclusion, we mention several further areas that are deeply involved with differential forms.

(i) Forms play a major role in any mathematical development of classical mechanics. (See, e.g., [38, 2].)

(ii) Differential forms, especially the invariant ones, are basic to the theory of Lie groups and algebras. In addition to Chapter 9, of this book, the reader might consult [20, 54].

(iii) Of course, the origins of differential forms are basically tied to differential geometry. There is an enormous literature, from the classical works of E. Cartan [17] and Levi-Civita [71] to the modern texts, such as [65, 104].

PROBLEMS AND PROJECTS

1. Let $\omega \in D^p(\mathbb{R}^n)$ be a closed p-form, $p > 0$. Suppose M^p is a smooth, orientable submanifold (without boundary) in $\mathbb{R}^n$. Show

$$\int_{M_p} \omega = 0.$$

2. Let M^2 be a surface with boundary, lying in $\mathbb{R}^3$, that is given by a smooth function $z = f(x, y)$ for (x, y) belonging to some region D (bounded in $\mathbb{R}^2$). For each (x, y), let N be the normal vector to M^2 at $(x, y, f(x, y))$ and $\gamma(x, y)$ be the angle between N and the z-axis. Interpret the absolute value of

$$\int_D \sec \gamma(x, y) \, dx \wedge dy.$$

3. Let $D^2 \subseteq \mathbb{R}^2$ be a bounded region, with boundary ∂D^2 diffeomorphic to S^1. Show that the absolute value of

$$\int_{\partial D^2} y \, dx$$

is the area in the region D^2.

4. Given a multilinear alternating function

$$\tau : S(M^n) \times \cdots \times S(M^n) \to S(M^n),$$

where $S(M^n)$ is the vector space of vector fields on M^n and the domain has i factors, the map

$$\tau^* : D^1(M^n) \to D^i(M^n)$$

may be defined in terms of $\tau^*(\omega) = \omega \circ \tau$, where $\omega : S(M^n) \to C^\infty(M^n)$ is the given 1-form (Proposition 5.1).

Construct a specific example in $\mathbb{R}^3$, with $i = 2$.

5. If the manifold is nonorientable, the theory of integration may be carried out with suitable modifications [23].

Alternatively, every nonorientable manifold M^n has a two-fold covering space (called the *orientation cover*). This may be constructed from equivalence classes of paths in M^n beginning at a fixed point, where two paths are equivalent if they have the same end point, and if one transverses one in a given direction, followed by the other in the opposite direction, one obtains a closed path around which the orientation, chosen at the beginning point, is *not* reversed. Consult [25, 75].

If $\tilde{M}^n$ is the orientation cover, $\pi: \tilde{M}^n \to M^n$ the projection map, and $\omega \in D^n(M^n)$, discuss the theory of integration that results by defining

$$\int_{M^n} \omega = \int_{\tilde{M}^n} D(\pi)\omega.$$

What meaning can be given to the volume of a compact nonorientable manifold?

6.   Use formula (∗) before Definition 5.9 to calculate the correct formulas for volumes in cylindrical and spherical coordinates in $\mathbb{R}^3$.

If $D \subseteq S^2$ is a nice region, such as a submanifold with boundary, find an integral formula for the area of D.

CHAPTER

6

Differential Operators

on Manifolds

Classically, a differential operator in Euclidean space consists of an expression involving sums of terms, each consisting of products of functions and partial derivatives, for example,

$$x^2 \frac{\partial}{\partial y} + xy \frac{\partial^2}{\partial x \, \partial y},$$

that when applied to functions (sufficiently smooth) yield other functions. Since partial derivatives take sums to sums and respect scalar products (a number times a function), such a differential operator is a linear operator from one vector space of functions to another. Furthermore, if a function vanishes identically in some open set, then all partial derivatives of that function must also vanish identically in the open set. Hence, the support of the result of a differential operator applied to a function will be no bigger than the original support of the function.

Most of the basic analysis of linear differential operators may be generalized immediately to manifolds, instead of simply Euclidean spaces. The basic facts are largely the same. While one can build this general theory in terms of piecing together differential operators on Euclidean spaces, it turns out to be much more elegant and convenient to give a general definition

171

in terms of vector bundles. This approach, which we shall adopt, avoids—at the beginning—any cumbersome details involving coordinate charts.

We shall then start the chapter with the very general definition. After that, we shall look in detail at various important cases and examples, including the Laplace operator on an arbitrary Riemann manifold (definition to follow). Among various things, we shall see then that the only harmonic functions on compact Riemann manifolds are constants. This completed, we must in the interests of thoroughness and consistency, relate differential operators with the usual notion in Euclidean space, in a sufficiently local sense. This amounts to a global version of a famous theorem of Peetre [89]. Following Hörmander [58], we shall also characterize differential operators of order $\leq m$, for some fixed m. An excellent general reference is [84].

At this point in the book, we need somewhat more real analysis than we have assumed up to this point. This includes some material about Lebesgue integration and Fourier transforms. We shall state, but not prove, everything that we are going to use and also give complete references.

Finally, we shall speak briefly on the important notions of ellipticity, the symbol of an operator, and the analytic index. (This latter is beautifully characterized, in purely topological terms, in the famous index theorem of Atiyah and Singer (see [6, 88]), which unfortunately lies entirely beyond the scope of this book.)

DIFFERENTIAL OPERATORS
ON SMOOTH BUNDLES

Let M^n be a differential manifold (of class C^r), and let (E, M^n, p) be a k-dimensional vector bundle over M^n. The bundle is called *differentiable* (of class C^s, $1 \leq s \leq r$) when E is a differentiable manifold, the maps $\psi_\alpha : p^{-1}(O_\alpha) \xrightarrow{\equiv} O_\alpha \times \mathbb{R}^k$ are diffeomorphisms, and p is a differentiable map, all of this class. Usually, we shall consider smooth or infinitely differentiable vector bundles over M^n. For example, if M^n is smooth, then we have seen (Proposition 2.4) that the tangent bundle is a smooth vector bundle over M^n. Similar remarks would apply to all the basic bundles constructed from the tangent bundle in Chapter 5. It is easy to see that if we start with a smooth (or differentiable) vector bundle (E, M^n, p) and a smooth (or differentiable) map $f : N^q \to M^n$, then the induced bundle

$$(f^{-1}E, N^q, p'),$$

from Definition 4.12, is also smooth (or differentiable). Using this fact, one may conclude that if (E, M^n, p) is smooth and $f : N^q \to M^n$ is any continuous map whose domain N^q is smooth, then $f^{-1}E$ is equivalent to a smooth vector bundle.

It will be convenient at this time to discuss complex vector bundles. These are virtually the same as the real vector bundles of Chapter 4 except that $p^{-1}(x)$ is always a vector space over the complex numbers and the overlap maps are given by linear transformations $g_{\alpha\beta}(x)$ that belong to the group of invertible complex matrices $\mathrm{Gl}(n;\mathbb{C})$. The only basic difference in the theories is that $\mathrm{Gl}(n;\mathbb{C})$ is connected; this means that complex vector bundles are automatically oriented, which makes the theory slightly simpler. The chief reason for this extension to the complex numbers is that we wish to use the methods of Fourier transforms in our main theorems, and these are rather more conveniently stated over the complex numbers. Of course associated with any k-dimensional vector space over the complex numbers there is an underlying $2k$-dimensional real vector space obtained by taking the same operation of addition, but allowing scalar products with real numbers only. This immediately gives a $2k$-dimensional real vector bundle whenever we have a k-dimensional complex vector bundle over a given space.

Our basic definition is now the following:

Definition 6.1 Given an n-dimensional smooth manifold M^n and two smooth, complex vector bundles over M^n, say

$$(E_1, M^n, p_1) \qquad \text{and} \qquad (E_2, M^n, p_2),$$

let $C^\infty(M^n, E_i)$ be the complex vector space of sections of the bundle (E_i, M^n, p_i), i.e., smooth maps $s: M^n \to E_i$ such that $p_i s = 1_{M^n}$.

A *linear differential operator* means a linear map of vector spaces

$$P: C^\infty(M^n, E_1) \to C^\infty(M^n, E_2)$$

such that

$$\mathrm{supp}\, P(s) \subseteq \mathrm{supp}\, s.$$

(Recall that $\mathrm{supp}\, s = \overline{\{x \in M^n \,|\, s(x) \neq 0\}}$.)

Remarks (1) Note that the definition is extremely elegant and simple; we shall need some work before we can conclude that these operators are locally generated by differentiation.

(2) The definition may of course be given verbatum over the real numbers as well.

(3) If $M^n = \mathbb{R}^n$ and E_1 and E_2 are trivial 1-dimensional vector bundles (say real vector bundles), and if $\alpha = (\alpha_1, \ldots, \alpha_n)$ is any multi-index (i.e., α_i is

a nonnegative integer), then for smooth f we may define a linear differential operator D^α by

$$D^\alpha f = \partial^{|\alpha|} f / \partial x_n^{\alpha_n} \cdots \partial x_1^{\alpha_1}.$$

This is because in any open set where f vanishes, $D^\alpha f$ vanishes as well.

More generally, if P is any polynomial over the ring $C^\infty(M^n, \mathbb{R})$ in n variables $u_1, \ldots, u_n$, then if we substitute $\partial/\partial x_i$ for u_i, the resulting polynomial gives a linear differential operator

$$P: C^\infty(M^n) \to C^\infty(M^n),$$

defined by $P(f) = P(\partial/\partial x_1, \ldots, \partial/\partial x_n)(f)$, where $C^\infty(M^n, \mathbb{R}) = C^\infty(M^n)$ is precisely the smooth sections of the trivial vector bundle over M^n.

Once again if s vanishes in an open set, every term of $P(s)$ (and hence $P(s)$ itself) vanishes on that open set. Then clearly supp $P(s) \subseteq$ supp s.

(4)   From a given differential operator, one may manufacture various others by standard constructions. For example, if $O \subseteq M^n$ is open and P is a linear differential operator on M^n, one defines

$$(P_O s)(x) = P(\phi s)(x),$$

where ϕs is the pointwise product of ϕ and s, for any smooth function ϕ whose support lies in O but that is identically 1 on a neighborhood of x. (It is trivial to verify that this definition is independent of the choice of ϕ.)

Similarly, one defines sums and products of operators on the appropriate bundles.

If $g(x)$ is a smooth function on M^n and P is a linear differential operator, then gP (or more precisely $(gP)(s)(x) = g(x)P(s)(x)$) is a linear differential operator.

To restrict a differential operator to a submanifold with appropriate subbundles will in general require some special assumptions.

The reader may easily construct, in various ways, new linear differential operators, in terms of old.

An operator of the form $P(\partial/\partial x_1, \ldots, \partial/\partial x_n)$, or $P(D)$ mentioned above, has a well-defined order. That order is the highest-order partial derivative that occurs in the expression. We propose now to extend this concept of order to the general case of a linear (partial) differential operator, as defined in Definition 6.1 above.

Definition 6.2   Let M^n be a smooth manifold, $x_0 \in M^n$. Let E_1, E_2 be two smooth vector bundles over M^n, and let

$$P: C^\infty(M^n, E_1) \to C^\infty(M^n, E_2)$$

be a linear differential operator, in the sense of Definition 6.1.

Then we say that P has *order m* at the point x_0 if m is the largest nonnegative integer such that there is some $s \in C^\infty(M^n, E_1)$ and some smooth function f defined in an open neighborhood of x_0 and vanishing at x_0 such that

$$P(f^m s)(x_0) \neq 0.$$

(Recall that $P(f^m s)(x_0)$ means apply P to the product at x_0.)

The *order* of P is the maximum of the orders of P at all the points of M^n.

Remarks (1) It is the easiest thing to check that this notion agrees with the usual definition of order in the case of linear differential operators defined on Euclidean space. As an example, take two trivial 1-dimensional vector bundles over $\mathbb{R}^1$. Let

$$P(g)(x) = \frac{d^m g}{dx^m}(x).$$

Then, it is immediate that

$$P(f^r s)(x) = \sum_{i+j=m} \binom{m}{i} \frac{d^i f^r}{dx^i} \frac{d^j s}{dx^j}(x) \qquad \text{and} \qquad \frac{d^i f^r}{dx^i} \bigg|_{x_0} = 0$$

for any x_0 and $i < r$. Thus, $P(f^r s)(x_0) = 0$ when $m < r$.

On the other hand, if $r = m$, set $s(x) = 1$, $f(x) = x$ and observe that

$$\frac{d^m(f^m s)}{dx^m} = \frac{d^m(f^m)}{dx^m} s = m! \neq 0.$$

(2) Even in the simplest cases, if it happens that M^n is not compact, the order of a linear differential operator may not exist. Take, as an example, $M^n = \mathbb{R}^1$, and choose functions

$$\phi_i \in C^\infty(\mathbb{R}^1)$$

with supp $\phi_i \subseteq [i, i+1]$ and $\phi_i(i + \frac{1}{2}) > 0$.

We then set

$$P(f)(x) = \sum_i \phi_i(x) \frac{d^i f}{dx^i}(x).$$

This is then trivially checked to be a linear differential operator for which the order is not defined, and which we shall refer to (informally) as an operator of infinite order.

We also note that our basic theorem (Theorem 6.1) will assure us that a linear differential operator will have an order locally, and thus that a linear differential operator defined on a compact manifold will have an order (finite).

(3)   The sum of two linear differential operators is trivially defined. When defined, the order of a sum of two operators will be the maximum of the orders of the summands.

If P is a linear differential operator of order j and f is a C^∞ function, both defined on M^n, then we have defined fP by the expression

$$f(x)P(s)(x).$$

Excluding the obvious cases where f vanishes, the order of $f \cdot P$ will also be j.

(4)   The theory of differential forms, from the previous chapter, yields an important example. Let M^n be a smooth manifold, and consider the smooth bundles

$$\wedge^i(T^*M^n)   \quad\text{and}\quad   \wedge^{i+1}(T^*M^n)$$

from Definition 5.3. For these bundles, the vector spaces of sections, that is, the differential forms of degrees i and $i + 1$, already have the names $D^i(M^n)$ and $D^{i+1}(M^n)$. The linear map given by exterior differentiation is then a linear differential operator, because if $\omega \in D^i(M^n)$ and O is an open set for which $\mathrm{supp}(\omega) \cap O = \varnothing$, then the formula

$$d(f(x_1, \ldots, x_n)\, dx_{n_1} \wedge \cdots \wedge dx_{n_i}) = (df) \wedge dx_{n_1} \wedge \cdots \wedge dx_{n_i},$$

with

$$df = \sum_k \frac{\partial f}{\partial x_k}\, dx_k,$$

shows immediately that $\mathrm{supp}(d\omega) \cap O = \varnothing$.

To calculate the order of d in full detail, according to Definition 6.2, we let f be a C^∞-function that vanishes at $a \in M^n$. We must then calculate $d(f^m\omega)$ at a. If we write

$$\omega = \sum_{r_1 < \cdots < r_i} f_{r_1, \ldots, f_i}\, dx_{r_1} \wedge \cdots \wedge dx_{r_i},$$

we have

$$d(f^m \cdot \omega) = d\left(\sum_{r_1 < \cdots < r_i} f^m f_{r_1, \ldots, r_i}\, dx_{r_1} \wedge \cdots \wedge dx_{r_i} \right)$$

$$= \sum_{r_1 < \cdots < r_i} d(f^m f_{r_1, \ldots, r_i}) \wedge dx_{r_1} \wedge \cdots \wedge dx_{r_i}.$$

But in addition we have

$$d(f^m f_{r_1,\ldots,r_i}) = \sum_k \frac{\partial}{\partial x_k}(f^m f_{r_1,\ldots,r_i})\, dx_k$$

$$= \sum_k \left(m\frac{\partial f}{\partial x_k} f^{m-1} f_{r_1,\ldots,r_i} + f^m \frac{\partial}{\partial x_k}(f_{r_1,\ldots,r_i}) \right) dx_k.$$

Clearly, if $m > 1$, $d(f^m \omega) = 0$ at a whenever f vanishes at a.

On the other hand, there are obviously functions that vanish at a, yet in a coordinate chart at a

$$\frac{\partial f}{\partial x_j}(a) \neq 0, \qquad \frac{\partial f}{\partial x_k}(a) = 0 \qquad \text{for} \quad k \neq j.$$

Then for $i < n$, if we choose ω so that the only nonzero term is some $f_{r_1,\ldots,r_i}$, with j not equal to any $r_1,\ldots,r_i$, and this $f_{r_1,\ldots,r_i}$ is constantly 1 in a neighborhood of a, then at $x = a$

$$d(f\omega) = \left(\frac{\partial f}{\partial x_j} \cdot 1 \mid 0 \right) dx_j \wedge dx_{r_1} \wedge \cdots \wedge dx_{r_i}$$

$$= \left(\frac{\partial f}{\partial x_j} \right) dx_j \wedge dx_{r_1} \wedge \cdots \wedge dx_{r_i} \neq 0.$$

Therefore, the order of d at any point is , as we might have suspected, precisely 1.

RIEMANN METRICS AND THE LAPLACIAN

We shall now make a digression to study one very important linear differential operator on manifolds, the Laplacian. The Laplacian is a linear differential operator from $D^i(M^n)$ to itself (or in case $i = 0$, from $C^\infty(M^n)$ to itself) that is defined in terms of a specific, given local metric structure on the manifold. It generalizes the usual Laplacian

$$\frac{\partial^2}{\partial x^2} + \frac{\partial^2}{\partial y^2}$$

in the case of $\mathbb{R}^2$ and the usual metric. It plays an important role in a variety of subjects from potential theory to differential geometry.

We shall need a few initial definitions to make these concepts clear.

Definition 6.3   Let V be a vector space over a field such as the real numbers. The *i-fold tensor product* on V is defined as follows:

Let S be the vector space generated by all i-tuples of elements of V,

$$(v_1, \ldots, v_i),$$

into which we introduce the equivalence relation, defined in terms of the elementary equivalence relations,

$$\alpha(v_1, \ldots, v_i) \sim (v_1, \ldots, \alpha v_j, \ldots, v_i) \tag{1}$$

for any $v_1, \ldots, v_i \in V$, with $1 \leq j \leq i$ and $\alpha \in \mathbb{R}$.

$$(v_1, \ldots, v_j', \ldots, v_i) + (v_1, \ldots, v_j'', \ldots, v_i) \sim (v_1, \ldots, v_j' + v_j'', \ldots, v_i). \tag{2}$$

The quotient space of S by this equivalence relation, the i-fold tensor product is denoted

$$\otimes^i V.$$

By convention $\otimes^0 V$ is the field, say, $\mathbb{R}$.

Remarks   (1)   Note that this definition differs from that of the i-fold exterior product only in the absence of an anticommutativity relation

$$(v_1, \ldots, v_j, \ldots, v_l, \ldots, v_i) \sim (v_1, \ldots, v_l, \ldots, v_j, \ldots, v_i).$$

(Compare with Definition 5.1.)

(2)   While all this works over any field, the principal application, involving the Riemann metric, concerns real vector spaces.

(3)   Just as in the case of exterior products and powers, we may immediately define the i-fold tensor product of a vector bundle (E, X, p). For each $x \in X$, the bundle over x consists of

$$\otimes^i (p^{-1}(x)).$$

The overlap maps on an element $\{(v_1, \ldots, v_i)\}$ are obtained by taking the original overlap map on each coordinate. (Compare Definition 5.3.)

(4)   If M^n is a smooth manifold,

$$\otimes^i (T^*M^n)$$

is called the *bundle of (contravariant) i-tensors* over M^n (compare Definition 5.3b).

Similarly,

$$\otimes^i (TM^n)$$

is the *bundle of i-tensors* that are *covariant* on M^n. In addition, by taking tensor products of both kinds of tensors, one defines mixed tensors, covariant in some variables and contravariant in others. The constructions, basic to differential geometry, are not needed in this text. Consult [47, 23].

The terminology, admittedly confusing, arises because of certain historical developments. If one starts with the cotangent bundle, one gets the bundle of contravariant tensors; if one starts with the tangent bundle, one gets a covariant tensor. DeRham cohomology (Definition 5.8) is a contravariant functor (maps go backwards). The historical basis has always proved strong; attempts [49] to change terminology have not been widely accepted.

Definition 6.4   A section (or cross section) of the bundle
$$\otimes^i (TM^n)$$
will be called an *i-fold covariant tensor* on M^n. If it is smooth, one may speak of a smooth *i*-fold tensor, and we generally assume smoothness without specific comment.

We pause momentarily for an example: $\otimes^1 V = V$, trivially. If dim $V = n$, choose a basis and use the Definition 6.3 with relations (1) and (2) to associate with each element of $\otimes^2 V$, an $n \times n$ matrix over $\mathbb{R}$, whose ijth entry is the coefficient of the product of the ith and jth basis vectors.

Therefore, $\otimes^2 (TM^n)$ consists of a smooth assignment of $n \times n$ matrices in $\otimes^2 (TM^n)_x$ to each point $x \in M^n$.

Definition 6.5   A *Riemann* (or sometimes Riemannian) *metric* on a (smooth) manifold M^n is a smooth 2-fold covariant tensor s on M^n such that

(a)   for each $x \in M^n$ and some basis for $(TM^n)_x$, $s(x) \in \otimes^2 (TM^n)$ is a symmetric matrix, i.e., $s(x) = (g_{ij}(x))$, with $g_{ji}(x) = g_{ij}(x)$;

(b)   for each nonzero n vector with respect to our basis in (a) above, say
$$y = \begin{pmatrix} y_1 \\ \vdots \\ y_n \end{pmatrix}, \qquad \text{we have} \quad y^t s(x) y > 0$$

(y^t means the transpose of y). This condition is often referred to by saying that $s(x)$ is *positive definite*.

Remarks   (1)   Clearly a finite sum of Riemann metrics is a Riemann metric, and in a single coordinate chart or neighborhood a Riemann metric exists. It therefore follows from a standard partition of unity argument (Definition 1.9) that there exists a Riemann metric on a manifold (satisfying the second axiom of countability).

(2)   If we write $(dx)^t = (dx_1, \ldots, dx_n)$, one frequently writes
$$ds = +\sqrt{(dx)^t s(x)\, dx}.$$

A Riemann metric defines a distance in the small by noting that $y^t s(x) y$ may be called the *length of y at, or near, x*. If γ is a smooth curve parameterized by, say, t, i.e., $x(t)$ describes the points of γ in M^n, then we define the length of γ to be

$$\text{l.u.b.} \sum_i \sqrt{(x(t_i) - x(t_{i-1}))^t s(x(t_i^*))(x(t_i) - x(t_{i-1}))},$$

where $x(t_i) - x(t_{i-1})$ refers to the vector between the points in a suitable coordinate chart, for some point $x(t_i^*)$, $t_{i-1} \leq t_i^* \leq t_i$.

It is easy to see that this expression must be

$$\int_\gamma ds$$

when γ is a smooth curve. One may verify without difficulty that this is all well defined. Consult any basic text in differential geometry, such as [104, 65].

(3) Given a Riemann metric on M^n, we may define the structure in TM^n by local maps

$$\psi_\alpha : p^{-1}(O_\alpha) \to O_\alpha \times \mathbb{R}^n$$

so that the overlap maps

$$\psi_\beta \circ \psi_\alpha^{-1}(x, y) = (x, g_{\alpha\beta}(x)y)$$

have the $g_{\alpha\beta}(x)$ all orthogonal matrices. Simply choose n nonvanishing vector fields on M^n in the neighborhood of a point that are orthogonal and of unit length in terms of the inner product

$$(w, v)_x = w^t s(x) v,$$

where s is the Riemann metric. Define ψ_α so as to send these orthogonal unit vectors into the standard basis for $\mathbb{R}^n$, $e_j = (0, \ldots, 0, 1, 0, \ldots, 0)$, 1 in the jth place. Extend by linearity. Then ψ_α (as well as ψ_α^{-1}) preserves the inner product in the sense that the inner product of any two vectors in $p^{-1}(x)$ is the same as the usual inner product of their images in $\mathbb{R}^n$.

It follows at once that $\psi_\beta \circ \psi_\alpha^{-1}$, as a map of second coordinates, preserves the usual inner product in $\mathbb{R}^n$; thus, the $g_{\alpha\beta}(x)$ are orthogonal.

In the language of bundle theory, we have *reduced* the group to the orthogonal group (see [103]).

A manifold with a given Riemann metric will be called a *Riemann manifold* (occasionally Riemannian manifold).

As an easy example, in case $M^n = \mathbb{R}^2$ one Riemann metric is given by choosing $s(x)$ to be constantly the identity matrix. In this case

$$ds = \sqrt{(dx)^2 + (dy)^2}.$$

But in polar coordinates, the metric is given by

$$s(r, \theta) = \begin{pmatrix} 1 & 0 \\ 0 & r^2 \end{pmatrix},$$

so that an element of distance is the usual

$$ds = \sqrt{(dr)^2 + (r\, d\theta)^2}.$$

This metric is not constant.

If $N^k \subseteq M^n$ is a submanifold and s is a Riemann metric (or merely an i-fold tensor), then s defines in an obvious way a Riemann metric on N^k. If we view the torus T as an inner tube in $\mathbb{R}^3$, we get one (rather nontrivial) Riemann metric. If we view T as a quotient space of $\mathbb{R}^2$ with the constant identity metric, we get a very different (simpler) Riemann metric.

Now, let us assume that M^n is an oriented Riemann manifold. Consider, for any given $x \in M^n$ and oriented basis for $(TM^n)_x$, the matrix $s(x) = (g_{ij}(x))$. Because it is a symmetric matrix, all of its eigenvalues (characteristic or proper values) are real numbers. Because

$$y^t s(x) y > 0$$

for nonzero y, if λ is an eigenvalue,

$$s(x)y = \lambda y,$$

and then $y^t s(x) y = y^t \lambda y = \lambda y^t y > 0$. Because $y^t y = \|y\|^2 > 0$, we conclude that all the eigenvalues are in fact positive real numbers.

Now, recalling that the determinant of a matrix is the product of its eigenvalues, we see that the determinant

$$g_{1,\dots,n}(x) = \begin{vmatrix} g_{11}(x) & g_{12}(x) & \cdots & g_{1n}(x) \\ \vdots & & & \\ g_{n1}(x) & & \cdots & g_{nn}(x) \end{vmatrix}$$

is a positive number.

We may therefore introduce the following notation:

$$e_{1,\dots,n}(x) = +\sqrt{\det(g_{ij}(x))}.$$

(Recall that we use the bars and det alternatively for the determinant.)

It is an easy verification in coordinate overlap transformations, via the Jacobian, that there is a well-defined differential form whose expression in our oriented local coordinate chart is

$$e_{1,\dots,n}(x)\, dx_1 \wedge \cdots \wedge dx_n.$$

This form is called the volume element or form associated with the Riemann metric. It is a natural choice for the differential form μ in our definition of integration on a manifold (Definition 5.9). If $B \subseteq M^n$ is an n-dimensional submanifold with boundary, one would define its volume to be

$$\int_B e_{1,\ldots,n}(x) \, dx_1 \wedge \cdots \wedge dx_n.$$

In Euclidean space, with the usual metric, one checks immediately that

$$e_{1,\ldots,n}(x) \equiv 1,$$

giving agreement with the ordinary notion of volume therein.

To summarize, we have indicated various consequences from the existence of a Riemann metric: (i) a theory of arc length, (ii) the reduction of the group necessary to define the overlap maps in the tangent bundle to the orthogonal group, (iii) the existence of a cannonical positive n-form on M^n to play the role of μ.

We now wish to define an *adjoint operation* on differential forms. In general, if $0 \leq p \leq n$ and $\omega \in D^p(M^n)$, then there is $*\omega \in D^{n-p}(M^n)$. This construction is linear in ω, and up to sign $**\omega$ will be ω.

We define this adjoint in terms of local representation as follows:

A real symmetric matrix $(g_{ij}(x))$ may be diagonalized by an orthogonal change in coordinates; i.e., there is an orthogonal transformation of $(TM^n)_x$ that results in a new basis for which $(g_{ij}(x))$ is a diagonal matrix. In this coordinate system,

$$e_{1,\ldots,n}(x) = +\sqrt{g_{11}(x)g_{22}(x) \cdots g_{nn}(x)}.$$

Because each $g_{ii}(x) \neq 0$ (since $(g_{ij}(x))$ is positive definite), we may change scale and find a new coordinate chart at x for which $g_{11}(x) = \cdots = g_{nn}(x) = 1$ and hence

$$e_{1,\ldots,n}(x) = 1.$$

We must emphasize that this is in general only true at the single point x in a specific coordinate chart. We define, for $i_1 < \cdots < i_p$,

$$*(dx_{i_1} \wedge \cdots \wedge dx_{i_p}) = \pm dx_{j_1} \wedge \cdots \wedge dx_{j_{n-p}},$$

where $i_1,\ldots,i_p,j_1,\ldots,j_{n-p}$ is a permutation of $1,\ldots,n$ and where we choose $+$ when this permutation is even and $-$ otherwise.

We extend this definition, at the point x, by linearity; i.e.,

$$*(f \, dx_{i_1} \wedge \cdots \wedge dx_{i_p}) = f(*(dx_{i_1} \wedge \cdots \wedge dx_{i_p}))$$

and

$$*(\omega_1 + \omega_2) = *\omega_1 + *\omega_2.$$

Of course for some other choice of coordinate chart at x the value of the form $*\omega$ must be calculated by the usual complicated change of coordinate functions.

If $\omega \in D^p(M^n)$, $*\omega$ is the $(n - p)$-form defined by taking $*\omega$ at each point.

It is immediate that if 1 is the constant function (or 0-form), then

$$*1 = e_{1,\ldots,n}(x)\, dx_1 \wedge \cdots \wedge dx_n,$$

the volume form. A somewhat more exciting fact would be

$$**\omega = (-1)^{np+p}\omega$$

for any $\omega \in D^p(M^n)$. To see this, we shall calculate in the coordinate chart for which we first made the definition. For simplicity, we set

$$e_{i_1,\ldots,i_p,j_1,\ldots,j_{n-p}} = \pm 1$$

with the same choice of signs as above, namely, $+1$ if the numbers are an even permutation of $1, \ldots, n$. If we let $\omega = dx_{i_1} \wedge \cdots \wedge dx_{i_p}$, as before, then

$$**\omega = *(e_{i_1,\ldots,i_p,j_1,\ldots,j_{n-p}}\, dx_{j_1} \wedge \cdots \wedge dx_{j_{n-p}})$$
$$= e_{i_1,\ldots,i_p,j_1,\ldots,j_{n-p}}(e_{j_1,\ldots,j_{n-p},i_1,\ldots,i_p}\, dx_{i_1} \wedge \cdots \wedge dx_{i_p}).$$

Now, we simply note that $p(n - p)$ simple permutations, or interchanges, take $j_1,\ldots,j_{n-p},i_1,\ldots,i_p$ into $i_1,\ldots,i_p,j_1,\ldots,j_{n-p}$. Therefore,

$$**\omega = (-1)^{p(n-p)}(e_{i_1,\ldots,i_p,j_1,\ldots,j_{n-p}})^2\omega$$
$$= (-1)^{p(n-p)}\omega,$$

as we had claimed. Note also that p and p^2 are either both even or both odd, so that

$$**\omega = (-1)^{pn+p}\,\omega.$$

We conclude that for any p, $0 \le p \le n$,

$$*: D^p(M^n) \to D^{n-p}(M^n)$$

is a linear isomorphism.

Yet another interesting property concerns the relationship with inner products. Suppose we are given two p-forms, in local form

$$f\, dx_{i_1} \wedge \cdots \wedge dx_{i_p} \qquad \text{and} \qquad g\, dx_{l_1} \wedge \cdots \wedge dx_{l_p}.$$

Then at any point x of this coordinate neighborhood we define the bracket of these forms by

$$(f\,dx_{i_1} \wedge \cdots \wedge dx_{i_p}, g\,dx_{l_1} \wedge \cdots \wedge dx_{l_p})e_{1,\ldots,n}(x)dx_1 \wedge \cdots \wedge dx_n$$
$$= f \cdot g(dx_{i_1} \wedge \cdots \wedge dx_{i_p}) \wedge (*(dx_{l_1} \wedge \cdots \wedge dx_{l_p})).$$

This is easily checked to extend to an inner product by linearity; the product of two terms above will be nonzero if $(l_1, \ldots, l_p)$ is a permutation of $(i_1, \ldots, i_p)$. The inner product, on differential forms on a compact smooth manifold, is obtained by integrating the inner product above over the entire manifold with respect to the volume element $e_{1,\ldots,n}(x)\,dx_1 \wedge \cdots \wedge dx_n$

Definition 6.6   (a)   We define an *antiderivative* δ,

$$\delta : D^p(M^n) \to D^{p-1}(M^n)$$

by $\delta\omega = (-1)^{np+n+1}*d*(\omega)$, where d is the usual exterior differentiation (as in Proposition 5.2). One trivially checks that $\delta^2 = 0$.
(b)   We define the Laplacian

$$\Delta : D^p(M^n) \to D^p(M^n)$$

by $\Delta = d\delta + \delta d$.

It is immediate that Δ is a linear differential operator (in the sense of Definition 6.1) with respect to the vector bundle

$$\wedge^p(T^*M^n)$$

for both domain and range $(0 \le p \le n)$. This is the basic Laplacian defined on an oriented Riemann manifold. One should verify that Δ on smooth functions or 0-forms is precisely the negative of the customary Laplacian, in Euclidean space, i.e.,

$$-\frac{\partial^2}{\partial x_1^2} - \cdots - \frac{\partial^2}{\partial x_n^2}.$$

While this definition is technically correct, it does not directly show the dependence of Δ on the given Riemann metric, but goes by the rather indirect route of the adjoint or * operation on forms. For computational purposes, it would be clearly more desirable to have a specific formula, which might be more complex than Definition 6.6b but would avoid the operator*. Such a formula is furnished neatly by the following proposition in the case $p = 0$ (i.e., smooth functions).

Proposition 6.1   Let u be a 0-form (smooth function) on the oriented Riemann manifold M^n. Then in any local coordinate chart oriented with respect to the given orientation, we have

$$\Delta u(x) = -\sum_{i,j} \frac{1}{e_{1,\ldots,n}(x)} \frac{\partial}{\partial x_i} \left[g^{ij}(x) e_{1,\ldots,n}(x) \frac{\partial u}{\partial x_j} \right].$$

Here, $(g^{ij}(x)) = (g_{ij}(x))^{-1}$.

Proof Suppose

$$*dx_j = \sum h_{ij} dx_1 \wedge \cdots \wedge \widehat{dx_i} \wedge \cdots \wedge dx_n$$

for suitable functions h_{ij}, and where the caret, as usual, means to omit the symbol lying under it.

Then

$$dx_k \wedge *dx_j = (-1)^{k-1} h_{kj} dx_1 \wedge \cdots \wedge dx_n.$$

But we also know that

$$dx_k \wedge *dx_j = (dx_k, dx_j) e_{1,\ldots,n}(x) dx_1 \wedge \cdots \wedge dx_n,$$

where (dx_k, dx_j) is the inner product evaluated on basic 1-forms, as in the remark preceding Definition 6.6. Referring there to the definition of $*$ and calculating in the special coordinates used to make the definition explicit, we have

$$(dx_k, dx_j) e_{1,\ldots,n}(x) dx_1 \wedge \cdots \wedge dx_n = dx_k \wedge (-1)^{j-1} dx_1 \wedge \cdots \wedge \widehat{dx_j} \wedge \cdots \wedge dx_n$$
$$= \delta_{jk} dx_1 \wedge \cdots \wedge dx_n$$
$$= \delta_{jk} e_{1,\ldots,n}(x) dx_1 \wedge \cdots \wedge dx_n,$$

where $\delta_{jk} = 1$ if $j = k$, 0 otherwise. Hence, in the special system of coordinates used to define $*$, $\{(dx_k, dx_j)\}$ is given by the identity matrix.

It is not hard to figure out what (dx_k, dx_j) is in the original coordinate notation. The effect of the linear coordinate change, given by a matrix A, on the Riemann metric written (g_{ij}) in the first basis, is that the metric in the new coordinates is $A^t(g_{ij})A$. On the other hand, if one takes the inner product in new coordinates and substitutes the result of applying first A^{-1} and then the original 1-forms (that is, reexpress the basic 1-forms in old coordinates), one concludes that the new inner product results from the old by left multiplication by A^{-1} and right multiplication by $(A^{-1})^t$.

Since the Riemann metric and the inner product form are both the identity matrix in the new, special coordinates used to define*, we conclude that in the original coordinates, we must have had $(g_{ij}) = (A^t)^{-1} A^{-1}$ and that the inner product in the old coordinates was AA^t. It follows at once that in the original, or old, coordinates the inner product of basic forms was $(g_{ij})^{-1} = (g^{ij})$.

We now compare our two formulas for $dx_k \wedge * dx_j$. We conclude that

$$(-1)^{k-1} h_{kj} = g^{kj} e_{1,\ldots,n}(x),$$

and thus $h_{ij} = (-1)^{i-1} g^{ij} e_{1,\ldots,n}(x)$. We then calculate

$$d^*\left(\sum_j h_j \, dx_j\right) = d \sum_j h_j {}^* dx_j$$

$$= d\left(\sum_{i,j} (-1)^{i-1} h_j g^{ij} e_{1,\ldots,n}(x) \, dx_1 \wedge \cdots \wedge \widehat{dx_i} \wedge \cdots \wedge dx_n\right)$$

$$= \sum_{i,j} \frac{\partial}{\partial x_i} (h_j g^{ij} e_{1,\ldots,n}(x)) \, dx_1 \wedge \cdots \wedge dx_n$$

$$= \left(\frac{1}{e_{1,\ldots,n}(x)} \sum_{i,j} \frac{\partial}{\partial x_i} (h_j g^{ij} e_{1,\ldots,n}(x))\right) (\text{volume form})$$

since the volume form is $e_{1,\ldots,n}(x) \, dx_1 \wedge \cdots \wedge dx_n$.

Because * takes the volume form to 1, we get

$$*d^*\left(\sum_i h_j \, dx_j\right) = \frac{1}{e_{1,\ldots,n}(x)} \sum_{i,j} \frac{\partial}{\partial x_i} (h_j g^{ij} e_{1,\ldots,n}(x)).$$

Checking the signs from Definition 6.6a, we have

$$\delta\left(\sum h_j \, dx_j\right) = -\frac{1}{e_{1,\ldots,n}(x)} \sum_{i,j} \frac{\partial}{\partial x_i} (h_j g^{ij} e_{1,\ldots,n}(x)).$$

Finally, we may calculate

$$\Delta u = \delta \, du = \delta\left(\sum_j \frac{\partial u}{\partial x_j} \, dx_j\right) = -\sum_j \frac{1}{e_{1,\ldots,n}(x)} \frac{\partial}{\partial x_i}\left(g^{ij} e_{1,\ldots,n}(x) \frac{\partial u}{\partial x_j}\right),$$

as desired.

For an alternate description of this formula, expressing the Laplacian on functions, see [65, note 14].

Definition 6.7 A form ω satisfying $\Delta\omega = 0$ is called *harmonic*. A zero-form that is harmonic is called a *harmonic function*.

To proceed further we need two technical lemmas for oriented Riemann manifolds.

Lemma 6.1 If f is a smooth function on a smooth manifold, and f^2 means the product of f with itself, then

$$\Delta(f^2) = 2f \, \Delta f - 2(\text{grad } f)f,$$

where

$$\operatorname{grad} f = \sum_i \left(\sum_j g^{ij} \frac{\partial f}{\partial x_j} \right) \frac{\partial}{\partial x_i}$$

(grad f is an operator on functions).

Proof.

$$\Delta(f^2) = -\sum_{i,j} \frac{1}{e_{1,\dots,n}(x)} \frac{\partial}{\partial x_i} \left(g^{ij} e_{1,\dots,n}(x) 2f \frac{\partial f}{\partial x_j} \right)$$

$$= -\sum_{i,j} \frac{1}{e_{1,\dots,n}(x)} \frac{\partial}{\partial x_i} \left(2f g^{ij} e_{1,\dots,n}(x) \frac{\partial f}{\partial x_j} \right)$$

$$= -\sum_{i,j} \frac{2f}{e_{1,\dots,n}(x)} \frac{\partial}{\partial x_i} \left(g^{ij} e_{1,\dots,n}(x) \frac{\partial f}{\partial x_j} \right) - \sum_{i,j} 2 \frac{\partial f}{\partial x_i} \frac{\partial f}{\partial x_j} g^{ij}$$

$$= 2f \, \Delta f - 2(\operatorname{grad} f)f.$$

Lemma 6.2 If $\mu = e_{1,\dots,n}(x) \, dx_1 \wedge \cdots \wedge dx_n$ is the volume form, f is a smooth function, and M^n is also compact,

$$\int_{M^n} (\delta df)\mu = 0.$$

Proof Recall that δdf is a zero form, or smooth function. We then have

$$\int_{M^n} (\delta df)\mu = \int_{M^n} (\delta df)*1 = \int_{M^n} *(\delta df)1 = \int_{M^n} *(\delta df).$$

Now, an easy check yields

$$*\delta d = d\delta*.$$

Hence, our integral is

$$\int_{M^n} d(\delta * f),$$

which vanishes by Corollary 5.1. The following proposition, essentially due to E. Hopf (see [65, 23] for references in this area), generalizes the well-known results about harmonic functions.

Proposition 6.2 Let M^n, $n > 0$, be a compact, connected, oriented, smooth Riemann manifold. Let f be a smooth real-valued function, and suppose

$$\Delta f \geq 0 \qquad \text{for all} \quad x \in M^n.$$

Then f is a constant.

Proof   By Lemma 6.2 and the fact that δ vanishes on 0-forms,

$$\int_{M^n} \Delta f = 0.$$

Because $\Delta f \geq 0$, this forces $\Delta f = 0$.

Now we use Lemma 6.1 also to get

$$0 = \int_{M^n} \Delta\left(\frac{f^2}{2}\right)\mu = \int_{M^n} (f\,\Delta f)\mu - \int_{M^n} (\text{grad } f \cdot f)\mu = 0 - \int_{M^n} (\text{grad } f \cdot f)\mu.$$

But now grad $f \cdot f$ is clearly the result of evaluating the quadratic form

$$y^t(g^{ij}(x))y$$

when $y^t = (\partial f/\partial x_1, \ldots, \partial f/\partial x_n)$.

If we can show that the quadratic form is positive definite, then our above calculation, i.e.,

$$0 = -\int_{M^n} (\text{grad } f \cdot f)\mu,$$

will force $(\partial f/\partial x_1, \ldots, \partial f/\partial x_n)$ to be the zero vector, and hence f to be a constant.

To complete the proof, choose a basis for which $(g_{ij}(x))$ is diagonal, i.e.,

$$g_{ij}(x) = \begin{pmatrix} g_{11}(x) & & 0 \\ & \ddots & \\ 0 & & g_{nn}(x) \end{pmatrix},$$

at least at a point x. Then, one can immediately write

$$g^{ij}(x) = \begin{pmatrix} (g_{11}(x))^{-1} & & 0 \\ & \ddots & \\ 0 & & (g_{nn}(x))^{-1} \end{pmatrix},$$

which is visibly symmetric and positive definite. This completes the proof.

Remarks   (1)   One may clearly reformulate this in the language of a *maximum principal*, as in the usual basic texts in complex analysis.

(2)   Harmonic forms on compact Riemann manifolds have been studied extensively. (See the text of Hodge [55].) The basic result is that there is a unique harmonic form in each deRham cohomology class (see Definition 5.8).

(3)   On noncompact Riemann manifolds, Proposition 6.2 is no longer true. There are classic examples of nonconstant harmonic functions on Euclidean space with the origin deleted, i.e., nonconstant f such that

$$\Delta f = \sum_{i=1}^{n} \frac{\partial^2 f}{\partial x_i^2} = 0.$$

(4)   More generally, one may study the differential equation

$$\Delta\omega = \sigma$$

for a preassigned form or function $\sigma \in D^p(M^n)$. For $p = 0$, consult any basic text on partial differential equations (such as [57]). For $p > 0$, consult the final chapter of de Rham [23].

CHARACTERIZATION OF LINEAR DIFFERENTIAL OPERATORS

We must now head towards the characterization of linear differential operators, as defined by Definition 6.1, in terms of local expressions involving partial differentiation. At this point, we must use some fundamental material from graduate-level real analysis that has become quite standard. This involves norms, as mentioned earlier, uniform convergence, and Fourier transforms. Because one must have spaces of functions that are complete under the norms in question, we should at this point work with the Lebesgue integral. The general references here are the books of Halmos [44], Saks [95], Rudin [94], and Narasimhan [84]. I shall review these topics briefly now, with specific references:

Norms   Recall the m-norm of a smooth function with compact support defined on an open set $U \subseteq \mathbb{R}^n$:

$$\|f\|_m - \sum_{|\alpha| \le m} \sum_{j=1}^{r} \frac{1}{\alpha!} \sup_{x \in U} |D^\alpha f_j(x)|.$$

Here $f(x) = (f_1(x), \ldots, f_r(x))$, $f : U \to \mathbb{R}^r$, and α is a multi-index (as in Theorem 1.3). These norms are clearly subadditive.

We observe that these norms also make sense when we work over the complex numbers, provided we reinterpret $|\,|$ to mean modulus and include all partial derivatives, of the appropriate orders, with respect to both the real and imaginary parts of the variables. See [119 Chap. 1] for a good concise good reference.

Uniform Convergence   We recall that a sequence of functions $f_n : \mathbb{R}^n \to \mathbb{R}^r$ converges uniformly if there is a function f such that, for each $\varepsilon > 0$, there is $N > 0$ such that if $n > N$

$$\|f_n - f\|_0 < \varepsilon.$$

Of course, the norm $\|\ \|_0$ is just the usual l.u.b. or sup of the absolute values.

We may generalize this immediately, to define C^k-*uniform convergence*, or uniform convergence in the norm $\| \ \|_k$. Specifically, let $O \subseteq \mathbb{R}^n$ be open and $f_n : O \to \mathbb{R}^r$, $f : O \to \mathbb{R}^r$ be C^k-functions. Then we say that f_n converges to f uniformly in C^k if for each $\varepsilon > 0$ there is $N > 0$ such that if $n > N$

$$\| f_n - f \|_k < \varepsilon$$

in O.

If it is necessary to specify the domain, one may write $\| \ \|_k^O$ for this norm.

We may also extend this to C^∞-uniform convergence by demanding that the above inequality hold for $\varepsilon > 0$ on O, for all positive integers k.

Note also that one may immediately extend these concepts to functions of the form

$$f : M^n \to \mathbb{R}^r$$

by taking the supremums over all sets in a cover of M^n by coordinates charts.

We note however that the cases where M^n is compact, or the f_n and f have compact support, are the most important.

In our study of linear differential operators, we shall need to extend these concepts to smooth sections of trivial vector bundles. Suppose we are given a smooth, trivial vector bundle over M^n

$$\tau : M^n \times \mathbb{R}^k \to M^n.$$

A section of this bundle is then given by a map

$$s(x) = (y, z).$$

But since $\tau s(x) = x$, $x = y$, and the section is completely determined by a function $z = f(x)$, where

$$f : M^n \to \mathbb{R}^k,$$

all the previous notions of norm and convergence must carry over to sections by application of the definitions to these functions f.

Later we shall touch on sections of general vector bundles, rather than trivial ones. We shall handle this in terms of a metric defined on the entire vector bundle (details later).

Fourier Transforms   To do a correct treatment of Fourier transforms, we should work in the framework of Lebesgue integration. This is an extension of ordinary integration in such a way that the L^p-spaces, which we recall below, are complete metric spaces. Although it might be possible to give a minimal treatment of Fourier transforms, without Lebesgue theory, this would constitute a considerable corruption of the area in its present state. For Lebesgue integration, consult [44, 94].

Let $O \subseteq \mathbb{R}^n$ be open, $p \geq 1$. $L^p(O)$ is the complex-valued Lebesgue measurable functions defined on O such that $|f|^p$ has a (finite) integral over O. (Here, $|\ |$ means modulus, i.e., the square root of the sum of the squares of the real and imaginary parts.) $L^p(O)$ is a vector space over the complex numbers with the usual addition and scalar multiplication by complex constants. In $L^p(O)$ one defines a norm by

$$\|f\|_{L^p} = \left(\int_O |f(x)|^p \, dx \right)^{1/p},$$

where we write dx as a shorthand for $dx_1 \cdots dx_n$ here. In the case $p = 2$, in addition to the norm there is an inner product

$$(f, g)_{L^2} = \int_O f(x)\overline{g(x)} \, dx,$$

where the bar means the complex conjugate. The standard Cauchy–Schwarz inequality assures us that the integral of the modulus of $f(x)\overline{g(x)}$ is no bigger than

$$\left(\int_O |f(x)|^2 \, dx \right)^{1/2} \left(\int_O (|g(x)|^2 \, dx \right)^{1/2}.$$

We also note that

$$\|f\|_{L^2} = ((f, f)_{L^2})^{1/2}.$$

Now, to define the Fourier transform, initially for some $f \in L^1(\mathbb{R}^n)$, we set

$$\hat{f}(\xi) = \frac{1}{(2\pi)^{n/2}} \int_{\mathbb{R}^n} f(x) e^{-i\langle x, \xi \rangle} \, dx,$$

where here, of course, we are using the following abreviations:

$$x = (x_1, \ldots, x_n), \qquad \xi = (\xi_1, \ldots, \xi_n), \qquad \langle x, \xi \rangle = x_1\xi_1 + \cdots + x_n\xi_n,$$

and

$$dx = dx_1 \cdots dx_n.$$

To get at a theory of Fourier transforms that is well suited to studying linear differential operators, we must focus on a subset of the smooth functions first clearly noticed by L. Schwartz. Define $\mathscr{S} \subseteq C^\infty(\mathbb{R}^n)$ to be the subset of all functions f so that for any α, $D^\alpha f$ tends to zero as $|x| \to \infty$ faster than the reciprocal of any power of $1 + |x|^2$. For example, $e^{-x^2} \in \mathscr{S}$, and any function with compact support is in $\mathscr{S}$. Clearly, if $f \in \mathscr{S}$, then $D^\alpha f \in \mathscr{S}$ for any multi-index α. It is also easy to check that the functions of $\mathscr{S}$ are dense in the spaces $L^p(\mathbb{R}^n)$.

It is well known that functions in $\mathscr{S}$ have many pleasant properties with respect to Fourier transforms.

(i)   If $f \in \mathscr{S}, \hat{f}$ exists and $\hat{f} \in \mathscr{S}$.

(ii)   If $f \in \mathscr{S}$,

$$f(x) = \frac{1}{(2\pi)^{n/2}} \int_{\mathbb{R}^n} \hat{f}(\xi) e^{i\langle x, \xi \rangle} \, dx.$$

This is the so-called Fourier inversion formula.

(iii) $\int_{\mathbb{R}^n} \hat{f}(x)g(x)\,dx = \int_{\mathbb{R}^n} f(x)\hat{g}(x)\,dx$

for $f, g \in \mathscr{S}$.

(iv)   The *convolution* $f * g$ is defined by

$$(f * g)(x) = \int_{\mathbb{R}^n} f(x - t)g(t)\,dt.$$

It exists if both functions are in $L^1(\mathbb{R}^n)$.

If $f \in L^1(\mathbb{R}^n)$ and $g \in \mathscr{S}$, then

$$\int_{\mathbb{R}^n} g(y)\hat{f}(y)e^{i\langle x, y \rangle}\,dy = -f * \tilde{g}(x),$$

where $\tilde{g}(x) = \hat{g}(-x)$.

Now, using the above properties, for the inner product $\int_{\mathbb{R}^n} f(x)\overline{g(x)}\,dx$, we have that whenever $f, g \in \mathscr{S}$

$$(f, g)_{L_2} = (\hat{f}, \hat{g})_{L_2},$$

and taking $f = g$ we conclude that $\|f\|_{L_2} = \|\hat{f}\|_{L_2}$ for $f \in \mathscr{S}$. This is a special case of the more important

(v)   Plancherel's theorem. If $f \in L^2(\mathbb{R}^n)$, then $\hat{f} \in L^2(\mathbb{R}^n)$ and

$$\|f\|_{L_2} = \|\hat{f}\|_{L_2}.$$

Returning to the convolution from (iv) above, we have

(vi)   if $f, g \in \mathscr{G}, f * g \in \mathscr{G}$ and

$$\widehat{(f * g)} = (2\pi)^{n/2}\hat{f} \cdot \hat{g}, \qquad \widehat{(f \cdot g)} = (2\pi)^{-n/2}\hat{f} * \hat{g}.$$

In short, up to suitable multiplication by a constant, Fourier transforms interchange convolution and multiplication.

As for differential operators, the most useful facts are that

(vii)   when $f \in \mathscr{S}$,

$$\widehat{(D^\alpha f)}(y) = i^{|\alpha|} y^\alpha \hat{f}(y) \qquad \text{and} \qquad (D^\alpha \hat{f})(y) = (\widehat{(-iy)^\alpha f(y)}).$$

Here, of course, $i = \sqrt{-1}$ and $|\alpha| = \alpha_1 + \cdots + \alpha_n$. Thus, as with the Laplace transform, this transform converts differentiation to polynomial multiplication.

Some of these properties, for example (vii) are simply routine calculations, while others are actually theorems. For more details, consult [57, 84, 119].

We now wish to head towards the theorem of Peetre [89], which characterizes linear differential operators. We will then follow with Hormander's characterization [58] of operators of order less than or equal to m. The first result may be given over the real numbers, while the second, closely tied to Fourier transform methods, sits more naturally in the language of complex vector spaces and bundles.

We begin by fixing the terminology. Let $O \subseteq \mathbb{R}^n$ be open and $C_0^\infty(O,s)$ be the space of smooth sections, with compact support, of the s-dimensional, trivial vector bundle (complex or real) over O. If $f \in C_0^\infty(O,s)$, $m \geq 0$, then $\|f\|_m$ is defined by the formula above. We denote the trivial bundle by τ_s and a linear differential operator by P. The following lemma would be obvious if P were a finite combination of functions and partial derivatives, but with P defined by Definition 6.1 it will take some proof.

Lemma 6.3   With the above terminology, suppose $a \in O$. Then there is an open neighborhood O_1 of a in O, a positive integer m, and a positive constant C such that if $f \in C_0^\infty(O_1 - a,s)$,

$$\|Pf\|_0 \leq C\|f\|_m.$$

Proof   We shall work by contradiction.

Let U_0 have compact closure in O and $a \in U_0$. If the lemma is false, there is an open U_1,

$$U_1 \subseteq \overline{U_1} \subseteq U_0 - a,$$

and some $u_1 \in C_0^\infty(U_1,s)$ with

$$\|Pu_1\|_0 > 2^2\|u_1\|_1.$$

Repeat this for the nonempty open set $U_0 - \overline{U_1}$. We then find an open U_2,

$$U_2 \subseteq \overline{U_2} \subseteq U_0 - \overline{U_1} - a,$$

and some $u_2 \in C_0^\infty(U_2,s)$ for which

$$\|Pu_2\|_0 > 2^{2 \circ 2}\|u_2\|_2.$$

In general, there are open sets U_k whose closures are disjoint and for which $\overline{U_k} \subseteq U_0 - a$, for each of which we have a $u_k \in C_0^\infty(U_k, s)$, with

$$\|Pu_k\|_0 > 2^{2k}\|u_k\|_k.$$

On the other hand, the functions u_k are all smooth and all have compact support in $U_0 - a$. We may then form

$$u = \sum_{k=1}^{\infty} \frac{1}{2^k\|u_k\|_k} u_k.$$

For any fixed positive integer l, this series converges in the norm $\|\ \|_l$ (just ignore the first l terms). Therefore, the series converges with all derivatives, and u is a smooth function with compact support lying within $\overline{U_0}$.

But because of the disjointness of the sets $\overline{U_k}$,

$$u \mid U_k = (1/2^k\|u_k\|_k)u_k \mid U_k.$$

Using Definition 6.1 for P, we see that

$$P(u \mid U_k) = (1/2^k\|u_k\|_k)(Pu_k).$$

By the earlier inequalities for Pu_k, we may select for each k some $a_k \in U_k$

$$|Pu_k(a_k)| > 2^{2k}\|u_k\|_k.$$

Then

$$|Pu(a_k)| = \left| \sum_l \frac{1}{2^l\|u_l\|_l} P(u_l(a_k)) \right| = \left| \frac{1}{2^k\|u_k\|_k} P(u_k(a_k)) \right| > 2^k.$$

But Pu is smooth on O, and hence bounded on the compact subset $\overline{U_0}$. This is not compatible with

$$|Pu(a_k)| > 2^k \qquad \text{for all } k.$$

This contradiction proves the lemma.

We now come to our first characterization theorem, here at first expressed locally over the real numbers.

Theorem 6.1 (Local Peetre Theorem) [89]   Let $O \subseteq \mathbb{R}^n$ be open and let

$$P : C^\infty(O, s) \to C^\infty(O, t)$$

be a linear differential operator from the sections of the trivial s-dimensional (real) vector bundle over O to those of the trivial t-dimensional bundle.

Let $O_1 \subseteq O$ have a compact closure contained in O. Let $M(s, t)$ be the vector space of linear maps from $\mathbb{R}^s$ to $\mathbb{R}^t$. Then there is an $m \geq 0$ such that for every multi-index α, $|\alpha| \leq m$, there are C^∞-maps

$$a_\alpha : O_1 \to M(s, t)$$

such that for any $f \in C^\infty(O_1, s)$, $x \in O_1$,

$$(Pf)(x) = \sum_{|\alpha| \leq m} a_\alpha(x)(D^\alpha f)(x).$$

Proof With an eye towards an eventual use of Lemma 6.2, let U be an open set in O, so that there is $C > 0$ and a positive integer m, with $\|Pf\|_0 \leq C\|f\|_m$, for all $f \in C_0^\infty(U, s)$. Assuming this about U, we make the following claim:

If $f \in C_0^\infty(U, s)$ is m-flat at $a \in U$ (i.e., $(D^\alpha f)(a) = 0$ for $|\alpha| \leq m$), then $(Pf)(a) = 0$.

We use Proposition 1.7 to find a sequence of functions

$$f_k \in C_0^\infty(U, s)$$

such that every f_k vanishes in some open neighborhood of a and

$$\lim_{k \to \infty} \|f_k - f\|_m = 0.$$

By our assumption about U we conclude that Pf_k converges to Pf uniformly in the open set U. Because P is a linear differential operator and f_k vanishes in a neighborhood of a, we also know that

$$(Pf_k)(a) = 0$$

for all k. Hence, $(Pf)(a) = 0$.

Now suppose $v_1, \ldots, v_s$ is a basis for $\mathbb{R}^s$ and $g \in C^\infty(U, s)$. We may write

$$g(x) = \sum_{i=1}^s g_i(x)v_i, \quad \text{with} \quad g_i \in C^\infty(U).$$

Then we may use Taylor's theorem (Theorem 1.3) to calculate

$$g(x) = \sum_{|\alpha| \leq m} \frac{1}{\alpha!}(x-a)^\alpha \left(\sum_{i=1}^s (D^\alpha g_i)(a)v_i \right) + f,$$

where f is an m-flat vector function at a.

Now, we use the above claim and the linearity of P to arrive at

$$(Pg)(a) = \sum_{|\alpha| \leq m} \sum_{i=1}^s \frac{(D^\alpha g_i)(a)}{\alpha!} P((x-a)^\alpha v_i)(a).$$

Referring to the definitions in Theorem 1.3, it is clear that $P((x - a)^{\alpha}v_i)(a)$ is a C^{∞}-function since a varies in O. We thus conclude that there is a C^{∞} map $a_{\alpha}: O \to M(s, t)$ such that for all $g \in C^{\infty}(U, s)$ and $x \in U$,

$$(Pg)(x) = \sum_{|\alpha| \leq m} a_{\alpha}(x)(D^{\alpha}g)(x). \qquad (*)$$

We may finally pass to the theorem in general. By Lemma 6.3 and the compactness of $\overline{O}_1$ there must be finitely many $x_1, \ldots, x_l \in \overline{O}_1$ and $C > 0$ and some positive integer m with

$$\|Pu\|_0 \leq C\|u\|_m$$

for all $u \in C_0^{\infty}(\overline{O}_1 - x_1 - \cdots - x_l, s)$. Recalling that Pu at a point depends only on u in a neighborhood of the point, and that when $x \in \overline{O}_1 - x_1 - \cdots - x_l$, one may find $\varphi \in C^{\infty}(O)$ that is 1 in a neighborhood of x but vanishes outside a compact neighborhood of x lying in $\overline{O}_1 - x_1 - \cdots - x_l$, we may calculate Pu by calculating $P(\varphi u)$; we deduce that formula $(*)$ is valid for all $g \in C^{\infty}(\overline{O}_1 - x_1 - \cdots - x_l; s)$ and $x \in \overline{O}_1 - x_1 - \cdots - x_l$.

Now both sides $(*)$ are defined and continuous over all of O_1. Therefore, they must have the same value for all points in O_1, completing the proof of the theorem.

Our next task, essentially trivial, is to translate this result into a global theorem for differential operators defined on a manifold.

Theorem 6.2   Let M^n be a smooth manifold, and let E_1 and E_2 be two vector bundles over M^n, with dim $E_i = s_i$.
   Let

$$P: C^{\infty}(M^n, E_1) \to C^{\infty}(M^n, E_2)$$

be a linear differential operator. Take $a \in M^n$. Then there is a coordinate neighborhood of a, say U, over which both bundles are trivial and a positive integer m such that in U

$$(Pf)(x) = \sum_{|\alpha| \leq m} a_{\alpha}(x)(D^{\alpha}f)(x),$$

for smooth maps $a_{\alpha}: U \to M(s_1, s_2)$.

Proof   Choose a small enough neighborhood of a that the bundles are trivial over it. In a coordinate neighborhood of a, choose a small open ball around a that lies in this original neighborhood. As an open ball in $\mathbb{R}^n$ is diffeomorphic to all $\mathbb{R}^n$, this small open ball will function as U.
   The theorem is then a direct application of Theorem 6.1 for this set U. Note that f is a vector-valued function of n variables, with values in $\mathbb{R}^{s_1}$,

Remarks   (1)   This theorem shows that a linear differential operator has a meaningful order locally. It follows immediately that over a compact manifold M^n a linear differential operator has an order.

(2)   We should emphasize that the condition on supports in Definition 6.1 is quite essential here. Without it a linear operator may have nothing to do with derivatives. For example, let $M^n = \mathbb{R}$ and E_1 and E_2 be two trivial 1-dimensional bundles. Define

$$(Ps)(x) = (x, \pi_2(s(0))),$$

where $\pi_2 : \mathbb{R} \times \mathbb{R} \to \mathbb{R}$ projects onto the second factor. (P) is clearly a section of E_2, and

$$
\begin{aligned}
P(\alpha_1 s_1 + \alpha_2 s_2)(x) &= (x, \pi_2(\alpha_1(s_1(0)) + \alpha_2(s_2(0)))) \\
&= (x, \alpha_1 \pi_2(s_1(0)) + \alpha_2 \pi_2(s_2(0))) \\
&= \alpha_1(x, \pi_2(s_1(0))) + \alpha_2(x, \pi_2(s_2(0))) \\
&= \alpha_1[(Ps_1)(x)] + \alpha_2[(Ps_2)(x)]
\end{aligned}
$$

because addition of the sections is addition in the second coordinate, i.e., in the vector space over x. Clearly, P is *not* a *differential operator*.

We now describe a different characterization due to Hormander. To state this in full generality we need a special notion of convergence of a sequence of sections.

Definition 6.8   (a)   Let (E, M^n, p) be a smooth complex vector bundle, $s_i \in C^\infty(M^n, E)$ a sequence of C^∞-sections of E, and $s \in C^\infty(M^n, E)$ a given fixed C^∞-section of E.

We say that s_i converges to s *locally uniformly* provided for any $a \in M^n$ there is a coordinate neighborhood U of a over which E is trivial such that, in U, s_i and $D^\alpha s_i$ (any α) converge uniformly to s and $D^\alpha s$, respectively

(b)   With two given smooth (complex) vector bundles E_1 and E_2, a linear map

$$L : C^\infty(M^n, E_1) \to C^\infty(M^n, E_2)$$

is *weakly continuous* if wherever $s_i \to s$ locally uniformly and $K \subseteq M^n$ is compact, then

$$L(s_i) \to L(s)$$

uniformly over K.

Naturally, any classically defined linear differential operator is weakly continuous. The term weakly refers to the fact that we have not topologized the vector spaces of cross sections $C^\infty(M^n, E_i)$, so that it does not make sense to ask whether the map L is continuous. This property will be used at a key place on the next theorem, when we need to take L inside the integral sign.

Theorem 6.3 (Hormander [58])   With the above notations, let

$$L: C^\infty(M^n, E_1) \to C^\infty(M^n, E_2)$$

be a weakly continuous linear map. Then a necessary and sufficient condition that L be a linear differential operator of order $\leq m$ is the following:
For any $s \in C^\infty(M^n, E_1)$, $x \in M^n$, and $f \in C^\infty(M^n)$, that is, f a smooth real function,

$$g(y)(x) = e^{-iyf(x)}[L(e^{iyf}s)(x)]$$

is a function in y from $\mathbb{R}$ to $\mathbb{C}^{\dim E_2}$ (dim E_2 being the dimension of the complex vector bundle E_2) that in each coordinate is a polynomial of degree $\leq m$.
Note that the expression in parentheses following L refers to the product of the function e^{iyf} and the section s, both evaluated at x.

Proof   Suppose that $L = d^m/dx^m$ and that we are in the 1-dimensional case. We may then calculate

$$e^{-iyf(x)}[L(e^{iyf}s)(x)] = e^{-iyf(x)} \sum_{j=0}^{m} \delta_j \frac{d^j s}{dx^j} \frac{d^{m-j}e^{iyf(x)}}{dx^{m-j}}$$

$$= e^{-iyf(x)} \sum_{j=0}^{m} \delta_j \frac{d^j s}{dx^j}$$

$$\times [c_1 e^{iyf(x)} iyf^{m-j-1}(x) + \cdots + c_{m-j} e^{iyf(x)}(iyf(x))^{m-j}],$$

where the δ_j and c_j are constants in y. This is visibly a polynomial of degree $\leq m$, as required.
The general case, where L is a linear differential operator of degree $\leq m$, that is (by our earlier theorems), locally,

$$(Lf)(x) = \sum_{|\alpha| \leq m} a_\alpha(x)(D^\alpha f)(x),$$

follows by an entirely similar computation.
Conversely, using the terminology of the statement of the theorem, and assuming that $g(y)(x)$ is a polynomial function of degree less than or equal to m in each coordinate, we may write

$$g(y)(x) = \sum_{l=0}^{m} g_l(f, s)y^l,$$

where each $g_l \in C^\infty(M^n, E_2)$.

Or more generally, at any given x, one might write

$$e^{-it(y_1 f_1 + \cdots + y_k f_k)} L(e^{it(y_1 f_1 + \cdots + y_k f_k)} s) = \sum_{l=0}^{m} g_l(y, f, s) t^l,$$

where $g_l(y, f, s) \in C^\infty(M^n, E_2)$ and t is now the variable. It is easy to check that each $g_l(y, f, s)$ is homogeneous of degree l in $y = (y_1, \ldots, y_k)$. It then follows from Taylor's theorem that the (components of) $g_l(y, f, s)$ are homogeneous polynomials of degree l in the variables $y_1, \ldots, y_k$.

Now, taking a coordinate chart, or a possible subset of a chart, we let O be a connected open set in M^n whose closure is compact, and which is homeomorphic (via a map $\phi : O \to \mathbb{R}^n$) to an open subset in $\mathbb{R}^n$.

If $\beta \in C_0^\infty(\phi(O))$, and thus $\beta \in \mathcal{G}$, we may write the basic Fourier formulas

$$\beta(u) = \frac{1}{(2\pi)^{n/2}} \int_{\mathbb{R}^n} \hat{\beta}(y) e^{i\langle u, y \rangle} \, dy$$

and

$$\hat{\beta}(y) = \frac{1}{(2\pi)^{n/2}} \int_{\mathbb{R}^n} \beta(x) e^{-i\langle x, y \rangle} \, dx.$$

Suppose $s \in C_0^\infty(M^n, E_1)$ and

$$\tilde{s} = \beta(\phi(x)) s(x).$$

We then get (for $\phi(x) = (\phi_1(x), \ldots, \phi_n(x))$)

$$\tilde{s}(x) = \frac{1}{(2\pi)^{n/2}} \int_{\mathbb{R}^n} \hat{\beta}(y) e^{i(y_1 \phi_1(x) + \cdots + y_n \phi_n(x))} s(x) \, dy_1 \cdots dy_n.$$

Because L is weakly continuous and because an integral (here in the variable y) may be approximated by finite sums, if we apply L to the integral on the right of the above formula, the result is the same as if we were to apply L inside the integral sign. Therefore, we may calculate

$$L(\tilde{s})(x) = \frac{1}{(2\pi)^{n/2}} \int_{\mathbb{R}^n} \hat{\beta}(y) L(s(x) e^{i(y_1 \phi_1(x) + \cdots + y_n \phi_n(x))}) \, dy_1 \cdots dy_n$$

$$= \frac{1}{(2\pi)^{n/2}} \int_{\mathbb{R}^n} \left[\sum_{l=0}^{m} \hat{\beta}(y) e^{i\langle y, \phi(x) \rangle} g_l(y, \phi, s) \right] dy_1 \cdots dy_n$$

(by taking $t = 1$ in the formula above, which defines $g_l(y, \phi, s)$).

But this last expression may clearly be rewritten

$$\frac{1}{(2\pi)^{n/2}} \sum_{l=0}^{m} \sum_{|\alpha|=l} h_{\alpha,l}(x) \int_{\mathbb{R}^n} \hat{\beta}(y) e^{i\langle y, \phi(x) \rangle} y^\alpha \, dy_1 \cdots dy_k$$

for suitable smooth functions $h_{\alpha,l}$, recalling that $g_l(y, \phi, s)$ is a homogeneous polynomial in y.

Using the relationship of Fourier transforms and derivatives (property (vii) in the review above), we finally arrive at the formula

$$L(\tilde{s})(x) = \sum_{l=0}^{m} \sum_{|\alpha|=l} k_{\alpha,l}(D^{\alpha}\beta)(\phi(x))$$

for suitable smooth $k_{\alpha,l}$, in our given open set. It is from this formula that we shall deduce our theorem.

   (i)   Observe that by weak continuity, one needs only show that

$$\text{supp}(L(s)) \subseteq \text{supp}(s)$$

when s has compact support, for one may consider the support of $\beta \cdot s$, with β a smooth function, $0 \le \beta(x) \le 1$, taking β to have compact support, but taking $\beta^{-1}(1)$ to be larger and larger sets.

   (ii)   If supp(s) is compact, choose β in our formula for $L(\tilde{s})$ above to be 1 on supp(s) and to vanish outside an arbitrary neighborhood of supp(s). One can then easily conclude that if $a \notin$ supp(s) there is a β and hence $\tilde{s} = \beta \cdot s$ such that $\tilde{s}$ is equal to s, and hence $L(\tilde{s})$ is equal to $L(s)$, yet $L(\tilde{s})(a) = 0$ (because if β vanishes in a neighborhood of a, our formula shows that $L(\tilde{s})$ also vanishes in that neighborhood).

   Therefore, if $a \notin$ supp(s), $L(s)(a) = 0$, i.e.,

$$\text{supp}(L(s)) \subseteq \text{supp}(s).$$

   (iii)   That L has order less than or equal to m follows from the formula for $L(\tilde{s})(x)$ by simply choosing s to be (locally) constant and β to be arbitrary, so that $L(\tilde{s})$ obviously has order at most m, near x. Note that near x any arbitrary section may be written as a finite sum of terms of the form

$$\beta(\phi(x))s(x),$$

where $s(x)$ is a constant section. This completes the proof.

Remarks   (1)   The need for the assumption of weak continuity can be seen by taking a fixed vector bundle for E_1 and E_2, say E, and choosing a basis (generally infinite) $\{v_\alpha\}$ for $C^\infty(M^n, E)$. Then a permutation of the basis vectors will offer a linear transformation from $C^\infty(M^n, E)$ to itself, which will usually not be a linear differential operator (in fact, such an operator will have an almost random effect on supports).

   (2)   The question of continuity of linear differential operators can be addressed directly only if the spaces $C^\infty(M^n, E)$ are topologized. In general, this poses difficulties, but in sufficiently special cases one may introduce a

topology. Suppose E has a metric such that restricted to $p^{-1}(x)$ it gives the usual topology on a finite-dimensional vector space. For $s \in C^{\infty}(M^n, E)$, one may define a continuous function

$$\rho_s(x) = \text{distance from } s(x) \text{ to the origin in } p^{-1}(x).$$

If in addition M^n is compact, set $\sigma(s) = \max_{x \in M^n} \rho_s(x)$. Then one can define a distance between sections s_1 and s_2 to be $\sigma(s_1 - s_2)$. In more general circumstances, constructing a natural metric is problematical. Various topologies are possible, but not directly related to metrics.

Suitable metrics on E might include a Riemann metric on E when $E = TM^n$.

THE SYMBOL; ELLIPTICITY

As a final topic, we wish to define the symbol of a linear differential operator and then address the notion of ellipticity. In the local case, that is, a linear partial differential operator L of order m defined over Euclidean space $\mathbb{R}^n$, the symbol may be constructed as follows: (1) ignore the terms of order less than m and (2) in each term, replace $\partial/\partial x_i$ by a symbol ξ_i, obtaining a form in the variables ξ_i, with coefficients being smooth functions. Higher derivatives, i.e., $\partial^k/\partial x_i^k$, are written as powers of ξ_i, i.e., ξ_i^k in this case.

If this form, in local case, is definite, that is to say, it vanishes only when all the variables are set equal to 0, we say that the linear differential operator is *elliptic*. Thus, in the classical two-dimensional case

$$\Delta = \frac{\partial^2}{\partial x^2} + \frac{\partial^2}{\partial y^2}$$

has symbol $\xi_1^2 + \xi_2^2$, and it is visibly elliptic. As another case,

$$3\frac{\partial^2}{\partial x^2} + 5\frac{\partial^2}{\partial y^2} - g\frac{\partial}{\partial x} + 4\frac{\partial}{\partial y}$$

has symbol $3\xi_1^2 + 5\xi_2^2$ and is elliptic. On the other hand,

$$2\frac{\partial^4}{\partial x^4} - 3\frac{\partial^4}{\partial y^4} + 3x\frac{\partial}{\partial x}$$

is *not* elliptic.

The term elliptic comes from the theory of quadratic forms as it relates to conic sections.

In order to treat the symbol in a global setting, one cannot expect to consider a single form, but rather something attached to—and reflecting the complexity of—the bundle. To this end, we set up the following machinery:

Suppose we are given a smooth manifold M^n and two smooth complex vector bundles over M^n, (E_i, M^n, p_i). T^*M^n is the (real) contangent bundle, with projection map π. Suppose

$$P: C^\infty(M^n, E_1) \to C^\infty(M^n, E_2)$$

is a linear differential operator of order m.

We consider the induced bundle

$$\pi^{-1}(E_1) = \{(y, x) \in T^*M^n \times E_1 \,|\, p_1(x) = \pi(y)\}.$$

Let $a \in M^n$, $\omega \in (T^*M^n)_a$. Let f be a smooth function in a neighborhood of a for which $f(a) = 0$ and $df(a) = \omega$ [as is easily constructed in $\mathbb{R}^n$ and then (locally) in M^n]. For $e \in p_1^{-1}(a)$, let s be a smooth section of (E_1, M^n, P_1) such that $s(a) = e$.

We set

$$\sigma_p(\omega, e) = P(f^m s)(a).$$

Notice that if $s(a) = 0$, $f^m s$ has a zero of order greater than m at a, making $P(f^m s)(a) = 0$; this shows independence of the choice of s. In a similar fashion, one shows independence of the choice of f. Observe that (ω, e) is a point of $\pi^{-1}(E_1)$.

Definition 6.9   The map

$$\sigma_p: \pi^{-1}(E_1) \to E_2$$

is called the *symbol* of the differential operator P. For each $x \in M^n$ and $\omega \in (T^*M^n)_x$, σ_p is a map from $p_1^{-1}(x)$ to $p_2^{-1}(x)$.

Definition 6.10   A linear differential operator P is *elliptic* if for each nonzero $\omega \in (T^*M^n)_a$, $a \in M^n$, the map σ_p is 1–1.

As an elementary example, consider the classical Laplacian in the plane,

$$\Delta = \frac{\partial^2}{\partial x^2} + \frac{\partial^2}{\partial y^2},$$

with respect to the (trivial) 1-dimensional vector bundle over $\mathbb{R}^2$. Then

$$\pi^{-1}(E_1)$$

is a trivial 1-dimensional real vector bundle over $\mathbb{R}^2 \times \mathbb{R}^2 = \mathbb{R}^4$.

Given $\omega \in \mathbb{R}^2$, $\omega = a_1 dx + a_2 dy$, choose $f(x, y) = a_1 x + a_2 y$. Set $s(x, y) = ((x, y), a)$. We calculate

$$\left(\frac{\partial^2}{\partial x^2} + \frac{\partial^2}{\partial y^2}\right)((a_1 x + a_2 y)^2 a) = \left(\frac{\partial^2}{\partial x^2} + \frac{\partial^2}{\partial y^2}\right)(a_1^2 a x^2 + 2a_1 a_2 axy + a_2^2 a y^2)$$

$$= 2a_1^2 a + 2a_2^2 a.$$

If a_1 and a_2 are not both zero, this is clearly a 1–1 map in the variable a, verifying that the operator is elliptic.

A more sophisticated example would be the following: Let M^n be a smooth manifold, and consider

$$\wedge^i T^* M^n,$$

the bundle of i-forms (Definition 5.3b). Consider smooth differential forms, i.e.,

$$D^i(M^n) = C^\infty(M^n, \wedge^i T^*(M^n)),$$

so that we have a linear differential operator

$$d: D^i(M^n) \to D^{i+1}(M^n)$$

of order 1. Let us calculate the symbol σ_d when $i = 0$.

Given $\omega \in (T^* M^n)_a$, $f \in C^\infty(M^n)$, with $(df)(a) = \omega$. Let $s \in D^0(M^n) = C^\infty(M^n)$, $s(a) = e$. Then by definition, in a suitable local coordinate chart, we may calculate

$$\sigma_d(\omega, e) = d(fs)(a) = \sum_{i=1}^{n} \frac{\partial(fs)}{\partial x_i}(a)\, dx_i$$

$$= \sum_{i=1}^{n} \left(\frac{\partial f}{\partial x_i}(a)s(a) + f(a)\frac{\partial s}{\partial x_i}(a)\right) dx_i$$

$$= e \cdot \omega + \sum_{i=1}^{n} f(a)\frac{\partial s}{\partial x_i}(a)\, dx_i.$$

Because we may choose (in our definition) any s with $s(a) = e$, we select $s(x)$ to be constantly e, making the last term zero.

Thus,

$$\sigma_d(\omega, e) = e\omega,$$

and $d: D^0(M^n) \to D^1(M^n)$ is elliptic.

As a final topic, we shall sketch the theory of the *analytic index* of an operator. With the same given data as before, for a elliptic differential operator P of positive order on a compact manifold, it is known [88] that the vector spaces

$$\ker P = \{f \in C^\infty(M, E_1) | P(f) = 0\}$$

and

$$\operatorname{coker} P = C^\infty(M^n, E_2)/\{g \in C^\infty(M^n, E_2) | g = Pf\}$$

are both finite dimensional.

It is then natural to look at the integer

$$i_a(P) = \dim(\ker P) - \dim(\operatorname{coker} P),$$

which is called the *analytic index* of P.

It had been known that $i_a(P)$ is not changed by small perturbations, and this suggested (to Gelfand) that there might be a purely topological description of i_a. This has lead to the theory of the topological index and the beautiful index theorem of Atiyah and Singer (see [6, 88]), which asserts that these two integers are the same.

The analytic index is known without the full index theorem in various interesting special cases. One may give an immediate interpretation of the Euler characteristic in analytic terms (the Euler characteristic of a triangulated manifold is equal to the number of vertices, minus the number of edges, plus the number of triangles, etc; see [5].)

PROBLEMS AND PROJECTS

1. Use Proposition 6.1 to derive the expressions for the Laplacian in polar, spherical, and cylindrical coordinates.

2. Is the linear map (for a Riemann manifold)

$$\delta : D^p(M^n) \to D^{p-1}(M^n)$$

a differential operator? If so, what would be its order?

3. Show that on a Riemann manifold the Laplacian Δ is an elliptic, linear differential operator. (One knows that the property of being elliptic does not depend on the choice of coordinate chart. The result is not difficult if one uses the normal coordinates, as in [65].)

4. If M^n is a compact Riemann manifold, $\omega \in D^p(M^n)$, then ω is harmonic ($\Delta\omega = 0$) if and only if both $d\omega = 0$ and $\delta\omega = 0$. (Using the inner product, check that $(\Delta\omega, \omega) = (d\omega, d\omega) + (\delta\omega, \delta\omega)$.)

5. Examine the theory of complex forms of type (p, q) on a complex or almost complex manifold M^{2n}, as well as various naturally defined linear differential operators there. Consult [88, 5].

6. Let M^n be a smooth manifold immersed or embedded in $\mathbb{R}^{n+k}$. Let $a \in \mathbb{R}^{n+k}$ be a fixed vector. For $x \in M^n$, let $f(x)$ be the inner product of the image of x under the immersion and a.

Then one has the following basic results from the differential geometry of submanifolds (consult pp. 339–340 of Vol. II [65]).

(a)   The mean normal curvature of M^n is zero if and only if $\Delta f = 0$ for every f as constructed above.

(b)   There is no compact, minimal submanifold in Euclidean space. (This may be deduced from Proposition 6.2).

7.   The formulas that express the relationship of differentiation to Fourier transforms [Remark (vii) preceding Lemma 6.3] point to a theory of non-integral order derivatives, because α need not be integral (that is made up of whole numbers) to take the Fourier transform of $y^\alpha \hat{f}(y)$. This is the starting point for *pseudodifferential* operators. For this very important area, consult [86, 58].

7

Infinite-Dimensional Manifolds

The basic idea of a differentiable manifold was to give the most reasonable generalization of Euclidean space, a generalization to which most of the traditional analysis and geometry that one carries out in basic mathematics would extend immediately. Infinite-dimensional manifolds may be viewed in a similar way. For example, starting with some fixed, infinite-dimensional vector space, such as Hilbert space or some Banach space, one can form a manifold in the traditional way, that is, by gluing together open sets by nice overlap functions. But the need for infinite-dimensional manifolds will arise in other ways. For example, given two manifolds in the sense of Chapter 1, say M^n and N^k, there is a substantial interest in the space $C(M^n, N^k)$ of continuous maps from M^n to N^k, as well as the various subspaces obtained by requiring certain amounts of differentiability for these maps. Under certain conditions, this should be (and we shall prove it is) an infinite-dimensional manifold. The fact that various function spaces are infinite-dimensional manifolds can then be exploited to draw certain conclusions about the topology of these function spaces. On top of this, the theory of infinite-dimensional manifolds will afford the opportunity to make explicit various connections between geometry and analysis, particularly operator theory.

Within the obvious limitations imposed by the fact that this must be done within a single chapter of the present book, we propose here to do the following:

(i) We recall the basic definitions and facts about Banach and Hilbert spaces.

(ii) We define an infinite-dimensional manifold, modeled on a given topological vector space.

(iii) We discuss the basic problems concerned with generalizing the standard concepts, such as the tangent space or a partition unity, to infinite-dimensional manifolds. We shall prove the existence of partitions of unity, in the case of Hilbert manifolds (i.e., manifolds modeled on Hilbert space) and outline various other results that are relevant there.

(iv) We shall prove a theorem (Theorem 7.2) to the effect that many function spaces made up of maps between ordinary manifolds are actually infinite-dimensional manifolds.

(v) We shall discuss the case of a general linear group acting on Hilbert space. We mention connectivity, as well as the special result of Jänich [61] on contractibility. The general result of Kuiper [67] is indicated, but we omit the proof, which is too detailed to be included here.

(vi) We give a brief survey of some of the literature in this ever expanding field.

TOPOLOGICAL VECTOR SPACES

Definition 7.1 Let V be a vector space over the real or complex numbers, so that we are given the usual operations of addition, $v_1 + v_2$, and scalar multiplication, λv_1. We shall call V a *topological vector space* if there is a topology on V for which the maps

$$\phi: V \times V \to V, \qquad \phi(v_1, v_2) = v_1 + v_2,$$

and

$$\psi: k \times V \to V, \qquad \psi(a, v) = av,$$

are continuous (here k is either the real or the complex numbers, endowed with the usual topology).

There are various important classes of topological vector spaces. For example:

Definition 7.2 A topological vector space V is *locally convex* if whenever we are given an open neighborhood of the origin 0, say U_1, then we can find a smaller open neighborhood of 0, say U_2, that is convex; i.e., if x, $y \in U_2$, $0 \le t \le 1$, then $tx + (1 - t)y \in U_2$.

The vector spaces $C^r(O)$ of C^r-functions on O may be topologized to give examples.

For the basic theory of such spaces, see [119, 57]. More important are the Banach spaces, whose definition we now recall.

Definition 7.3   Let V be a vector space. A *norm* on V is a function $V \to \mathbb{R}$, usually written $\|x\|$, for each $x \in V$, such that

(a) $\|x\| \geq 0$; in addition $\|x\| = 0$, if and only if $x = 0$;
(b) $\|\lambda x\| = |\lambda| \|x\|$, for $x \in V$, $\lambda \in k$;
(c) $\|x + y\| \leq \|x\| + \|y\|$, for any $x, y \in V$.

Note that if V has a norm then we may define a metric on V by

$$\rho(x, y) = \|x - y\|.$$

We may then immediately define a topology on V by taking the sets

$$B_\varepsilon(x) = \{y \in V \mid \rho(x, y) < \varepsilon\},$$

that is to say, the open balls, as a basis.

Definition 7.4   A topological vector spaces whose topology arises from a norm is called a *Banach space*, if it is complete with respect to the norm or metric; that is to say, given a sequence $\{x_n\}$ in V that is Cauchy (if $\varepsilon > 0$, there is $N > 0$ so that $m, n > N$ implies $\rho(x_m, x_n) < \varepsilon$), then the sequence converges to some $x \in V$ (if $\varepsilon > 0$, there is $M > 0$ so that $\rho(x, x_n) < \varepsilon$ whenever $n > M$).

Of course, we understand that $\rho(x, y)$ is defined to be $\|x - y\|$ in this context.

We now discuss various examples.

Examples *1.* The set of real numbers with $\|x\| = |x|$ is trivially a Banach space. More generally, consider $\mathbb{R}^n$ with the norm

$$\|x\| = \sqrt{x_1^2 + \cdots + x_n^2}$$

for $x = (x_1, \ldots, x_n)$.

There are other norms which yield the same topology.

2. $C[a, b]$, the space of continuous functions on the closed interval of real numbers $[a, b]$, with norm

$$\|f\| = \max_{a \leq x \leq b} |f(x)|.$$

3. Hilbert space. This is most easily described as l^2, the set of all sequences of real numbers

$$(a_1, a_2, a_3, \ldots.)$$

such that

$$\sum_{i \geq 0} a_i^2$$

converges. If we abbreviate the sequence by (a_i), we define

$$\|(a_i)\| = \sqrt{\sum_{i \geq 0} a_i^2}.$$

Strictly speaking, this is a separable Hilbert space (in the sense that l^2 has a countable dense subset). To show that l^2 is complete, one takes a limit of a Cauchy sequence in some fixed place and thereby fabricates the limit sequence. l^2 is then a Banach space.

We note that l^2 has various other representations. For example, if $a < b$, $L^2([a, b])$ is the set of Lebesgue measurable functions defined on $[a, b]$ whose squares are integrable. Similarly, one may define $L^2(-\infty, \infty)$.

We also remark that l^2 has fundamentally more structure than a Banach space, in that there is an inner product from which the norm is derived. We simply set

$$((a_i), (b_i)) = \sum_{i \geq 0} a_i b_i.$$

The sequence on the right is easily checked to converge, because it follows from the usual Cauchy–Schwarz inequality that

$$\sum_{i=1}^{m} |a_i b_i| \leq \sqrt{\sum_{i=1}^{m} a_i^2} \sqrt{\sum_{i=1}^{m} b_i^2} \leq \|(a_i)\| \|(b_i)\|.$$

This inner product follows all the usual laws, and furthermore, one may define the norm in terms of the inner product via

$$\|(a_i)\| = \sqrt{((a_i), (a_i))}.$$

In similar fashion, one may do Hilbert spaces over the complex numbers.

4. Let B_1 and B_2 be Banach spaces and let f be a linear transformation, i.e.,

$$f : B_1 \rightarrow B_2.$$

Suppose that f is bounded in the sense that there is $K > 0$ with

$$\|f(x)\| \leq K\|x\|$$

for all $x \in B_1$. (It is clear from context which norm is meant by $\| \, \|$.)

I claim that such an f is necessarily continuous. Because B_1 and B_2 are metric spaces it will suffice to show that whenever $\lim_n x_n = x$, then

$\lim_n f(x_n) = f(x)$. But we easily calculate

$$\lim_n \|f(x_n) - f(x)\| = \lim_n \|f(x_n - x)\| \leq \lim_n K\|x_n - x\| = K \lim_n \|x_n - x\| = 0,$$

from which $\lim_n f(x_n) = f(x)$ follows at once.

The converse of this assertion, that if f is continuous K exists, is not difficult to check.

We define

$$\|f\| = \text{g.l.b.}\{K \mid \|f(x)\| \leq K\|x\| \quad \text{for all} \quad x \in B_1\}.$$

This norm makes the space of linear transformations f into a Banach space, $\text{Lin}(B_1, B_2)$.

Observe that the closed graph theorem from basic functional analysis implies that if f is linear, continuous, $1-1$, and onto, then f is an isomorphism of Banach spaces (in the sense that it has a two-sided inverse). Furthermore, the linear isomorphisms are open in $\text{Lin}(B_1, B_2)$.

5. Many topological vector spaces are not complete, for example, the subspace of $C[0, 1]$ consisting of the differentiable functions, or the subspace of $L^1[0, 1]$ (the Lebesgue integrable functions) that is the Riemann integrable functions.

Now we wish to extend to this general context some basic ideas of differentiable calculus. We take two open subsets of our Banach spaces

$$U_1 \subseteq B_1 \quad \text{and} \quad U_2 \subseteq B_2.$$

Suppose $g: U_1 \to U_2$ is a continuous function.

Definition 7.5 We say that g is *differentiable* at $x \in U_1$ with derivative $g' \in \text{Lin}(B_1, B_2)$ if for all $h \in B_1$ of sufficiently small norm the function

$$\varepsilon(h) = g(x + h) - g(x) - g'(x)h$$

satisfies

$$\lim_{\|h\| \to 0} \|\varepsilon(h)/\|h\| \| = 0.$$

Note that g' in this notation consists of operator multiplication by the (constant) derivative operator at the point x. g' consists, as x varies, of a map from U_1 to $\text{Lin}(B_1, B_2)$.

It is not difficult to check the usual linearity and the chain rule. The higher-order derivatives may be similarly defined.

Our next goal is to introduce the notation of an infinite dimensional manifold modeled after some fixed, given topological vector space, rather than just $\mathbb{R}^n$, and to discuss the basic elementary properties. This works easily, and it constitutes, by and large, a routine generalization of the basic theory that we have discussed up to this point. The first point where the new theory is more difficult involves localization and partition of unity. There arise here questions of the existence of suitable smooth functions, in the spirit of Urysohn's lemma from basic point-set topology. We shall prove that these constructions are indeed possible in the case of infinite-dimensional manifolds modeled after Hilbert space l^2. In full generality, some of these questions have not yet been resolved.

ELEMENTS OF INFINITE-DIMENSIONAL
MANIFOLDS

Definition 7.6   Let B be a given topological vector space (usually a Banach space). A Hausdorff space M is called a *manifold modeled on B*, if for every $x \in M$ there is some open $O_x \subseteq M$, with $x \in O_x$, and a homeomorphism ϕ_x of O_x onto an open subset of B.

In general, there will be many O_x and ϕ_x for each x in M.

Note that the definition differs slightly from our earlier definition of a manifold in that ϕ_x is not required to be a homeomorphism onto all of B, but just onto an open subset of B. In general, we shall not be able to specify the range of ϕ_x.

Clearly, every (finite-dimensional) manifold, in the sense of Chapter 1, is a manifold modeled on $\mathbb{R}^n$ for some n. If B is a topological vector space, any open subset of B is a manifold modeled on B. If M^n is a finite-dimensional manifold and B is a Banach space, it is easy to check that $M^n \times B$ is a manifold, in the sense of Definition 7.6, modeled on the (obvious Banach) space $\mathbb{R}^n \times B$. For our general theory, no countability condition is necessary on M.

If we wish to define an infinite-dimensional differentiable manifold, it is more convenient to work with norms (compare Definition 7.5), so we shall then restrict attention to manifolds modeled on a Banach space (which we shall now refer to as *Banach manifolds*).

Definition 7.7   Let B be a given Banach space. M is called a *smooth manifold modeled on B* if M is a manifold modeled on B, as in Definition 7.6 above, such that if $O_x \cap O_y \neq \phi$, the composition

$$\phi_x(O_x \cap O_y) \xrightarrow{(\phi_x^{-1} \mid \phi_x(O_x \cap O_y))} O_x \cap O_y \xrightarrow{\phi_y} \phi_y(O_x \cap O_y)$$

is a smooth map in the sense that it has continuous derivatives, of any order, obtained from iterating Definition 7.5. (Similarly, C^r-manifolds may be defined for any positive integer r. A C^1-manifold might be called a differentiable manifold modeled on B.)

Just as before, we may define maps.

Definition 7.8 Let M and N be manifolds modeled on Banach spaces. Let $f: M \to N$ be a continuous map. Then we say that f is *smooth* (or *differentiable*) if for each $x \in O_x \subseteq M$ with homeomorphism $\phi_x: O_x \xrightarrow{\cong} \phi_x(O_x)$ and $y \in O_y$ with $\phi_y: O_y \xrightarrow{\cong} \phi_y(O_y)$, where $f(x) = y$, then in a sufficiently small neighborhood of x (so that the following composition is defined),

$$\phi_x(O_x) \xrightarrow{\phi_x^{-1}} M \xrightarrow{f} N \xrightarrow{\phi_y} \phi(O_y)$$

is smooth (or differentiable).

We note that, just as before, we can define two covers of a manifold by open sets, as in Definition 7.6, to be equivalent, by requiring that the mixed overlap maps are smooth. Alternatively, one might require that a family of open sets, as in Definition 7.6, be maximal.

Several of our earlier notions go through in a trivial manner, although on occasion some care is needed. For example:

(i)   One defines a submanifold as in Chapter 1. One requires that each point has a neighborhood homeomorphic to $O_1 \times O_2$ where O_1 is open in B_1 and O_2 is open in B_2. The big manifold is to be modeled on $B_1 \times B_2$, and the submanifold is described locally in terms of $O_1 \times$ pt.

A submanifold is easily checked to be a manifold.

(ii)   A diffeomorphism (or less often "isomorphism") of infinite-dimensional manifolds is a smooth map with a smooth, two-sided inverse.

(iii)   An embedding $\phi: M \to N$ is a smooth 1–1 map that is a diffeomorphism of M onto a submanifold of N.

(iv)   An immersion is a smooth map which is locally an embedding.

We note that it is also possible to formulate notions of immersion and embedding in terms of the tangent space (as we remark below). This would be more in line with the presentation of Chapter 2.

Definition 7.9   Given a smooth manifold M modeled on a Banach space B, choose for each $x \in M$ a coordinate neighborhood O_x endowed with a homeomorphism ϕ_x onto an open set in B, i.e.,

$$\phi_x: O_x \xrightarrow{\cong} \phi_x(O_x) \subseteq B.$$

Consider the sets $O_x \times B$, and define an equivalence relation in their union by specifying that if $(u, v_1) \in O_x \times B, (u, v_2) \in O_y \times B$, then (u, v_1) is equivalent to (u, v_2) if and only if

$$(\phi_y \circ \phi_x^{-1})'(v_1) = (v_2).$$

(Here, the prime refers to the derivative. The derivative is to taken at u.)

Define the quotient space to be TM, and define the map

$$\pi : TM \to M$$

by $\pi\{(u, v)\} = u$. We write $(TM)_x = \pi^{-1}(x)$.

In the same fashion as before, we may define the differential of a smooth map between smooth manifolds. One might then characterize immersions f by requiring that at every point the differentials

$$(df)_x : (TM)_x \to (TN)_{f(x)}$$

have a left inverse. The proof of the equivalence of the two appraoches is not difficult. It involves generalizing the inverse function theorem (our Theorem 1.1) to Banach spaces.

HILBERT MANIFOLDS; PARTITION OF UNITY

We now look into the question of generalizing the technique of the partition of unity to infinite-dimensional manifolds (compare Theorem 1.4). We shall do this for separable Hilbert manifolds. Of course, such a separable manifold is a fortiori a paracompact space, and hence partitions of unity exist by the basic theorems of point-set topology. However, we are interested in smooth manifolds, and hence in the construction of smooth partitions of unity. To achieve this one can proceed as follows:

Lemma 7.1 Let X be a (separable) Hilbert space, such as l^2, and let $B_\varepsilon(x)$ be an open ball of radius $\varepsilon > 0$ about $x \in l^2$. Then there is a smooth function

$$\phi : l^2 \to \mathbb{R}$$

such that $\phi(y) = 0$ if $y \notin B_\varepsilon(x)$, yet $\phi(y) > 0$ if $y \in B_\varepsilon(x)$.

Proof An element of l^2 is a sequence (a_i) such that

$$\sum_{i=0}^{\infty} a_i^2$$

converges. This number is (by definition) the square of the norm of (a_i), i.e., $\|(a_i)\|^2$.

If $(h_i) \in l^2$, we form (with an eye towards eventually using Definition 7.5)

$$\varepsilon((h_i)) = \|(a_i + h_i)\|^2 - \|(a_i)\|^2 - 2\sum (a_i h_i)$$
$$= ((a_i + h_i), (a_i + h_i)) - \|(a_i)\|^2 - 2((a_i), (h_i)),$$

where we recall the inner product on l^2 is

$$((a_i), (b_i)) = \sum_{i \geq 0} a_i b_i.$$

Therefore,

$$\varepsilon((h_i)) = \|(a_i)\|^2 + 2((a_i), (h_i)) + \|(h_i)\|^2 - \|(a_i)\|^2 - 2((a_i), (h_i)) = \|(h_i)\|^2.$$

Clearly, if we let (h_i) approach 0 in norm,

$$\lim_{\|(h_i)\| \to 0} (\varepsilon((h_i)) / \|(h_i)\|) = 0.$$

In other words, the linear function

$$(x_i) \mapsto 2\sum_{i \geq 0} a_i x_i,$$

which we may also write

$$(x_i) \mapsto 2((a_i), (x_i)),$$

is the derivative of $\| \ \|^2$ at (a_i).

One may continue to show without difficulty that $\| \ \|^2$ is a smooth function.

Now, just as we did at the section on partitions of unity in Chapter 1, we construct a smooth function

$$\chi : \mathbb{R} \to \mathbb{R}$$

so that

$$\chi(x) = 0 \quad \text{if } |x| \geq \varepsilon^2, \qquad \chi(x) > 0 \quad \text{if } |x| < \varepsilon^2.$$

We set $\phi((a_i)) = \chi(\|a_i\|^2)$. To complete the proof, note that ϕ is smooth and bounded, and ϕ satisfies the conditions required by this lemma.

We remark that for arbitrary Banach spaces there is no easy candidate for such a smooth (or just C^1-) function ϕ.

Lemma 7.2 Let M be a separable metric space that is a *Hilbert manifold*, that is to say, a manifold modeled on l^2. Suppose that $\{O_n\}$ is a countable, locally finite cover by open sets, so that for each n, we have

$$O_n = B_{r_1}(x_1) \cap \cdots \cap B_{r_n}(x_n) \cap (M - \overline{B_{s_1}(y_1)}) \cap \cdots \cap (M - \overline{B_{s_m}(y_m)}).$$

Here the x_i's and y_j's are points in M, the r_i and s_j are positive real numbers, and the bar means the closure. (For various O_n, the x_i, r_i, s_j, and y_j are generally different.)

Then there is a partition of unity subordinate to this cover.

Proof Using the previous lemma, select smooth ϕ_i, $1 \le i \le n$, such that

$$\phi_i(z) > 0 \qquad \text{if} \quad z \in B_{r_i}(x_i)$$

and

$$\phi_i(z) = 0 \qquad \text{if} \quad z \notin B_{r_i}(x_i).$$

We also select smooth ψ_j such that

$$\psi_j(z) > 0 \qquad \text{if} \quad z \notin \overline{B_{s_j}(y_j)}$$

and

$$\psi_j(z) = 0 \qquad \text{if} \quad z \in \overline{B_{s_j}(y_j)}.$$

This latter type of function is easily constructed from the previous lemma applied to a suitable family of open balls covering the complement of $\overline{B_{s_j}(y_j)}$ (check).

Define the product function

$$\tilde{f}_n(x) = \phi_1(x) \cdots \phi_n(x) \cdot \psi_1(x) \cdots \psi_m(x).$$

Clearly, $\tilde{f}_n(x) > 0$, if and only if $x \in O_n$ and $\tilde{f}_n$ is visibly smooth.

Finally, we set

$$f_n(x) = \tilde{f}_n(x) / \sum_n \tilde{f}_n(x),$$

to get the desired partition of unity.

Lastly we need

Lemma 7.3 If M is a separable metric space and $\{U_n\}$ is a countable open cover, then there is a countable, locally finite refinement consisting of sets of the form needed in Lemma 7.2.

Proof Because a separable metric space satisfies the second axiom of countability, we can find a countable refinement of our original open cover by open balls, say B_n. This cover might not be locally finite, but we shall find a cover by sets of the form in Lemma 7.2, say O_n, so that $O_n \subseteq B_n$, for each n, and the O_n are a locally finite cover.

Set $O_1 = B_1$. Let B_n be the ball of radius $\rho_n > 0$ about the point x_n, to fix the terminology.

Set

$$r_{i,n} = \max(\rho_i - 1/n, 0), \qquad 1 \le i \le n - 1.$$

Set

$$O_n = B_n \cap (M - \overline{B_{r_{1,n}}(x_1)}) \cap \cdots \cap (M - \overline{B_{r_{n-1,n}}(x_{n-1})}).$$

Clearly $O_n \subseteq B_n$.

To see that the O_n's cover M, set, for a given $x \in M$,

$$j = \min_{x \in B_i} (i).$$

If $x \notin O_j$, then

$$x \in (M - B_j) \cup \overline{B_{r_1,j}(x_1)} \cup \cdots \cup \overline{B_{r_{j-1,1,j}}(x_{j-1})}.$$

But this means that either

$$x \in \overline{B_{r_1,j}(x_1)} \subseteq B_{\rho_1}(x_1) = O_1 \qquad \text{or} \qquad x \in \overline{B_{r_2,j}(x_2)} \subseteq B_{\rho_2}(x_2),$$

etc. This would contradict the definition of j. We conclude that with our definition of j, for a given x, we must have $x \in O_j$. Thus, $\{O_n\}$ is a cover.

Finally, if $x \in O_i$, select $\varepsilon > 0$ such that

$$x \in B_\varepsilon(x) \subseteq O_i \subseteq B_i = B_{\rho_i}(x_i).$$

If $B_\varepsilon(x)$ meets only finitely many B_j, we are done (since the cover is locally finite at x for $O_j \subseteq B_j$). In general, observe that there is $\varepsilon' > 0$ such that

$$x \in B_{\varepsilon'}(x) \subseteq \overline{B_{\rho_i - 1/n}(x_i)} = \overline{B_{r_{i,n}}(x_i)}$$

for all sufficiently large n. But then it follows from the definition that $B_{\varepsilon'}(x)$ does not meet O_n for all sufficiently large n. Hence, the cover $\{O_n\}$ is locally finite.

Summarizing the previous three lemmas, we have

Theorem 7.1 Let M be a manifold modeled on l^2, which is separable metric. Let $\{B_\alpha\}$ be an open cover. Then there is a countable, locally finite, open cover $\{O_n\}$, refining the cover $\{B_\alpha\}$, and there exists a smooth partition of unity subordinate to this cover $\{O_n\}$.

The problems associated with doing this for an arbitrary, infinite-dimensional manifold are twofold. First, what meaning can be given to the term smooth? Second, can one find "nice" smooth functions of the sort required in Lemma 7.1?

FUNCTION SPACES

We now propose to look at infinite-dimensional manifolds that arise as function spaces.

Definition 7.10   Let M^n be a smooth Riemann manifold. That is, for any $x \in M^n$ vectors $v_1, v_2 \in (TM^n)_x$, the tangent space at x, we have a continuous, symmetric, positive definite inner product defined by

$$(v_1, v_2)_x = v_1^t(g_{ij}(x))v_2.$$

We set

$$\|v\|_x = \sqrt{(v, v)_x}.$$

If X is any compact Hausdorff space, we set

$$C(X, M^n) = \{f \mid f: X \to M^n, \quad f \text{ continuous}\}.$$

We introduce a metric on $C(X, M^n)$ by putting

$$d(f, g) = \text{l.u.b.}_{x \in X} \rho(f(x), g(x)),$$

where $\rho(a, b)$ is the greatest lower bound of the lengths of all smooth curves in M^n from a to b. (See Definition 6.5 and sequel for the definition of the length of a smooth curve on M^n as an integral.)

Remarks   (1)   It is easy to see that d makes $C(X, M^n)$ into a metric space. We shall view $C(X, M^n)$ as a topological space whose topology comes from this metric; i.e., the open balls are taken to be a basis for the topology.

(2)   We observe that a homotopy class of maps (continuous) from X to M^n is the same as a pathwise connected component in the space $C(X, M^n)$, for a homotopy $F: X \times I \to M^n$ easily translates to a path in $C(X, M^n)$.

Definition 7.11   For a given map $f \in C(X, M^n)$, we may define a tangent space to $C(X, M^n)$ at that map, written

$$T_f C(X, M^n),$$

as follows.

Let $\pi: TM^n \to M^n$ be the tangent bundle map. Then $T_f C(X, M^n)$ is the set of all

$$\tilde{f}: X \to TM^n$$

such that $\pi \circ \tilde{f} = f$.

We may obviously make $T_f C(X, M^n)$ into a *linear space*, because if we set for each x

$$(\tilde{f}_1 + \tilde{f}_2)(x) = \tilde{f}_1(x) + \tilde{f}_2(x),$$

then $\pi(\tilde{f}_1 + \tilde{f}_2)(x) = f(x)$, and similarly for $\alpha \tilde{f}$, via $(\alpha \tilde{f})(x) = \alpha(\tilde{f}(x))$.

We introduce a *norm* into $T_f C(X, M^n)$ by setting

$$\|\tilde{f}\|_f = \operatorname*{l.u.b.}_{x \in X} \|\tilde{f}(x)\|_{f(x)}.$$

Here, we recall that $\|\tilde{f}(x)\|_{f(x)}$ is the norm in $(TM^n)_{f(x)}$, defined by the Riemann metric.

Finally, it is trivial to verify that $T_f C(X, M^n)$ is a complete metric space with respect to this norm (the proof being much the same as the proof that $C([a, b])$ is a complete metric space, i.e., uniform limits of continuous functions are continuous functions). In other words, $T_f C(X, M^n)$ is a (real) Banach space.

We now may prove a theorem which assures us that in a certain case function spaces are infinite-dimensional manifolds.

Theorem 7.2 If X is a compact Hausdorff space and M^n is a smooth Riemann manifold, then $C(X, M^n)$ is a (smooth) manifold modeled on a real Banach space, in this case $T_f C(X, M^n)$. (The choice of f does not matter here.)

Proof We must show that every point in $C(X, M^n)$ has an open neighborhood that is homeomorphic to an open subset of $T_f C(X, M^n)$.

First we note that on a Riemann manifold every point has a sufficiently small neighborhood such that every point in that neighborhood may be joined to our original point by a unique smooth curve (geodesic) of minimal length. This follows from basic facts about ordinary differential equations (see [22, 52]).

Second, because X is compact, we may easily find an $\varepsilon > 0$ such that every ball of radius ε in $f(X)$, given $f \in C(X, M^n)$, is small enough in the sense of the previous paragraph. In fact, one sees in the same fashion that any two points in $f(X)$ whose distance is less than some fixed $\varepsilon > 0$ may be joined by a unique geodesic of minimal length.

Hence, if we take $g \in C(X, M^n)$, with

$$d(f, g) = \operatorname*{l.u.b.}_{x \in X} \rho(f(x), g(x)) < \varepsilon,$$

then for any $x \in X$ we have a unique geodesic of minimal length joining $f(x)$ and $g(x)$. Call this geodesic σ_x, so that $\sigma_x(0) = f(x)$, $\sigma_x(\alpha) = g(x)$ for some positive α. The derivative of σ_x with respect to its variable, when that variable equals 0, is a tangent vector to σ_x at $f(x)$. If σ_x is parameterized by are length, then this is a nonzero vector at $f(x)$, say v_x.

We then define $\phi(f)(g)(x)$ to be the vector having the same direction as v_x and having length equal to $\rho(f(x), g(x))$. We then have defined a map

$$\phi(f): \{g \in C(X, M^n) \mid d(f, g) < \varepsilon\} \to T_f C(X, M^n).$$

One may easily check directly from the definitions that

$$\|\phi(f)(g)\|_f = d(f, g).$$

Proceeding in the opposite sense, suppose that we are given $\tilde{f} \in T_f C(X, M^n)$ of small norm; then it follows (from the basic theory of ordinary differential equations) that there is a unique geodesic through $f(x)$ that is tangent to $\tilde{f}(x)$, and well defined for a distance at least $\|\tilde{f}(x)\|_{f(x)}$. If we follow this geodesic from the point $f(x)$ for a distance of $\|\tilde{f}(x)\|_{f(x)}$, we arrive at a point that we call

$$\psi(f)(\tilde{f})(x).$$

This process clearly defines a map

$$\psi(f): (\text{an open neighborhood of } \tilde{f} \in T_f C(X, M^n)) \to C(X, M^n).$$

There is no difficulty checking the continuity of these maps ϕ and ψ. In general, one may drop the f from the notation without risk of confusion.

Finally, we claim that these maps are inverse to one another. For example, we may calculate $\psi \circ \phi(g)(x)$. In our earlier terminology, we have

$$(\psi \circ \phi)(g)(x) = \psi\left(\frac{\rho(f(x), g(x))}{\|v_x\|} v_x\right),$$

that is, ψ is applied to a vector of length $\rho(f(x), g(x))$ pointing in the direction of v_x. v_x is tangent to the (unique) geodesic tangent to v_x, that is σ_x, which is the (unique) geodesic of minimal length from $f(x)$ to $g(x)$. If we apply ψ, which means proceeding along this geodesic a distance equal to the distance from $f(x)$ to $g(x)$, naturally we arrive at $g(x)$. Therefore, $(\psi \circ \phi)(g) = g$.

On the other hand, we may calculate $(\phi \circ \psi)(\tilde{f})(x)$ by noting that $\psi(\tilde{f})$ gives the result of following the geodesic tangent to $\tilde{f}(x)$ a distance of $\|\tilde{f}(x)\|_{f(x)}$, so that when we apply ϕ we have a vector identical to $\tilde{f}(x)$.

One may show by a tedious but straightforward argument that the coordinate overlaps are smooth.

Remarks We have only scratched the beginning in this area; the interested, reader is urged to consult various other sources, such as [28, 29, 69, or the more specialized 68, 35, 66].

THE UNITARY GROUP

As a final topic in infinite dimensional manifolds, we discuss the general linear and unitary groups acting on Hilbert space. Let H be a separable Hilbert space, such as l^2, over the real or complex numbers. An *operator* means simply a linear map

$$g: H \to H.$$

Define (as we have already mentioned)

$$\|g\| = \operatorname*{l.u.b.}_{0 \neq x \in H} \|g(x)\|/\|x\|.$$

If $\|g\|$ exists, we say that g is a *bounded* linear operator; it follows easily that g is then continuous.

We write End(H) for the bounded linear operators from H to itself; it is clearly an algebra over the real or complex numbers, according as we consider real or complex Hilbert spaces, because the composition of two bounded linear operators is clearly a bounded linear operator. End(H) is clearly a normed linear space.

We introduce the following terminology:

$$\mathrm{GL} = \mathrm{GL}(H) = \{g \in \mathrm{End}(H) | \exists\, g^{-1} \in \mathrm{End}(H), \quad g^{-1}g = gg^{-1} = 1\}$$

(here of course 1 denotes the identity operator);

$$\mathrm{U} = \mathrm{U}(H) = \{g \in \mathrm{GL} | \|gx\| = \|x\| \quad \text{for all} \quad x \in H\}.$$

It is clear that GL and U generalize $\mathrm{Gl}(n; \mathbb{R})$ (or $\mathrm{Gl}(n; \mathbb{C})$) and $\mathrm{O}(n)$ (or $\mathrm{U}(n)$), when the Hilbert space is merely $\mathbb{R}^n$ (or $\mathbb{C}^n$). In addition, GL and U inherit topologies as subsets of the normed linear space End(H), and they then become, as is easily checked, topological groups (Definition 4.1) that act as transformation groups (Definition 4.2) on H.

GL and U have many interesting properties, such as the following:

(1)   It is well known, from results in functional analysis, that these groups are connected (see [67]).

(2)   There is a retraction from GL onto U; that is, there is a continuous map

$$r: \mathrm{GL} \to \mathrm{U}$$

such that if i denotes the inclusion of U in GL, then $r \circ i = 1$.

(3)   GL is an open subset of End(H). GL is thus an infinite-dimensional manifold modeled on the normed linear space End(H).

(4)   Kuiper [67] has shown that GL (and hence U) have the homotopy type (remarks following Proposition 4.3) of a single point; that is, they are

contractible spaces. This striking result contrasts strongly with the situation for the ordinary general linear group,

$$\text{Gl}(n; \mathbb{R}) \subseteq \text{End}(\mathbb{R}^n).$$

For example, $\text{Gl}(1; \mathbb{R})$ has the homotopy type of two points. $\text{Gl}(2; \mathbb{R})$ has the homotopy type of two circles. In general, $\text{Gl}(n; \mathbb{R})$ is a sophisticated manifold, with, for example, rather nontrivial deRham cohomology (as in Definition 5.8).

Kuiper's theorem is both deep and technical. To get an idea of the methods used in this work, we shall now give a preliminary theorem of this type, which was first proved by Jänich [61], and which is actually one of the lemmas used in Kuiper's work. For the basic ideas about homotopy, refer to Definition 4.15.

Theorem 7.3   Let H_0 and H_1 be two infinite-dimensional closed subspaces of H such that

(a)   every element of H is a sum of an element of H_0 and an element of H_1, and
(b) H_0 and H_1 are mutually orthogonal; i.e., if $x \in H_0$ and $y \in H_1$, then $(x, y) = 0$.

Let $V = \{g \in \text{GL} \,|\, g \,|\, H_0 = 1_{H_0}$ and $g(H_1) = H_1\}$. In other words, an element of V is the identity on H_0 and takes H_1 to itself.
    Then the inclusion map $V \subseteq \text{GL}$ is homotopic to the constant map that sends all of V to the identity 1.

Proof H_0 may be decomposed into an infinite sum

$$H_0 = H_2 + H_3 + \cdots$$

of mutually orthogonal, infinite-dimensional subspaces. For example, if $e_1, e_2, \ldots$ is an orthonormal basis for H_0 (mutually perpendicular elements of length 1), let S_n be the subset of these basis vectors e_i such that i has a prime decomposition into $n - 1$ factors, $n \geq 2$. Put $e_1 \in S_2$, by convention, and let H_j be the subspace of H_0 spanned by the e_j belonging to S_n.
    Note that we then get an orthogonal decomposition

$$H = H_1 + H_0 = H_1 + H_2 + H_3 + \cdots.$$

Given a bounded linear map $f: H \to H$, f is clearly determined by what it does on each H_i.
    Let $\pi_n: H \to H_n$ be the projection onto H_n, i.e., the map that ignores components of basic vectors not in H_n. Then f is visibly determined by an

infinite matrix of elements

$$f_{ij} = \pi_j(f\,|\,H_i),$$

where each f_{ij} is a linear map from H_i to H_j. Of course not every such infinite matrix will yield a bounded linear operator f.

Now if $g \in V$, as defined in the statement of the present theorem,

$$g_{ij} = 0 \quad \text{if} \quad i \neq j, \qquad g_{ii} = 1 \quad \text{if} \quad i > 1, \qquad g_1 = g\,|\,H_1,$$

shortening g_{11} to g_1. It will suffice to exhibit a sequence of homotopies from the given map g, through elements of GL, ending with the map $l: V \to$ GL defined by $l(g) = 1$.

Observe that g is the same as h described as

$$h_{ij} = 0, \quad i \neq j; \qquad h_{11} = g_{11} = g_1; \qquad h_{ii} = g_1^{-1}g_1, \qquad i > 1.$$

(i)　We define a homotopy from h to h', where h' is given by the infinite matrix

$$h'_{ij} = 0, \quad i \neq j; \qquad h'_{11} = g_1; \qquad h'_{2i,2i} = g_1^{-1}; \qquad h'_{2i+1,2i+1} = g_1.$$

To see this, note that as θ ranges from 0 to $\pi/2$,

$$\begin{pmatrix} \cos\theta & -\sin\theta \\ \sin\theta & \cos\theta \end{pmatrix}\begin{pmatrix} g_1 & 0 \\ 0 & 1 \end{pmatrix}$$

goes from

$$\begin{pmatrix} g_1 & 0 \\ 0 & 1 \end{pmatrix} \quad \text{to} \quad \begin{pmatrix} 0 & -1 \\ g_1 & 0 \end{pmatrix}.$$

Thus, as θ goes from 0 to $\pi/2$,

$$\begin{pmatrix} \cos\theta & -\sin\theta \\ \sin\theta & \cos\theta \end{pmatrix}\begin{pmatrix} g_1 & 0 \\ 0 & 1 \end{pmatrix}\begin{pmatrix} \cos\theta & \sin\theta \\ -\sin\theta & \cos\theta \end{pmatrix}$$

goes from

$$\begin{pmatrix} g_1 & 0 \\ 0 & 1 \end{pmatrix} \quad \text{to} \quad \begin{pmatrix} 1 & 0 \\ 0 & g_1 \end{pmatrix}.$$

Finally,

$$\begin{pmatrix} \cos\theta & -\sin\theta \\ \sin\theta & \cos\theta \end{pmatrix}\begin{pmatrix} g_1 & 0 \\ 0 & 1 \end{pmatrix}\begin{pmatrix} \cos\theta & \sin\theta \\ -\sin\theta & \cos\theta \end{pmatrix}\begin{pmatrix} g_1^{-1} & 0 \\ 0 & 1 \end{pmatrix}$$

represents a homotopy from

$$\begin{pmatrix} 1 & 0 \\ 0 & 1 \end{pmatrix} = \begin{pmatrix} g_1 g_1^{-1} & 0 \\ 0 & 1 \end{pmatrix} \quad \text{to} \quad \begin{pmatrix} g_1^{-1} & 0 \\ 0 & g_1 \end{pmatrix}.$$

If we apply this to each pair of subscripts (except the first) in our decomposition of H, we have our homotopy from h to h'. (The reader should check that during this homotopy the maps are all bounded or continuous linear operators. It is routine to convert this deformation, parameterized by 0 to $\pi/2$, to a usual homotopy parameterized by 0 to 1.)

(ii)   We define a homotopy from $h': V \to GL$ to another map $h'': V \to GL$ described as

$$h''_{ij} = 0, \qquad i \neq j; \qquad h''_{2i-1,2i-1} = g_1 g_1^{-1}, \qquad i \geq 1; \qquad h''_{2i,2i} = 1, \qquad i \geq 1.$$

Actually h'' is just the identity $(g_1 g_1^{-1} = 1)$, so this will complete the proof.

First note that as θ ranges from 0 to $\pi/2$,

$$\begin{pmatrix} \cos\theta - 1 & \sin\theta - 1 \\ 1 - \sin\theta & \cos\theta - 1 \end{pmatrix} \begin{pmatrix} g_1^{-1} & 0 \\ 0 & 1 \end{pmatrix}$$

describes a homotopy from

$$\begin{pmatrix} 0 & -1 \\ g_1^{-1} & 0 \end{pmatrix} \qquad \text{to} \qquad \begin{pmatrix} -g_1^{-1} & 0 \\ 0 & -1 \end{pmatrix}.$$

Therefore,

$$\begin{pmatrix} \cos\theta - 1 & \sin\theta - 1 \\ 1 - \sin\theta & \cos\theta - 1 \end{pmatrix} \begin{pmatrix} g_1^{-1} & 0 \\ 0 & 1 \end{pmatrix} \begin{pmatrix} \cos\theta - 1 & 1 - \sin\theta \\ \sin\theta - 1 & \cos\theta - 1 \end{pmatrix}$$

is a homotopy from

$$\begin{pmatrix} 1 & 0 \\ 0 & g_1^{-1} \end{pmatrix} \qquad \text{to} \qquad \begin{pmatrix} g_1^{-1} & 0 \\ 0 & 1 \end{pmatrix}.$$

Finally, as θ ranges from 0 to $\pi/2$,

$$\begin{pmatrix} \cos\theta - 1 & \sin\theta - 1 \\ 1 - \sin\theta & \cos\theta - 1 \end{pmatrix} \begin{pmatrix} g_1^{-1} & 0 \\ 0 & 1 \end{pmatrix} \begin{pmatrix} \cos\theta - 1 & 1 - \sin\theta \\ \sin\theta - 1 & \cos\theta - 1 \end{pmatrix} \begin{pmatrix} g_1 & 0 \\ 0 & 1 \end{pmatrix}$$

is a homotopy from

$$\begin{pmatrix} g_1 & 0 \\ 0 & g_1^{-1} \end{pmatrix} \qquad \text{to} \qquad \begin{pmatrix} 1 & 0 \\ 0 & 1 \end{pmatrix}.$$

Applying this to each pair of subscripts, starting from the first, we get a homotopy from h' to h'', the identity. Once again, one converts to the range of parameter 0 to 1 and verifies that all maps in question are bounded.

Recapitulating, applying the homotopy in (i) yields a map $h': V \to GL$; following by a homotopy in (ii) yields the identity map. Hence, h is homotopic to l, $l(g) = 1$, as desired.

Remarks (1)   Theorem 7.3 is entirely superseded by the more general and difficult theorem of Kuiper [67], which asserts that GL is contractible. It follows that any map $f : V \to GL$ is homotopic to a constant map.

(2)   The assumption that H_1 is infinite dimensional can be avoided, because one can "take away" an infinite dimensional part of H_0 to "give to H_1" and still leave an infinite dimensional H_0 (check details).

(3)   These results, such as Theorem 7.3 or Kuiper's theorem, are important for the modern theory of Hilbert manifolds. Some of the other developments here are the following:

(a)   A Hilbert manifold (that satisfies reasonable point-set topological conditions) must have a trivial tangent bundle.

(b)   The results of Eells and Elworthy [29] that any reasonable Hilbert manifold is equivalent to an open subset of l^2 contrast strongly with the case of finite-dimensional manifolds, where such a result is unthinkable, even without the word open. A manifold modeled on $\mathbb{R}^1$, for example S^1, is not homeomorphic to any subset of $\mathbb{R}^1$ (check).

(c)   The theorem (see [29]) about the uniqueness of the differentiable structure on a Hilbert manifold contrasts strikingly with the results of Milnor [76] on the plurality of differentiable structures on simple manifolds such as S^7.

For some applications of infinite dimensional manifolds see [1] or [28].

The interested reader should consult the literature here. We offer a few possible directions in which one might proceed in the problems and projects that follow.

PROBLEMS AND PROJECTS

1.   The group $O(n + 1)$ acts as a transformation group on S^n. Furthermore, $O(n + 1)$ has the same homotopy type as $Gl(n + 1; \mathbb{R})$ (because a matrix is a product of an orthogonal and a positive definite matrix, and the latter are contractable; compare [103]).

Since GL is contractible, one would thus expect that

$$S^\infty = \left\{ (a_i) \in l^2 \,\middle|\, \sum_{i \geq 0} a_i^2 = 1 \right\}$$

is also contractible.

Prove that S^∞ is contractible directly. (Hint: Construct a homotopy from the identity map to the shift map

$$\sigma : l^2 \to l^2, \qquad \sigma(a_i) = (0, a_1, a_2, \ldots).$$

Deform this latter map $S^\infty \to S^\infty$, which does not contain $(1, 0, 0, \ldots)$, into a constant map.)

2. Discuss the embeddings and immersions of finite-dimensional smooth manifolds in l^2. Can an immersion be approximated, with arbitrary closeness, by embeddings?

3. Consult the paper of Eells and Elworthy [29], as well as the references therein, on the subject of the structure of Hilbert manifolds.

4. Consult the paper of Kuiper and Burghelea [68] on homotopy equivalences of Hilbert manifolds.

5. One area of great importance concerns maps, between say Banach manifolds, that extend certain kinds of operators that have been well analyzed in functional analysis. Consult the work of Elworthy and Tromba [35].

6. Kuiper's result [67] about the contractability of the general linear group of l^2 does not extend to other topological vector spaces. The answer is known to be affirmative in some cases and negative in some other cases. References may be found in [35].

7. There are important connections between certain spaces of operators and the classification of vector bundles (as in our Chapter 4). Consult the basic work of Jänich [61].

8. Since GL is contractible, several authors have studied a restricted linear group GL_c, consisting of operators that differ from the identity by a compact operator (i.e., an operator that maps bounded sets into sets whose closure is compact). GL_c has interesting topological structure; consult, for example, the work of Koschorke [66].

CHAPTER

8

Morse Theory and Its Applications

It has been known for a long time that a topological space may often be characterized by the algebra of continuous functions on it. If the space is metric, one might recover both the space and the metric from some enriched structure on the algebra of continuous real functions, such as a suitable norm. (For all this, consult [37].) It is therefore not surprising that one can learn much about a smooth manifold from the smooth real-valued functions defined on it. But it was M. Morse who first realized that the information that a smooth function (sufficiently nice) reveals about a manifold is in a certain sense concentrated in the critical points. Morse (see [82, 77]) first called attention to the importance of nondegenerate critical points and the numerical invariant called the index, which completely characterizes the local behavior near that point. In addition, the number of critical points of various indexes relates to the topology of a manifold by means of the Morse inequalities. But the critical points play an even more direct role, in that a sufficiently isolated critical point signals the addition of a cell to the decomposition of the manifold, in the sense of Remark (ii) following Theorem 4.1. This shows how a manifold is put together, as a cell complex, in terms of the critical points of a sufficiently well-behaved function. Furthermore, one may deduce as a consequence of Sard's theorem (Theorem 3.1) that these well-behaved functions (nondegenerate critical points) are actually common.

There is another facet of Morse's work, which concerns geodesics (intuitively, curves of minimal length) on a manifold that has a Riemann metric.

226

Morse originally studied the variations (small changes) of paths on a manifold, and he called the theory the calculus of variations in the large. If we think of paths as some infinite-dimensional manifold, then the length function is something we wish to minimize if we wish to find a geodesic. More generally, one seeks stationary values of the length function. For example, if we look at a torus (Fig. 8.1) there are two stationary paths from a to b, in the sense that small perturbations, with endpoints fixed, result in greater arc length. However, one path has obviously shorter length.

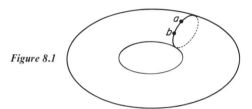

Figure 8.1

Morse did a great deal here, and we can only touch on a few high points. We show how the suitable spaces of broken geodesics are manifolds (of finite dimension) and indicate how the actual (unbroken) geodesics are critical values of the energy function.

There are two points in this chapter where we shall need some material that we have not required before. To be specific, we need

(i)   some algebraic topology, such as rational homology, Betti numbers, and the Euler characteristic, when we get to the Morse inequalities;

(ii)   some more details about Riemann manifolds when we get to look at geodesics.

At the appropriate points, we shall outline, without proof, all the basic facts. In both cases, there are now many excellent references available. For (i) one may consult [25, 33, 102], while for (ii) one has [17, 65], or the excellent but brief summary in [77].

NONDEGENERATE CRITICAL POINTS

We begin by recalling the basic concept of a critical point (compare Chapter 3).

Definition 8.1   Let M^n be a smooth, n-dimensional manifold, and let $f: M^n \to \mathbb{R}$ be a smooth function. Then $x \in M^n$ is a *critical point* if df is *not* onto at x. Because the range of df is a 1-dimensional vector space at x, x is a critical point precisely when df is the zero map at x.

In other words, $x \in M^n$ is a critical point if there is a coordinate chart

$$\phi_\alpha : U_\alpha \to \mathbb{R}^n,$$

$x \in U_\alpha$, such that all first partial derivatives of $f \circ \phi_\alpha^{-1}$ vanish at $\phi_\alpha(x)$.

A real number y such that $y = f(x)$, where x is some critical point, is called a *critical value*.

Naturally, a critical point means degenerate behavior of the first partial derivatives at that point. A critical point is called nondegenerate when the second derivatives are better behaved. To be specific, we have

Definition 8.2 If $x \in M^n$ is a critical point for a smooth function $f : M^n \to \mathbb{R}$, as in Definition 8.1, and the *Hessian matrix* at x

$$(\partial^2(f \circ \phi_\alpha^{-1})/\partial x_i \, \partial x_j)$$

is nonsingular, then we say that x is a *nondegenerate critical point* for f.

For example, let S^2 be the unit sphere about the origin in $\mathbb{R}^3$, and let f assign to any point its third (or z) coordinate; i.e.,

$$f(x, y, z) = z.$$

In this case, the north and south poles are quickly checked to be nondegenerate critical points.

On the other hand, if $f : S^3 \to \mathbb{R}$ is a constant, every point is a degenerate critical point, since the Hessian matrix is identically zero.

We observe that that the Definition 8.2 does not depend on the choice of coordinate chart. For we may calculate

$$\partial(f \circ \phi_\alpha^{-1})/\partial x_j = \partial(f \circ \phi_\beta^{-1} \circ \phi_\beta \circ \phi_\alpha^{-1})/\partial x_j,$$

which may be written out, in matrix notation, as

$$\left(\frac{\partial(f \circ \phi_\beta^{-1} \circ \phi_\beta \circ \phi_\alpha^{-1})}{\partial x_1} \cdots \frac{\partial(f \circ \phi_\beta^{-1} \circ \phi_\beta \circ \phi_\alpha^{-1})}{\partial x_n} \right)$$

$$= \left(\frac{\partial(f \circ \phi_\beta^{-1})}{\partial x_1} \cdots \frac{\partial(f \circ \phi_\beta^{-1})}{\partial x_n} \right) J_{\phi_\beta \circ \phi_\alpha^{-1}},$$

where $J_{\phi_\beta \circ \phi_\alpha^{-1}}$ is the Jacobian (necessarily nonsingular for a coordinate overlap map in a differentiable manifold). This is simply the formula for the Jacobian of a composite map. We use x_i freely, the meaning being clear from context.

To obtain the rows of the Hessian matrix for $f \circ \phi_\alpha^{-1}$, we may simply differentiate this with respect to x_j, getting

$$\left(\frac{\partial^2(f \circ \phi_\beta^{-1})}{\partial x_j \, \partial x_1} \cdots \frac{\partial^2(f \circ \phi_\beta^{-1})}{\partial x_j \, \partial x_n}\right) J_{\phi_\beta \circ \phi_\alpha^{-1}}$$

$$+ \left(\frac{\partial(f \circ \phi_\beta^{-1})}{\partial x_1} \cdots \frac{\partial(f \circ \phi_\beta^{-1})}{\partial x_n}\right) \frac{\partial}{\partial x_j} (J_{\phi_\beta \circ \phi_\alpha^{-1}}).$$

But at a critical point this last term will vanish, so that the Hessian transforms according to multiplication by the (nonsingular) Jacobian. Therefore, the question of whether the Hessian is nonsingular is independant of the choice of coordinate chart at x.

At a nondegenerate critical point, the Hessian is a nonsingular real symmetric matrix, so that we may define a bilinear form H by setting

$$H(f)(x)[v_1; v_2] = v_1^t(\partial^2(f \circ \phi_\alpha^{-1})/\partial x_i \, \partial x_j)v_2$$

for any two n-dimensional column vectors v_1 and v_2 (compare the material concerning the Riemann metric in Chapter 5). Naturally, a change of co-ordinate chart will give an equivalent form, and it will not effect any of the concepts which we now plan to study.

It is well known from linear algebra that if M is a symmetric $n \times n$ matrix there is an orthogonal matrix A (i.e., $A^t = A^{-1}$) such that the matrix

$$B = A^t M A$$

is diagonal ($b_{ij} = 0$ whenever $i \neq j$). With our additional assumption about the Hessian

$$(\partial^2(f \circ \phi_\alpha^{-1})/\partial x_i \, \partial x_j),$$

at a nondegenerate critical point, we can find orthogonal A such that

$$A^t \, (\partial^2(f \circ \phi_\alpha^{-1})/\partial x_i \, \partial x_j) \, A$$

is zero off the diagonal (i.e., $i \neq j$) and nonzero in every diagonal entry. These diagonal elements are clearly the eigenvalues, i.e., the roots of the characteristic polynomial of the Hessian. With this in mind, we make the following basic definition.

Definition 8.3   A symmetric bilinear form represented by a matrix M has *index i* if M has precisely i negative eigenvalues.

We say that M (or the form) has *nullity k* if k of the eigenvalues are zero.

Remarks x is a nondegenerate critical point for f precisely when f is a critical point but the nullity of $H(f)(x)$ is zero.

Our first goal is the famous lemma of Morse, which describes the local behavior of a smooth function, in a neighborhood of a nondegenerate critical point, in terms of the index.

Proposition 8.1 (Morse Lemma)   Let M^n be a smooth manifold, $f: M^n \to \mathbb{R}$ a smooth function, and $x_0 \in M^n$ a nondegenerate critical point.
    Then there is a coordinate chart O_α containing x_0, given by

$$\phi_\alpha : O_\alpha \xrightarrow{\approx} \mathbb{R}^n$$

with $\phi_\alpha(x_0) = 0$, and such that

$$f(\phi_\alpha^{-1}(u)) = f(\phi_\alpha^{-1}(0)) - u_1^2 - \cdots - u_i^2 + u_{i+1}^2 + \cdots + u_n^2,$$

where $u = (u_1, \ldots, u_n) \in \mathbb{R}^n$ and i is the index of f at x_0.

Proof   First notice that if $g: \mathbb{R}^n \to \mathbb{R}$ is a smooth function defined in a convex neighborhood of 0 and $g(0) = 0$, then

$$g(u_1, \ldots, u_n) = \int_0^1 \frac{d}{dt} g(tu_1, \ldots, tu_n)\, dt = \int_0^1 \sum_{s=1}^n \left(\frac{\partial g}{\partial u_s}(tu_1, \ldots, tu_n) \right) u_s\, dt.$$

Therefore, we have an expression

$$g(u_1, \ldots, u_n) = \sum_{s=1}^n u_s \cdot g_s(u_1, \ldots, u_n)$$

for a suitable g_s. Observe also that $g_s(0) = \partial g / \partial u_s(0)$, the derivative at 0.
    Second, observe that if the desired expression for f exists, f has index i. Therefore, we need only verify that if $f: \mathbb{R}^n \to \mathbb{R}$, $f(0) = 0$, and 0 is a non-degenerate critical point of index i for f, then in a suitable coordinate neighborhood of 0, f has the form

$$f(u) = -u_1 - \cdots - u_i^2 + u_{i+1}^2 + \cdots + u_n^2.$$

By our earlier remark, we write g for f and get

$$g(u) = \sum_{s=1}^n u_s g_s(u_1, \ldots, u_n)$$

with $g_s(0) = \partial g(0)/\partial u_s = 0$ (since 0 is a critical point).
    We apply our earlier remark once again to $g_s(u)$ and write

$$g_s(u) = \sum_{t=1}^n u_t h_{ts}(u_1, \ldots, u_n)$$

for suitable smooth functions h_{ts}.

Substituting, we have

$$g(u) = \sum_{s,t=1}^{n} u_s u_t h_{ts}(u).$$

We replace (if necessary) $h_{st}(u) = \frac{1}{2}(h_{st}(u) + h_{ts}(u))$. We then see that $g(u)$ may be expressed by such a formula with the matrix of functions $(h_{st}(u))$ symmetric. Furthermore, because 0 is a nondegenerate critical point, one trivially calculates that $(h_{st}(u))$ is nonsingular at 0, and thus nonsingular in a neighborhood of 0.

Now, at $u = 0$ it is well known that there is a change of coordinates so that $(h_{st}(0))$ is a diagonal matrix. The diagonal entries are the eigenvalues of $(h_{st}(0))$. By a change of scale along these axes, if necessary, we may assume that $(h_{st}(0))$ is a diagonal matrix with distinct real eigenvalues at 0. Summarizing, there is a coordinate neighborhood of 0 for which (h_{st}) is a matrix with distinct real eigenvalues at 0. Thus, (h_{st}) has distinct real eigenvalues in some (possibly smaller) neighborhood of 0 in this new coordinate neighborhood.

Now, if a matrix has distinct, nonzero, real eigenvalues at a point, the linear space on which the matrix operates has a basis of eigen- (or characteristic) vectors of unit length corresponding to these eigenvalues. Furthermore, since we may assume our matrix is symmetric, the transformation of the matrix to this basis, or diagonal form, is the same transformation as that of the bilinear form

$$\sum_{s,t=1}^{n} u_s u_t h_{st}(u)$$

to a diagonal form (see any linear algebra text on bilinear forms).

Thus, there is a continuous coordinate change in a neighborhood of 0 that writes our form in terms of a basis of eigenvectors at each point of the neighborhood, and hence in this coordinate chart

$$g(u) = -a_1(u)u_1^2 - \cdots - a_i(u)u_i^2 + a_{i+1}(u)u_{i+1}^2 + \cdots + a_n(u)u_n^2,$$

with all $a_j(u)$ positive. The final coordinate change to our desired form is then a trivial change of scale.

Remarks We observe that the change of coordinate chart in the Morse lemma is in general nonlinear.

It is also possible to give an inductive proof of the Morse lemma, modeled after the classical diagonalization procedures for forms (see [77]).

Corollary 8.1 The nondegenerate critical points of a smooth function are isolated.

Proof By Proposition 8.1, we need only check that a function

$$g(u) = -u_1^2 - \cdots + u_{i+1}^2 + \cdots + u_n$$

has the property that dg vanishes only at $u = 0$.

Corollary 8.2 If f is a smooth function on a compact smooth manifold with all critical points nondegenerate, then f has only finitely many critical points.

Proof If x is not a critical point, x has an open neighborhood without critical points. Every critical point has a coordinate neighborhood, of the form of Proposition 8.1, in which there is precisely one critical point. Because M^n is compact, M^n has a finite cover of open neighborhoods of these kind, and hence M^n has a finite number of critical points.

We observe that every function on a compact manifold will have a maximum and a minimum, yielding a lower bound on the number of critical points.

We now wish to study the number of nondegenerate critical points, as an indication of the topological complexity of a manifold. To this end we make the following definition.

Definition 8.4 Let M^n be a smooth manifold (compact) and $f: M^n \to \mathbb{R}$ a smooth function. If all the critical points of M^n are nondegenerate, then we say that f is a *Morse function*.

Remarks It is easy to see that the Morse functions are dense in a suitable function space topology. One sets things up in such a way that the differential lies in one subspace and the Hessian in another linearly independent subspace. The result is an easy consequence of the Thom transversality lemma (see Theorem 3.2), applied to the submanifold of those points whose differential is zero.

Definition 8.5 Let f be a Morse function on the compact smooth manifold M^n. Then for each i, $0 \le i \le n$, we set

$\mu_i(M^n, f) = $ the number of critical points of f that have the the index i.

From Corollary 8.2, $\mu_i(M^n, f)$ is a well-defined nonnegative integer, which we shall abreviate (when there is no danger of confusion) by μ_i.

Our goal is to obtain estimates of these numbers μ_i in terms of known topological invariants of the manifold M^n. To this end we must review some basic algebraic topology, specifically, the rational homology groups of the manifold M^n. The best references here are, in my opinion, [25, 33, 32].

HOMOLOGY AND MORSE INEQUALITIES

Summary (1) We recall that, following Theorem 4.2, we described the notion of a cell complex. Such a space is a union of cells

$$e^n = \{x \in \mathbb{R}^{n+1} \mid \|x\| \le 1\}$$

for various n having the property that e^n touches cells of lower dimension, e^i ($i < n$), along its boundary only. (The boundary of e^n, written ∂e^n, is the set $\{x \mid \|x\| = 1\}$; it is a sphere.)

A more specific kind of cell complex is a simplicial complex, which is a union of simplexes. An n-simplex Δ^n is the convex hull (or the smallest convex set containing) $n + 1$ independent points in $\mathbb{R}^n$. For example, a 1-simplex is a closed interval. It is not difficult to check that an n-simplex Δ^n is homeomorphic to an n-cell e^n.

Now, a *simplicial complex* is a union of simplexes for which any pair is either disjoint or intersects in a common subsimplex of each.

For example, the following is a simplicial complex with three 1-dimensional simplexes, or 1-simplexes, and three 0-simplexes.

It is clearly homeomorphic to the circle S^1. On the other hand,

is not a simplicial complex, because the two 1-simplexes intersect in two 0-simplexes, rather than a single subsimplex of each.

The boundary of a tetrahedron is a simplicial complex that is homeomorphic to S^2, etc.

Note that a simplicial complex is actually a cell complex.

(2) Given a simplicial complex K, let $\mathscr{S}(K, i)$ be the set of ordered i-dimensional simplexes, i.e., an element of $\mathscr{S}(K, i)$ is an i-dimensional simplex Δ^i with an order selected on its vertices. $C_i(K)$ is the free abelian group generated by $\mathscr{S}(K, i)$; that is, an element of $C_i(K)$ is a finite sum of integers times elements of $\mathscr{S}(K, i)$, i.e.,

$$n_1 \Delta_1^i + \cdots + n_k \Delta_k^i.$$

One adds and subtracts coordinatewise: that is, one adds and subtracts the coefficients in front of each Δ_j^i. If $i < 0$, $C_i(K) = 0$ by convention. $C_i(K)$ is called the *i-dimensional chains* of K.

If the vertices of K are written $v_0, v_1, \ldots, v_N$, then an i-simplex of K is completely specified by writing its vertices in some order, say $(v_{n_1}, \ldots, v_{n_i})$. Naturally, not every list of i vertices is the vertices of an i-dimensional simplex (indeed K may not even have an i-dimensional simplex), but if there is one, this notation specifies it uniquely.

We define ∂_j by the formula

$$\partial_j(v_{n_1}, \ldots, v_{n_j}) = \sum_{r=1}^{j} (-1)^r (v_{n_1}, \ldots, \hat{v}_{n_r}, \ldots, v_{n_j}),$$

where, as before, the caret means to omit. Because the chains are a free abelian group, ∂_j extends at once to yield a homorphism (called the *boundary homomorphism*)

$$\partial_j : C_j(K) \rightarrow C_{j-1}(K).$$

The groups $C_j(K)$ and homomorphisms ∂_j, written $\{C_j(K), \partial_j\}$, are the *simplicial chain complex* of K.

A map $f : K \rightarrow L$ between simplicial complexes is called *simplicial* when f sends each simplex of K into a simplex of L of no greater dimension, and f restricted to any given simplex is linear. For example, f must send vertices to vertices, 1-simplexes either to a vertex or to another 1-simplex in a linear fashion etc.

One defines

$$f_* : C_j(K) \rightarrow C_j(L)$$

by setting $f_*(\Delta^j) = f(\Delta^j)$ in case $f(\Delta^j)$ is a j-dimensional simplex of L, 0 otherwise. This map extends to a homomorphism f_* in an obvious way. One might verify that $\partial_j' f_* = f_* \partial_j$, where ∂_j' refers to the boundary homomorphism in L.

(3)   One may easily calculate from the definition that $\partial_j \circ \partial_{j+1} = 0$, so that the image of the homomorphism ∂_{j+1} is a subgroup of the kernel of ∂_j, i.e., those elements in $C_j(K)$ which are mapped to zero by ∂_j. Write $\text{Im}(\partial_{j+1})$ for this image and $\text{ker}(\partial_j)$ for the kernel of ∂_j. We define the jth homology group to be the quotient group

$$H_j(K) = \text{ker}(\partial_j)/\text{Im}(\partial_{j+1}).$$

Since all groups here are abelian, the quotient exists without question.

Now, if $f : K \rightarrow L$ is a simplicial map, we defined in (2) above a map

$$f_* : C_j(K) \rightarrow C_j(L).$$

One then easily defines a homomorphism

$$H_j(f): H_j(K) \to H_j(L).$$

It is routine to see that if f and g are both simplicial

$$H_j(g \circ f) = H_j(g) \circ H_j(f) \quad \text{and} \quad H_j(1_K) = 1_{H_j(K)},$$

where 1 refers to the identity map on whatever this subscript is.

(4) If A is any abelian group, one may replace the integers by A in (2) above defining the chain group with coefficients in A,

$$C_j(K; A).$$

The elements here are

$$n_1 \Delta_1^j + \cdots + n_k \Delta_k^j,$$

where now the n_i belongs to A rather than just the integers.

Proceeding exactly as in (3), one defines the jth *homology with coefficients in A*,

$$H_j(K; A).$$

Note that $H_j(K)$ is precisely $H_j(K; A)$, where A is taken to be Z, the integers.

In this chapter we shall only need to study the homology with coefficients in the groups of rational numbers Q, i.e., $H_j(K; Q)$. It has the advantage that each group $H_j(K; Q), j \geq 0$, is a vector space over the rational numbers.

(5) All smooth manifolds are known to be simplicial complexes (see [16, 117]), so that our simplicial homology theory suffices for the purposes we have in mind. It is important, nevertheless, to observe that the basic definitions of homology may be consistently extended to arbitrary topological spaces. This is the important singular homology (see [33, 32]).

(6) If M^n is a compact, smooth n-dimensional manifold, we have defined the deRham cohomology (Definition 5.8)

$$H^i(M^n).$$

This is known to be isomorphic to the dual vector space to $H_i(M^n; \mathbb{R})$, where the abelian group of coefficients is chosen to be the real numbers $\mathbb{R}$.

We now need to state some of the basic properties of these groups, properties that are useful in calculating these groups in important examples and useful in estimating numbers of critical points. These properties are true for any fixed abelian group of coefficients, but we write them for $H_j(K)$ for simplicity. Proofs may be found in the standard references (such as [25, 33, 102]).

(a) If X is a single point (0-complex);

$$H_j(X) = 0, \quad \text{the trivial group,} \quad \text{if} \quad j > 0.$$

On the other hand, $H_0(X; A) = A$ the coefficient group.

(b) If $f, g: K \to L$ are homotopic maps (Definition 4.15), then $H_j(f) = H_j(g)$ for all $j \geq 0$. That is to say, the homomorphisms from $H_j(K)$ to $H_j(L)$ defined by homotopic maps are actually equal. (If one does not use singular homology theory, one would also assume that the maps are simplicial.)

(c) If $X \subseteq Y$ is a subspace (or a subcomplex), one may define *relative* homology groups $H_j(Y, X)$. In all the cases we treat here it suffices to put

$$H_j(Y, X) = H_j(Y/X) \quad \text{when} \quad j > 0,$$

where Y/X is the quotient space of Y with X shrunk to a point.

If $U \subseteq X$ is a set whose closure is actually contained in the interior of X, then one has the important excision property

$$H_j(Y, X) = H_j(Y - U, X - U).$$

If U is a subcomplex of the simplicial complex X, and X is a subcomplex of the simplicial complex Y, then the assumption on the closure of U may be weakened.

(d) There is a sequence of groups and homomorphisms

$$\cdots H_{j+1}(Y, X) \to H_j(X) \to H_j(Y) \to H_j(Y, X) \to \cdots$$

that is exact in the sense that at any given group the image of the homomorphism coming from the left is precisely equal to the kernel of the homomorphism going out to the right. (Recall that the kernel is the subgroup of elements that are mapped to zero.)

More generally, if $X \subseteq Y \subseteq Z$, there is an exact sequence

$$\cdots \to H_{j+1}(Z, Y) \to H_j(Y, X) \to H_j(Z, X) \to H_j(Z, Y) \to \cdots.$$

These properties permit one to deduce many facts about the homology groups of spaces, and even to calculate these groups in certain known cases.

Examples **1.** Recall that X and Y have the same homotopy type if there are continuous maps $f: X \to Y$ and $g: Y \to X$ whose compositions are homotopic to the identity (see remarks before Lemma 4.1).

If X and Y have the same homotopy type, we use (b) above to conclude

$$H_j(g \circ f) = H_j(1_X) = 1_{H_j(X)} \quad \text{and} \quad H_j(f \circ g) = H_j(1_Y) = 1_{H_j(Y)},$$

which implies that the groups $H_j(X)$ and $H_j(Y)$ are isomorphic for all $j \geq 0$.

That is, spaces of the same homotopy type have isomorphic homology groups.

2. If X and Y are homeomorphic, then X and Y have isomorphic homology groups. (Recall that homeomorphic spaces trivially have the same homotopy type.)

3. If $X \vee Y$ means X and Y joined together at a common vertex, $j > 0$,

$$H_j(X \vee Y) = H_j(X) \oplus H_j(Y).$$

[One may look at the exact sequence (d) above for the space $X \vee Y$ and subspace X. The inclusion map $X \subseteq X \vee Y$ has a left inverse, and this makes $H_j(X)$ a summand of $H_j(X \vee Y)$.]

4. Spaces like $e^n = \{x \in \mathbb{R}^n \,|\, \|x\| \leq 1\}$, $\mathbb{R}^n$, or more generally any convex subset of $\mathbb{R}^n$ have the same homology groups of a single point [(a) above], because they all have the homotopy type of a single point. [Compare Remark (2) after Definition 4.15.]

5. If $n > 0$ and $j > 1$, then we have an isomorphism

$$H_j(e^n, S^{n-1}) = H_{j-1}(S^{n-1}),$$

while $H_1(e^n, S^{n-1})$ is either 0, if $n > 1$, or otherwise the coefficient group. (Recall $S^{n-1} \subseteq e^n$ is the subset of points at distance 1 from the origin.)

To check this, we consider the exact sequence

$$\cdots \to H_{j+1}(e^n) \xrightarrow{j_*} H_{j+1}(e^n, S^{n-1}) \xrightarrow{d_*} H_j(S^{n-1}) \xrightarrow{i_*} H_j(e^n) \to \cdots.$$

By Example 4 the outermost two groups are zero ($j \geq 1$). But then the image of the homomorphism j_* is zero, so that the kernel of d_* is zero by the exact sequence property (often called exactness). Thus, d_* is 1–1. Since the kernel of i_* is all of $H_j(S^{n-1})$, we immediately get that d_* is onto, proving that d_* is an isomorphism.

The assertion about $H_1(e^n, S^{n-1})$ is quite similar to the above, but it requires an analysis of the homomorphism

$$H_0(S^{n-1}) \to H_0(e^n),$$

which is not difficult to do or may be easily tracked down in the literature referred to above.

6. Recall the suspension SX of a space X (Definition 4.18). We claim that for $j > 0$ there is an isomorphism

$$H_j(X) = H_{j+1}(SX).$$

That is, S is a geometric construction that effectuates a shift in dimension in homology. Recall that we may break SX up onto two pieces,

$$SX = C_+X \cup C_-X,$$

where $C_+X \cap C_-X = X$ and C_+X, as well as C_-X, is a cone over X. Each C_+X or C_-X is contractible (to its vertex), so each has vanishing homology groups in positive dimensions.

Consider the exact sequence

$$\cdots \to H_{j+1}(C_+X) \to H_{j+1}(SX) \to H_{j+1}(SX, C_+X) \to H_j(C_+X) \to \cdots .$$

The vanishing of the outer groups, along with exactness, implies at once that

$$H_{j+1}(SX) = H_{j+1}(SX, C_+X).$$

Applying the excision property [(c) above] to remove the interior of C_+X from both spaces SX and C_+X, we conclude that this latter group is isomorphic to

$$H_{j+1}(C_-X, X).$$

(In this case, even if the spaces are not simplicial complexes, the hypothesis for the excision may be shown to be unnecessary.)

Lastly, a look at the exact sequence

$$\cdots \to H_{j+1}(C_-X) \to H_{j+1}(C_-X, X) \overset{d_*}{\to} (H_j(X) \to H_j(C_-(X) \to \cdots ,$$

coupled with the fact that C_-X has the homotopy type of a point, shows us that

$$H_{j+1}(C_-X, X) = H_j(X).$$

Putting these isomorphisms together gives us the desired claim. This property 6 is on occasion referred to as the suspension theorem.

7. We now make some computations of homology, with, for example, integer coefficients.

If X is a point, or contractible space (or even pathwise connected space), then property (a) of homology groups says that $H_0(X) = Z$, the integers. It is very easy to check that when X is k points with the discrete topology,

$$H_0(X) = Z \oplus \cdots \oplus Z, \qquad k \text{ summands,}$$

and

$$H_i(X) = 0, \qquad \text{for } i > 0.$$

In particular, since the 0-sphere is just two points, we have

$$H_i(S^0) = \begin{cases} Z \oplus Z, & i = 0, \\ 0, & \text{the trivial group,} \qquad \text{if } i > 0. \end{cases}$$

For $S^1 = SS^0$, the exact sequence

$$\cdots \to H_1(C_+S^0) \to H_1(S^1) \to H_1(S^1, C_+S^0) \to H_0(C_+S^0) \to H_0(S^1) \to \cdots$$

shows that

$$H_1(S^1) = H_1(S^1, C_+S^0).$$

Using excision as above,

$$H_1(S^1, C_+S^0) = H_1(C_-S^0, S^0).$$

Referring to the exact sequence for $i: S^0 \subseteq SS^0$, we conclude from exactness that

$$H_1(C_-S^0, S^0) = \ker(i_*), \qquad \text{where} \quad i_*: H_0(S^0) \to H_0(C_-S^0).$$

It follows from the definitions of (simplicial) homology groups (check!) that the map

$$i_*: Z \oplus Z \to Z$$

is the map $i_*(a, b) = a + b$, and thus $\ker(i_*) = Z$. It follows at once that $H_1(S^1) = Z$. For $j > 1$, the same argument as in 6 above shows that

$$H_j(S^1) = H_{j-1}(S^0) = 0, \qquad j > 1.$$

These arguments extend immediately to an inductive proof that

$$H_i(S^n) = \begin{cases} Z & \text{if } i = 0 \text{ or } n \\ 0 & \text{otherwise.} \end{cases}$$

Definition 8.6 Let X be a topological space, and as above $H_i(X; Q)$ the ith homology group with coefficients in the rational numbers. It is a vector space over Q. We define the ith *Betti number* to be the vector space dimension, if it is finite, i.e.,

$$\beta_i(X) = \dim_Q H_i(X; Q).$$

More generally, when $X \subseteq Y$,

$$\beta_i(Y, X) = \dim_Q H_i(Y, X; Q).$$

(Recall that in all cases at hand $H_i(Y, X; Q) = H_i(Y/X; Q), i > 0$.) Usually, we shall write dim for $\dim_Q$.

Definition 8.7 Let X be a topological space for which all Betti numbers are defined and all but a finite number are zero. For example, X might be a simplicial complex with a finite number of simplexes, i.e., a finite simplicial complex.

We then define the *Euler characteristic* of X,

$$\chi(X) = \sum_{i=0}^{\infty} (-1)^i \beta_i(X),$$

and similarly $\chi(Y, X)$.

Remarks The $\beta_i(X)$ and $\chi(X)$ have many interesting properties. For example,

(1) If X is a compact differentiable manifold, then $\chi(X) = 0$ if and only if X admits a continuous, nowhere vanishing tangent vector field (i.e., a nowhere zero section of the tangent bundle). For a proof, see [103].

(2) If X is a connected finite simplicial complex that is a topological group, $\chi(X) = 0$. For this consequence of the Lefschetz fixed point theorem, consult [25].

Our aim is now to get the Morse inequalities, relating the μ_i (Definition 8.5) to the β_i (Definition 8.6). We begin with a lemma.

Lemma 8.1 Let M^n be a compact smooth n-dimensional manifold, and let $f: M^n \to \mathbb{R}$ be a smooth function. For any real number c we set

$$M_c^n = \{x \in M^n | f(x) \leq c\} \qquad \text{and} \qquad M_{c-}^n = \{x \in M^n | f(x) < c\}.$$

Suppose x_0 is a nondegenerate critical point and that the index of x_0 is i. Let O_α be a coordinate neighborhood of x_0 in which f has the form given by the Morse lemma. Suppose that outside O_α f has no critical point x_1 such that $|f(x_1) - f(x_0)| < \varepsilon$, for some given $\varepsilon > 0$.
Then, writing $f(x_0) = c$, we have

$$H_m(M_c^n, M_{c-}^n; Q) = \begin{cases} Q & \text{if} \quad m = i \\ 0 & \text{otherwise.} \end{cases}$$

Proof By the Morse lemma (Proposition 8.1), we have in the coordinate neighborhood O_α

$$f(x) = f(x_0) - x_1^2 - \cdots - x_i^2 + x_{i+1}^2 + \cdots + x_n^2,$$

with $x = (x_1, \ldots, x_n)$. Clearly it is no loss of generality to assume $f(x_0) = c = 0$.

For a point z outside O_α, with $f(z) = 0$, we know from our hypotheses that z is not critical. In particular, df does not vanish at z. Now, we may give M^n a Riemann metric and (according to the theory developed in Chapter 6) identify the space of 1-forms at any point, via this metric, with the dual the tangent space at the same point. Or in other words, a 1-form may be considered to be $\omega(v) = (v, \omega)$ for some tangent vector ω, with the inner product given by the metric. In this way, we may associate a continuous, nonvanishing vector field with df at points where f is not critical. (Check!)

Now, where f is not critical, this vector is trivially not tangent to $M_c^n - M_{c-}^n$. Therefore, we may continuously deform M_c^n into M_{c-}^n at all points where f is not critical simply by moving along those vectors in the direction of M_{c-}^n. Thus, with the assumptions of our present lemma, one sees at once that outside the coordinate chart O_α M_0^n may be deformed into M_{0-}^n, and we may actually assume that outside O_α both M_0^n and M_{0-}^n are deformed into the same set, for example, by pushing along these vectors dual to df till arriving in some $M_{-\delta}^n$.

Now we use excision [property (c) of homology] to remove the (new deformed) part of M_0^n outside of O_α, so that

$$H_m(M_0^n, M_{0-}^n; Q) = H_m(M_0^n \cap O_\alpha, M_{0-}^n \cap O_\alpha; Q).$$

We must analyze this latter group.

First we observe that the identity map from $M_0^n \cap O_\alpha$ to itself may be deformed into a map which has the last $n - i$ coordinates zero by setting

$$F(x_1, \ldots, x_n; t) = (x_1, \ldots, x_i, (1 - t)x_{i+1}, \ldots, (1 - t)x_n).$$

If we call $B^i = \{x \in O_\alpha | x_{i+1} = \cdots = x_n = 0\}$, we see at once that F yields a homotopy equivalence between $M_0^n \cap O_\alpha$ and B^i, the maps being given by $p(x_1, \ldots, x_n) = F(x_1, \ldots, x_n, 1)$ and $q(x_1, \ldots, x_i, 0, \ldots, 0) = (x_1, \ldots, x_i, 0, \ldots, 0)$, i.e., $q = 1_{B^i}$.

Second, one checks at once that this construction yields a homotopy equivalence from $M_{0-}^n \cap O_\alpha$ to $B^i - (0, \ldots, 0)$, a ball minus the origin.

Thus, using the basic fact that homology groups are invariant under homotopy equivalences,

$$H_m(M_0^n \cap O_\alpha, M_{0-}^n \cap O_\alpha; Q) = H_m(B^i, B^i - (0, \ldots, 0); Q).$$

Now, deforming radially out from the origin, it is clear that the pair of spaces $(B^i, B^i - (0, \ldots, 0))$, which is short for $(B^i - (0, \ldots, 0)) \subseteq B^i$, has the same homotopy type of

$$\left(B^i, \left\{(x_1, \ldots, x_i) \in B^i \,\middle|\, \sum_{l=1}^{i} x_l^2 \geq \tfrac{1}{2}\right\}\right).$$

But clearly,

$$\left(B^i \middle/ \left\{(x_1, \ldots, x_i) \in B^i \,\middle|\, \sum_{l=1}^{i} x_l^2 \geq \tfrac{1}{2}\right\}\right) = S^i,$$

because the quotient of any disk by its boundary is a sphere (check).

Finally, when $m > 0$,

$$H_m(M_0^n \cap O_\alpha, M_{0-}^n \cap O_\alpha; Q) = H_m(B^i, B^i - (0, \ldots, 0); Q) = H_m(S^i; Q)$$

because $H_m(Y, X) = H_m(Y/X)$ for $m > 0$. If Y is path connected and X is nonempty, then $H_0(Y, X; Q) = 0$. If X happens to be empty, $H_0(Y, X; Q) = Q$. (Check any of our references.) Thus, the assertion is also true when $m = 0$, regardless of what i is, completing the proof.

Lemma 8.2   Let M^n be a compact, smooth n-dimensional manifold and let $f: M^n \to \mathbb{R}$ be a smooth Morse function. Then the number of critical points x of index i for which $f(x) = c$ equals

$$\dim H_i(M_c^n, M_{c-}^n; Q).$$

Proof   Suppose $i > 0$. We know by the Morse lemma (Corollary 8.1) that there are only a finite number of critical points. Away from critical points, M_c^n may be deformed into M_{c-}^n. But the argument of Lemma 8.1 shows that at each such critical point x the quotient space has the homotopy type of an i-dimensional sphere. Therefore, M_c^n/M_{c-}^n has the homotopy type of a collection of i-dimensional spheres, one for each critical point of index i and value c, all joined together with other spheres at a common point (check). We note that a collection of spheres joined at a point is commonly referred to as a wedge of spheres, and written $S^i \vee \cdots \vee S^j$.

But by property (c) of homology groups,

$$H_i(M_c^n, M_{c-}^n; Q) = H_i(S^i \vee \cdots \vee S^j; Q),$$

while by Example 2 in our discussion of homology groups, the homology of a wedge of spaces is the direct sum of the homology groups of each of the individual spaces.

Since we have seen that

$$H_i(S^i; Q) = Q,$$

we conclude that

$$H_i(M_c^n/M_{c-}^n; Q) = Q \oplus \ldots \oplus Q$$

with as many summands as i-spheres, that is, as many summands as critical points of index i and value c.

Lastly, the case $i = 0$ is handled by observing that near such a critical point, M_{c-}^n is empty and M_c^n is just a single point, so that by excision we are considering the group H_0 applied to distinct points (as many as such critical points). This completes the proof, because we have seen in 7 above that the group

$$H_0(l \text{ distinct points in } M^n; Q) = Q \oplus \cdots \oplus Q,$$

with l summands.

Lemma 8.3 Let M^n be a compact, smooth n-dimensional manifold, and let $f: M^n \to \mathbb{R}$ be a smooth function. Let $a < b$ be real numbers, and suppose that no y, with $a \leq y \leq b$, is a critical value except c, with $a < c < b$. Suppose that the set of critical points in $f^{-1}(c)$ is finite.

Then there is a continuous deformation (i.e., homotopy of the identity map) of M_b^n onto M_c^n, and M_a^n onto M_{c-}^n, and thus for each i an isomorphism

$$H_i(M_b^n, M_a^n; Q) \approx H_i(M_c^n, M_{c-}^n; Q).$$

Proof This is a standard exercise in the spirit of Lemma 8.1. We define a vector field on $M_b^n - M_{a-}^n - f^{-1}(c)$ that is perpendicular to the submanifolds on which f is constant (where perpendicular is in the sense of some chosen Riemann metric). Naturally the vectors will go to zero as one approaches a critical point.

Use the fundamental theorem of differential equations to pass to a unique family of curves tangent to these vectors. Of course, this theory only works locally, but in our compact case it is not difficult to check that one obtains a unique family of smooth curves on all of $M_b^n - M_{a-}^n - f^{-1}(c)$.

The deformations result from pushing along these curves. It is easiest to deform M_b^n to M_c^n. One then observes that one may similarly deform M_{c-}^n to M_a^n and then, by reversing the parameter, obtain the desired deformation. Putting these together and using the invariance of homology groups under a homotopy equivalence, one gets the desired isomorphism.

The reader should verify details here (see [77, 27] for a more detailed exposition).

The following technical lemma is needed to relate the number of critical points, of various indices, to the topology of M^n.

Lemma 8.4 Let $X_0 \subseteq \cdots \subseteq X_n$ be an increasing family of spaces (or simplicial complexes). Referring to Definition 8.6 and Definition 8.7, we have the following relations on Betti numbers and the Euler characteristic:

$$\beta_i(X_n, X_0) \leq \sum_{r=1}^{n} \beta_i(X_r, X_{r-1}); \qquad \chi(X_n, X_0) = \sum_{r=1}^{n} \chi(X_r, X_{r-1}).$$

Proof If $X \subseteq Y \subseteq Z$, we have the exact sequence of groups and homomorphisms

$$\cdots \to H_i(Y, X; Q) \xrightarrow{i} H_i(Z, X; Q) \xrightarrow{j} H_i(Z, Y; Q) \to \cdots.$$

Since Q is a field, each group is actually a vector space and each map is (easily checked to be) a linear transformation. The basic theorem on linear transformations says that the dimension of the domain equals the sum of the dimensions of the image and the kernel (or null space). For the case at hand, we have (with coefficients understood to be in Q)

$$\begin{aligned} \dim H_i(Z, X) &= \dim(\ker(j)) + \dim(\mathrm{Im}(j)) \\ &= \dim(\mathrm{Im}(i)) + \dim(\mathrm{Im}(j)) \\ &\leq \dim(H_i(Y, X)) + \dim(H_i(Z, Y)). \end{aligned}$$

This property, $\beta_i(Z, X) \leq \beta_i(Y, X) + \beta_i(Z, Y)$, is on occasion referred to as the subadditivity of Betti numbers.

On the other hand, it is a well-known property of the Euler characteristic (see [25, 33]) that

$$\chi(Z, X) = \chi(Y, X) + \chi(Z, Y).$$

(Check or compare with references above.)

The two claims in our lemma will then follow by a straightforward induction.

We are now in a position to tackle the Morse inequalities.

Theorem 8.1   Let M^n be smooth and compact, and let $f : M^n \to \mathbb{R}$ be a smooth Morse function. Then we have

(a) $\beta_i(M^n) \leq \mu_i(M^n; f)$.
(b) $\chi(M^n) = \sum_{l=0}^{\infty} (-1)^l \mu_l(M^n, f)$, where the right-hand side is finite (Corollary 8.2).
(c)   For each integer $l > 0$, we have

$$\beta_l(M^n) - \beta_{l-1}(M^n) + \cdots \pm \beta_0(M^n) \leq \mu_l(M^n, f) - \mu_{l-1}(M^n, f) + \cdots \pm \mu_0(M^n, f).$$

Proof   Let $c_1 < c_2 < \cdots < c_n$ be the critical values of f, and choose real numbers a_i, $1 \leq i \leq m + 1$, so that

$$a_1 < c_1 < a_2 < c_2 < \cdots < a_m < c_m < a_{m+1}.$$

Now according to Lemma 8.2,

$$\mu_i(M^n, f) = \sum_{k=1}^{m} \dim H_i(M^n_{c_k}, M^n_{c_k -}; Q).$$

But according to Lemma 8.3, we may rewrite this as

$$\mu_i(M^n, f) = \sum_{k=1}^{m} \dim H_i(M^n_{a_{k+1}}, M^n_{a_k}; Q).$$

By Lemma 8.4, this right side is greater than or equal to

$$\dim H_i(M^n_{a_{m+1}}, M^n_{a_1}; Q).$$

If we choose $a_{m+1} > \max f$ and $a_1 < \min f$, this is then equal to

$$\dim H_i(M^n, \phi; Q) = \dim H_i(M^n; Q) = \beta_i(M^n),$$

which completes the proof of (a).

Part (b) follows at once from our second formula above, for $\mu_i(M^n, f)$, combined with the second part of Lemma 8.4, in which we choose the M_{a_k} for the increasing family of spaces.

To prove (c), one first shows that the function

$$S_l(Y, X) = \beta_l(Y, X) - \beta_{l-1}(Y, X) + \cdots \pm \beta_0(Y, X)$$

is subadditive. Applying that fact to the sequence of spaces

$$\phi = M^n_{a_1} \subseteq \cdots \subseteq M^n_{a_{m+1}} = M^n,$$

one easily obtains the last claim.

Remarks These different parts of Theorem 8.1 are not at all unrelated. For example, it is not difficult to get (a) or (b) from (c). Consider, for instance, (c) in the case of two adjacent values of l.

Theorem 8.1 may be used to get estimates of the topology of a manifold or to show the impossibility of various situations with critical points. In some cases, there is actually enough information to determine the homology groups. As a good example, we shall consider complex projective n-space CP^n (see Definition 2.3). If

$$\{(z_1, \ldots, z_{n+1})\} \in CP^n,$$

it is no loss of generality to assume

$$\sum_{i=1}^{n+1} \|z_i\|^2 = 1,$$

where $\|z\|^2 = z\bar{z}$, for multiplication by any nonzero constant does not change the equivalence class of a point.

Now, we may define a real function by

$$f(\{z_1, \ldots, z_{n+1}\}) = \sum_{i=1}^{n+1} i \|z_i\|^2.$$

To analyze the critical points, we need coordinate charts. The jth coordinate chart is

$$U_j = \{\{(z_1, \ldots, z_{n+1})\} \in CP^n \,|\, z_j \neq 0\}.$$

For a point in U_j, multiply by $\|z_j\| z_j^{-1}$ and take the real and imaginary parts of the remaining coordinates, calling them x_i and y_i. We have

$$\|z_i\|^2 = x_i^2 + y_i^2, \qquad \|z_j\|^2 = 1 - \sum_{i \neq j} (x_i^2 + y_i^2).$$

It is clear that we have described a homeomorphism onto an open ball in $\mathbb{R}^{2n}$ giving the desired coordinate charts.

In the specific chart U_j, we calculate

$$f((z_1, \ldots, z_{n+1})) = \sum_{i=1}^{n+1} i\|z_i\|^2 = j\left(1 - \sum_{i \neq j} (x_i^2 + y_i^2)\right) + \sum_{i \neq j} i(x_i^2 + y_i^2)$$

$$= j + \sum_{i \neq j} (i - j)(x_i^2 + y_i^2).$$

It is then immediate that the only critical point is the equivalence class of

$$(0, \ldots, 0, 1, 0, \ldots, 0),$$

with 1 in the jth place. It is obviously nondegenerate, and the index is twice the number of i's that are less than j. This may be determined by inspection, since f is actually already in the form of the Morse lemma (Proposition 8.1).

Looking at these critical points for each j, we see that the possible indexes are every even number from 0 to $2n$. By Theorem 8.1, the odd Betti numbers must be equal to zero, and the even ones are less than or equal to 1. But, by part (b) of Theorem 8.1,

$$\chi(CP^n) = \sum_{l=0}^{\infty} (-1)^l \mu_l(CP^n, f) = \sum_{l=0}^{n} \mu_{2l}(CP^n, f) = n + 1.$$

Because

$$\beta_i(M^n) \leq \mu_i(M^n, f) \qquad \text{and} \qquad \chi(CP^n) = \sum_{l=0}^{\infty} (-1)^l \beta_l(CP^n),$$

we have no choice but that

$$\beta_{2l}(CP^n) = u_{2l}(CP^n, f) = 1,$$

for $0 \leq l \leq n$, and

$$\beta_{2l+1}(CP^n) = 0.$$

Therefore, we conclude that

$$H_{2l}(CP^n; Q) = Q, \qquad 0 \leq l \leq n;$$
$$H_{2l}(CP^n; Q) = 0, \qquad \text{if } l > n;$$
$$H_{2l+1}(CP^n; Q) = 0.$$

CELL DECOMPOSITIONS
FROM A MORSE FUNCTION

The approach described in this above example can be improved if one can show how a Morse function describes the structure of a cell complex on the manifold in question, rather that simply giving relations about the homology groups (such as the Betti numbers and Euler characteristic). We now state and sketch the proof of the theorem that gives a cellular structure on a smooth (compact) manifold, in terms of a Morse function. The proof is really a geometric argument that parallels the homology group argument in Lemmas 8.1 and 8.2. Recall that two spaces X and Y have the same homotopy type if there are continuous maps $f: X \to Y$ and $g: Y \to X$, with $g \circ f \simeq 1_X$ and $f \circ g \simeq 1_Y$. Recall also (from the remarks following Theorem 4.1 or Definition 8.5) that a cell complex is a union of cells e^n each of which may meet cells of lower dimension along its boundary only.

Theorem 8.2   Let M^n be a compact, smooth n-dimensional manifold and let $f: M^n \to \mathbb{R}$ be a smooth Morse function. Then M^n has the homotopy type of a finite cell complex, with one cell of dimension l for each critical point of index l.

Sketch of Proof   We give three basic steps.

(i) Suppose $f(x) = c$ and x is the sole critical point in $M^n_{c+\varepsilon} - M^n_{c-\varepsilon}$, $\varepsilon > 0$, with ε sufficiently small. Then $M^n_{c+\varepsilon}$ has the homotopy type of $M^n_{c-\varepsilon}$ with one cell, whose dimension equals the index of x, attached.

One uses the same method as Lemma 8.1, to get the appropriate cell attached. The uniqueness of this critical point means that all of $M^n_{c+\varepsilon}$ is homotopically equivalent to $M^n_{c-\varepsilon}$, plus this one cell.

(ii) If $x_1, \ldots, x_k$ are the distinct critical points, with $f(x_1) = \cdots = f(x_k) = c$, one argues as in Lemma 8.2 to conclude that $M^n_{c+\varepsilon}$ has the same homotopy type as $M^n_{c-\varepsilon}$ with k cells attached. The dimensions are the indices of the respective critical points.

(iii) One executes step (ii) repeatedly, passing through the (finite number of) critical points in the order of increasing critical values. The only technical difficulty is that cells of lower dimension are possibly attached then to cells of higher dimension, whereas in order to get a cell complex one must form successive unions where each new cell meets the earlier cells, of lower dimension, along the boundary. This potential difficulty is removed by cellular approximation and the construction of certain homotopies (compare Part 1, Section 3, of [77]).

Remarks (1) The origins of Theorem 8.2 are not completely clear. While clearly known to Bott, Milnor, Thom, etc., it was not involved in Morse's original work, which in fact predated the notion of a cell complex (as currently perceived, and elucidated by Whitehead [113]).

(2) It is debatable whether Theorem 8.1 or 8.2 is more powerful. In a cell complex, with many cells in adjacent dimensions, there are general technical difficulties involved with calculating the homology groups and Betti numbers. Thus, Theorem 8.1 may possibly give an easier access to the relation between critical points and the topology of the manifold.

It is instructive to compare the analysis of CP^n that follows Theorem 8.1 with what one might conclude via Theorem 8.2. Recall that we had a Morse function

$$f : CP^n \to \mathbb{R}$$

with $n + 1$ critical points, which have all the even numbers from 0 to $2n$ as their indices. Theorem 8.2 then shows that CP^n has the homotopy type of a cell complex

$$e^0 \cup e^2 \cup \cdots \cup e^{2n},$$

with one cell in every even dimension through $2n$. Referring to the basic definition of homology groups for cell complexes, we get, for any abelian group A as coefficients, the chain groups

$$C_i(CP^n, A) = \begin{cases} A & \text{if } i = 2k, \quad k \le n, \\ 0 & \text{otherwise.} \end{cases}$$

It follows immediately that

$$H_i(CP^n, A) = C_i(CP^n, A),$$

which is A for even i, running from 0 to $2n$, and 0 otherwise. Then we easily see that $\chi(CP^n) = n + 1$.

We point out lastly that in this exceptional case, CP^n actually has the structure of such a cell complex rather than just having the homotopy type of such a complex (see [102]).

Remarks Many of these results are not limited to the compact case; sometimes one needs only assume that the critical points and values do not have a cluster point or lie within some compact set. In general one may extract more general results from all of the basic references, such as [77, 90].

APPLICATIONS TO GEODESICS

We now wish to touch upon Morse's work on the space of paths on a manifold and geodesics. Because our goal is simply to introduce the reader to this material, we shall sketch the relevant material from differential geometry in its most local form. This will be in the spirit of the most classical works, such as [71, 17]. A more modern treatment would begin with the inner product on the tangent bundle, which is given by the Riemann metric, and then deduce all the local formulas from general geometric properties. Unfortunately, that would be beyond the space and time requirements that we have here.

A geodesic is, in a local sense, a curve of shortest distance between two points on a manifold. Or, alternatively, one could speak in terms of minimizing the length integral. More generally, one looks at stationary values of the length integral. But there are several alternative points of view. On a Riemann manifold M^n, there is a well-defined notion of parallel translation of a vector, in the tangent space TM^n, along a smooth curve, and this notion preserves the inner product; in this sense, a geodesic is a curve in which the velocity vectors are (apart from length) parallel translations of one another. Yet another point of view deals with covariant differentiation, which is a measure of the "acceleration" of a vector field. A geodesic in this context means that the covariant derivative of the tangent vector field to the curve vanishes.

All these points of view lead to the same local formulas, which we shall now take as our definitions.

Definition 8.8 Let M^n be a smooth manifold with a Riemann metric (Definition 6.5) that in a coordinate chart U about some point x_0 has the expression $(g_{ij}(x))$. We denote the matrix inverse of this by $(g^{ij}(x))$, and we then define the *Christoffel tensor* by the formula

$$\Gamma^l_{ij} = \frac{1}{2} \sum_k g^{kl} \left(\frac{\partial g_{jk}}{\partial x_i} + \frac{\partial g_{ik}}{\partial x_j} - \frac{\partial g_{ij}}{\partial x_k} \right),$$

where of course the x_i's are the coordinate axes in U, and the functions are understood to be taken at $x = (x_1, \ldots, x_n)$.

Definition 8.9 On the smooth Riemann manifold M^n, consider a smooth path

$$s : [a, b] \to M^n.$$

s is called a *geodesic* if for every c, $a < c < b$, and every coordinate chart U about $s(c)$ in which we have

$$s(t) = (s_1(t), \ldots, s_n(t))$$

we have the equations

$$\frac{d^2 s_k}{dt^2} + \sum_{i,j=1}^{n} \Gamma_{ij}^k(s_1, \ldots, s_n) \frac{ds_i}{dt} \frac{ds_j}{dt} = 0.$$

Here k runs from 1 to n, and we have suppressed the variable t for simplicity.

Remarks Consult [77, 65, 104] for an elaboration of the alternative approaches indicated above and a deduction of this formula from variants of the definition. We note that the differential equation in Definition 8.9 is invariant under coordinate transformations.

Examples *1.* In $\mathbb{R}^n$, with the usual metric, $(g_{ij}(x)) = 1$, the geodesics are straight line segments.
2. On the sphere S^2, with the traditional metric, geodesics are portions of great circles.
3. In the large, on a general M^n one cannot expect to join two points with a geodesic, even if M^n is connected. Consider

$$M^2 = \mathbb{R}^2 - (0, 0)$$

with the metric $(g_{ij}(x)) = 1$, and take as our two points the points $(-1, 0)$ and $(+1, 0)$. However, one can join points by a path that is piecewise made from geodesics.

The question of existence, in the local sense, is clarified by the following two propositions.

Proposition 8.2 Let M^n be a smooth manifold with a Riemann metric, $x \in M^n$. Abbreviate $T_x = (TM^n)_x$ for the tangent space at x.

Then there is a coordinate neighborhood of x, say U, and some $\varepsilon > 0$ such that if

$$y \in U \quad \text{and} \quad \vec{v} \in T_y, \quad \text{with} \quad \|\vec{v}\| < \varepsilon,$$

then there is a unique geodesic s,

$$s : (-a, a) \to M^n,$$

for some $a > 0$, with $s(0) = y$ and

$$ds/dt|_{t=0} = \vec{v}.$$

(Recall that ds/dt makes sense in a coordinate neighborhood by precisely the same formulas as in Euclidean space.)

Proof This is a direct consequence of the standard existence and uniqueness theorem from ordinary differential equations, as applied to Definition 8.9 (see [22, 52]).

We observe that for small enough ε, a may be chosen large, because whenever s is a geodesic and $c > 0$, then $\tilde{s}(t) = s(ct)$ is a geodesic.

Proposition 8.3 With the notation of the above proposition, there is an open neighborhood V of x and an $\varepsilon > 0$ such that

(a) any two points of V may be joined by a unique geodesic of length less than ε, and

(b) this geodesic depends smoothly on the two end points, which, in light of the above proposition, means effectively that the derivative at the first point depends smoothly on the second point.

Proof Consider an open coordinate neighborhood W of $(x, 0)$ in the tangent TM^n, with naturally $(x, 0) \in W$. W is homeomorphic to an open ball in $\mathbb{R}^{2n}$, but we think of W as pairs (y, v), where $v \in (TM^n)_y$.

Write $s_y(v)$ for the unique geodesic given by Proposition 8.2, if necessary in a smaller neighborhood. Define a map

$$f : W \to M^n \times M^n$$

by the formula

$$f(y, v) = (y, s_y(v)(1))$$

(where we may assume $s_y(v)$ to be defined at 1, were our neighborhood sufficiently small).

It is easy to check that the Jacobian of this map is nonsingular (in fact it is triangular), and hence, by invoking the inverse function theorem (Theorem 1.1), f is a diffeomorphism on a possibly smaller neighborhood. The proposition follows immediately.

Remarks (1) The map that sends v to $s_y(v)(1)$, at fixed y, is often referred to as the exponential,

$$\exp : O \to M^n,$$

where O is a suitable small neighborhood of $(y, 0)$ in $T_y = (TM^n)_y$. The terminology comes from the fact that if M^n is a suitable space (or group) of matrices, exp is actually expressed by a convergent power series of matrices with coefficients $1/n!$. This should be compared with the exponential map for Lie groups in the next chapter, which is actually defined without reference to a specific Riemann metric.

(2) Proposition 8.2 has various refinements. For example, V may be chosen so that the unique geodesic between a pair of points lies entirely in V. Consult [65, 77], where specific references to the literature may be found.

We now wish to show how Morse theory may be applied to study geodesics on a smooth Riemann manifold. Of course, the space of continuous, or smooth, or piecewise smooth curves on a manifold M^n is surely infinite dimensional. In order to apply our work here to study geodesics or just curves of shortest length, some sort of restriction on the type of path, is necessary in order to reduce the problem to Morse theory on a (finite-dimensional) manifold. The result will be to display geodesics as the critical points of the length function on a suitable, related, finite-dimensional manifold.

Definition 8.10 Let M^n be a smooth Riemann manifold, and let a and b be two points of M^n lying in the same path-connected component.

(a) $\Omega(M^n; a, b)$ is the set of all continuous, piecewise smooth paths s from a to b, parameterized by the unit interval. That is, there are t_i, $0 = t_0 \leq t_1 \leq \cdots \leq t_k = 1$, with s smooth on each interval $t_{i-1} < t < t_i$.

(b) If s is a path, let λ_s denote the arc length function, which one may write as

$$\lambda_s(t) = \int_0^t \left\| \frac{ds}{dt} \right\| dt,$$

and define a distance function on paths

$$d(s_1, s_2) = \underset{0 \leq t \leq 1}{\text{l.u.b.}}\ \rho(s_1(t), s_2(t)) + \sqrt{\int_0^1 \left(\frac{d\lambda_{s_1}}{dt} - \frac{d\lambda_{s_2}}{dt} \right)^2 dt}.$$

$\Omega(M^n; a, b)$ may be regarded as topologized, as a metric space with this metric.

(c) If $0 \leq t_i \leq t_{i+1} \leq 1$, we may define an *energy function*

$$E_{t_i, t_{i+1}}(s) = \int_{t_i}^{t_{i+1}} \left\| \frac{d\lambda_s}{dt} \right\|^2 dt.$$

This is easily checked to be continuous in s with respect to the topology defined in (b).

(d) We define a subspace

$$\Omega(M^n; a, b, t_0, \ldots, t_k)_c \subseteq \Omega(M^n; a, b)$$

to consist of all continuous, piecewise smooth paths that are smooth geo-desics when restricted to the intervals, $t_{i-1} < t < t_i$, and for which $E_{0,1}(s) < c$. Here, we assume $c > 0$ and $0 = t_0 \leq t_1 \leq \cdots \leq t_k = 1$.

We may now see how to reduce the problem of geodesics to that of critical points of functions on finite-dimensional manifolds.

Proposition 8.4 Let M^n be a compact, smooth manifold with a given Riemann metric. With the above definitions, suppose $c > 0$ is chosen so that there is some path in $\Omega(M^n; a, b)$ with $E_{0,1}(s) < c$.
Then for all sufficiently fine subdivisions of $[0, 1]$, say $0 = t_0 \leq t_1 \leq \cdots \leq t_k = 1$,

$$\Omega(M^n; a, b, t_0, \ldots, t_k)_c$$

is a nonempty, smooth, finite-dimensional manifold.

Proof We combine Proposition 8.3 with elementary facts about com-pactness. It is clear that for a sufficiently fine subdivision each geodesic $s([t_i, t_{i+1}])$ is unique. Furthermore, it depends on the endpoints differentiably.
Now, given a $s \in \Omega(M^n; a, b, t_0, \ldots t_k)_c$, we associate with s a point in the $(k-1)$-fold Cartesian product of M^n with itself by the map

$$\phi(s) = (s(t_1), \ldots, s(t_{k-1})).$$

(Recall $s(t_0) = a$ and $s(t_k) = b$ are fixed.) This exhibits our space as diffeo-morphic to an open subset of

$$M^n \times \cdots \times M^n.$$

Lastly, to see that it is nonempty, note that one may approximate a path in $\Omega(M^n; a, b)$ by one in some subset $\Omega(M^n; a, b, t_0, \ldots, t_k)$, and the energy function is continuous. (Check this!)

Further Remarks (1) Pursuing this line of thought, one may show that the energy function $E_{0,1}$ is smooth on the manifold that is constructed in Proposition 8.8 and that the unbroken geodesics are the critical points of $E_{0,1}$.
(2) The method of Theorem 8.2 may be invoked to estimate the numbers of geodesics. For example, in certain cases one may show that the number of geodesics is finite, in other cases not.
As a simple example, let S be a sphere of radius r in $\mathbb{R}^{n+1}$. If $a, b \in S$, $a \neq b$, but a and b are not antipodal (that is $a \neq -b$, if the center is the origin), then there is only one geodesic of length $\leq \pi r$, and only one of length $> \pi r$ but $\leq 2\pi r$, running from a to b. But if a and b are antipodal, there are infinitely many geodesics of length equal to πr.

All this is developed, at considerable length, in the beautiful work of M. Morse [77, 82].

(3)   By piecing together various finite-dimensional approximations, it is possible to get an infinite cell complex for the full path space $\Omega(M^n; a, b)$. This is better approached by the fiber space techniques first developed by Serre [99]. This is an interesting subject whose techniques unfortunately are beyond what we have available in this book.

(4)   Apart from its intrinsic beauty, Morse theory has practical applications. One of the most striking is Bott's famous periodicity theorem [11]. This theory amounts to a description of the homotopy type of the infinite orthogonal and unitary groups, the limits of $O(n)$ or $U(n)$.

We mention one case of Bott's work. Let $U(n)$ be the unitary group, of $n \times n$ complex matrices whose inverses are the conjugates of their transposes. $SU(n)$ will be the subgroup consisting of those matrices in $U(n)$ whose determinant is precisely 1. Bott shows that the space of all (minimal) geodesics in $SU(2n)$ that connect the identity matrix with its negative is homeomorphic to the (complex) Grassman manifold of (complex) n-dimensional subspaces of $\mathbb{C}^{2n}$. He then can combine an index argument with a cellular decomposition (as in Theorem 8.2) to analyze the homotopy structure of $SU(2n)$. To study this in more depth, consult [11, 77].

PROBLEMS AND PROJECTS

1.   Let M^2 be the n-holed torus. Consider M^2 as lying in $\mathbb{R}^3$, either horizontally or vertically, as in Fig. 8.2. Study the critical values and indices for the height function (orthogonal projection onto the vertical axis) in both cases.

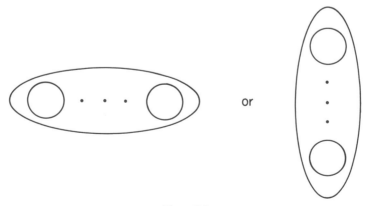

or

Figure 8.2

2. Use the Morse inequalities (Theorem 8.1) to exhibit some impossible combinations of critical values and indices for Morse functions on various elementary manifolds.

3. It is a well-known result that if M^k admits a nowhere vanishing tangent vector field, then $\chi(M^k) = 0$ (see [103]). Which n-holed toruses (see problem 1) admit nowhere vanishing tangent vector fields?

4. One classic example of basic Morse theory is the following Theorem of G. Reeb [77]:

Let M^n be a compact, smooth n-dimensional manifold, and let $f : M^n \to \mathbb{R}$ be a smooth Morse function with exactly two critical points. Then M^n is homeomorphic to S^n.

Clearly the two critical points are a minimum and a maximum, and the critical values are, say, m and M, with $m < M$. Show that $f^{-1}([m, m + \varepsilon])$ and $f^{-1}([M - \varepsilon, M])$ are two closed n-dimensional balls. The method of Lemma 8.3 shows that this is true for any ε, $0 < \varepsilon < M - m$. Choosing $\varepsilon = \frac{1}{2}(m + M)$, we see that M^n is the union of two n-dimensional balls, attached along their common boundary.

It is not hard to show that such a space, the union of two balls, must be homeomorphic to S^n. (Construct a latitude preserving 1–1, continuous map from M^n onto S^n.)

It is not true that such a homeomorphism must be a diffeomorphism. In fact, Milnor's famous example of 7-dimensional manifolds that are homeomorphic to, but not diffeomorphic to, S^7 makes use of this result (see [77]).

5. Analyze $\mathbb{R}P^n$ by constructing a Morse function

$$f(\{(x_1, \ldots, x_{n+1})\}) = \sum_{i=1}^{n+1} i x_i^2,$$

where $\sum_{i=1}^{n+1} x_i^2 = 1$.

Find critical points and indices and *estimate* the $\beta_i(\mathbb{R}P^n)$, using the method of Theorem 8.1 and the example that follows. (The precise determination of the β_i and χ are not a direct consequence of this method.)

CHAPTER

9

Lie Groups

The theory of Lie groups is somewhat older than much of the material that we have covered in this book. Lie groups exist naturally in many areas of mathematics where natural group structures may be found on certain manifolds. Examples are the Euclidean space $\mathbb{R}^n$ (as a group under addition) or the general linear group $Gl(n; \mathbb{R})$ (see Example 5 following Definition 1.1); these are naturally manifolds and groups at the same time. The basic theory was first developed by Lie, Engels, Cartan, and others in the late 19th century. These authors took a local and analytic (in the sense of convergent power series) view of the subject. Subsequent work of many mathematicians showed that the analytic hypotheses are not necessary (see [80] for a complete discussion), although the details there are so complex that most authors (including me) are forced to restrict attention to the (unnecessary) analytic definitions.

Our goals here consist of a concise introduction to the basic theory, including Lie subgroups and Lie algebras, the exponential map, etc. Time does not permit any classification theory. We do discuss integration on Lie groups, and we apply this to representations. The final topic is homogeneous spaces and analytic group actions.

BASIC THEORY OF LIE GROUPS

A Lie group is a group object in the category of analytic manifolds and analytic maps. To say this precisely, we adopt the following formal definition:

Definition 9.1 Let G be an analytic manifold in the sense of Definition 1.8. Suppose that G admits a group structure, written xy, satisfying the usual laws (associativity, existence of identity element, and existence of inverses).

If the map

$$\mu: G \times G \to G$$

defined by $\mu(x, y) = xy^{-1}$ is analytic, we say that G is a *Lie group*.

Remarks (1) Some authors make distinctions between analytic groups and Lie groups, in terms of connectivity. Most of what we study will be either connected, or have a finite number of connected components. Such distinctions offer no advantage here.

(2) Setting $x = e$, the identity element, we see that the map $y \to y^{-1}$ is analytic. Therefore, the map $(x, y) \to xy$ will also be analytic.

(3) Here are some examples of Lie groups.

(a) A discrete group is a 0-dimensional Lie group. 0-dimensional manifolds are analytic, as are maps between them.

(b) $\mathbb{R}^n$ with the usual addition. In this case, the map μ is given by

$$\mu((x_1, \ldots, x_n), (y_1, \ldots, y_n)) = (x_1 - y_1, \ldots, x_n - y_n).$$

Note that $x_i - y_i$ is a polynomial (and hence a trivial power series).

(c) $\mathrm{Gl}(n; \mathbb{R})$, the $n \times n$ matrices having inverses under matrix multiplication. The inverse of a matrix may be expressed in each place as a quotient of two polynomials, given by determinants. It is easy to see that locally this is analytic, starting from facts like $1/(1 - x) = \sum_{i=0}^{\infty} x^i$. On the other hand, the product of two matrices is expressed by polynomials in the entries. It is then clear that μ is analytic.

(d) An open subgroup of a Lie group is clearly a Lie group. In particular, the connected component of the identity in a Lie group is also a Lie group. Trivially, the Cartesian product of a finite collection of Lie groups is a Lie group.

(e) The spheres S^1 and S^3 are Lie groups, the respective multiplications being addition of angles and multiplication of quaternions of norm 1. Consult [20, 103] for further details.

Definition 9.2   A *homomorphism* of Lie groups is an analytic map

$$f: G_1 \rightarrow G_2$$

that is a homomorphism of the underlying groups ($f(xy) = f(x)f(y)$).

Remarks   (1)   Examples are the inclusion of a subgroup, the map $x \rightarrow e^{2\pi i x}$, which is a homomorphism from $\mathbb{R}$ to S^1, and the determinant, which is a homomorphism from $\mathrm{Gl}(n; \mathbb{R})$ to the group of nonzero real numbers under multiplication, in the other words, $\mathrm{Gl}(1; \mathbb{R})$.

(2)   Unlike the case of differentiable manifolds, where the underlying topological spaces of nondiffeomorphic manifolds may be homeomorphic, a given topological group can support at most one Lie group structure. A continuous homomorphism from one Lie group to another is always analytic. More about this later in the chapter.

We now look at the basic translation maps, their effect on tangent vectors, and the basic notion of an invariant vector field.

Definition 9.3   Let G be a Lie group, $y \in G$. We define the *left translation* by y,

$$\mathscr{L}_y: G \rightarrow G,$$

by $\mathscr{L}_y(x) = yx$ and, similarly, right translation by $\mathscr{R}_y(x) = xy$. The maps are easily checked to be analytic, and

$$\mathscr{L}_{y^{-1}} \circ \mathscr{L}_y = \mathscr{L}_e = 1_G$$

(and similarly $\mathscr{R}_{y^{-1}} \circ \mathscr{R}_y$).

Referring to Definition 2.6 and subsequent material, we abreviate

$$g_x = (TG)_x,$$

the tangent space to G at x, and write

$$d\mathscr{L}_y: g_x \rightarrow g_{yx} \quad \text{and} \quad d\mathscr{R}_y: g_x \rightarrow g_{xy}$$

for the differentials or mappings on the tangent spaces defined by $\mathscr{L}_y$ and $\mathscr{R}_y$

Finally, given a vector field $s: G \rightarrow TG$, s is called *left invariant* if

$$d\mathscr{L}_y(s(x)) = s(yx)$$

for all $x, y \in G$ (and similarly for right invariant).

Remarks   It is clear that the translation maps $\mathscr{L}_y$ give a specific way of translating the tangent space at one point to the tangent space at another point on a Lie group G. In this sense, the tangent space at any point may be recovered from that at the identity in a specific way. In the sense of Chapter 4, this means that the tangent bundle to G is a trivial vector bundle.

But this also suggests that the tangent space at the identity g_e should have some special importance. In fact it is the underlying vector space to the Lie algebra of G, as we shall soon see.

In order to get a deeper understanding of vector fields, especially invariant vector fields, we shall need to look at another (equivalent) description of a vector field on a manifold. This will enable us to get at the bracket product. Because of the applications we have in mind here, we shall carry this out in the analytic case.

Definition 9.4 Let M^n be an analytic manifold, and let $\mathscr{A}(x)$ be the set of (real) analytic functions defined in some neighborhood of x (similar to Definition 1.11). $\mathscr{A}(x)$ is clearly an algebra over the real numbers (with addition, multiplication, and scalar multiplication), because sums and products of functions may be defined on the intersection of their domains.

A mapping $L:\mathscr{A}(x) \to \mathbb{R}$, from $\mathscr{A}(x)$ to the real numbers, is called a *derivation* if

(a) $L(af + bg) = aL(f) + bL(g)$, where $a, b \in \mathbb{R}$, and $f, g \in \mathscr{A}(x)$.
(b) $L(fg) = (L(f))g(x) + (L(g))f(x)$, where fg refers to the product of f and g.

The set of derivations at x, visibly a vector space, is written $\mathscr{D}_x$.

As an example, let $M^n = \mathbb{R}^n$, $x = 0$, and set

$$L(f) = \partial f(0, \ldots, 0)/\partial x_1.$$

Then (a) is immediate, and (b) is the usual Leibnitz rule at (0)
To analyze the structure of $\mathscr{D}_x$, we have the following result.

Proposition 9.1 Let $x \in M^n$ and $x \in O_\alpha$, a coordinate chart with given homeomorphism $\phi_\alpha : O_\alpha \to \mathbb{R}^n$. Assume $\phi_\alpha(x)$ is the origin $0 \in \mathbb{R}^n$. Then for each $f \in \mathscr{A}(x)$ we have, for any derivation L,

$$L(f) = \sum_{i=1}^n \left(\frac{\partial(f \circ \phi_\alpha^{-1})}{\partial x_i}(0, \ldots, 0) \right) L(\pi_i \circ \phi_\alpha),$$

where $\pi_i : \mathbb{R}^n \to \mathbb{R}$ is the projection on the ith coordinate axis.
It follows from this that the map

$$\phi:(TM^n)_x \to \mathscr{D}_x,$$

defined by the formula or rule

$$(\alpha_1, \ldots, \alpha_n) \overset{\phi}{\mapsto} \left(f \to \sum_{i=1}^n \alpha_i \frac{\partial(f \circ \phi_\alpha^{-1})}{\partial x_i}(0) \right)$$

(where $(\alpha_1, \ldots, \alpha_n)$ is the coordinates of a tangent vector in terms of the basis given by ϕ_α^{-1}, and the right-hand parenthesis describes a derivation), is an isomorphism of n-dimensional vector spaces.

Proof ϕ_α being a diffeomorphism, it is no loss of generality to assume that O_α is $\mathbb{R}^n$ and ϕ_α is the identity map; this will simplify the notation.

Given $f \in \mathscr{A}(0)$, we write, by the Taylor's series (by Theorem 1.3),

$$f(x) = \sum_{|\alpha| \le 1} \frac{1}{\alpha!} D^\alpha f(0)x^\alpha + \sum_{|\alpha| = 2} x^\alpha g_\alpha(x),$$

where $g_\alpha(x)$ is a suitable analytic function obtained by factoring the x^α out of the terms of degree bigger than 1, which contain x^α, for each such α.

Observe that L annihilates constants (from the conditions in Definition 9.4) and also vanishes on quadratic (or higher) terms such as $x^\alpha g_\alpha(x)$, $|\alpha| = 2$ [by condition (b)]. Therefore, we apply L to both sides of the above equation and have

$$(Lf) = \sum_{|\alpha| = 1} D^\alpha f(0)L(x^\alpha) = \sum_{i=1}^{n} \left(\frac{\partial f}{\partial x_i}\right)(0)L(\pi_i).$$

In other words, the n numbers $L(\pi_i)$ suffice to describe the derivation in terms of the usual first partial derivatives. This proves our first assertion.

It is now trivial to check that the first partials evaluated at 0 are linearly independent and form a basis for the vector space $\mathscr{D}_x$, clearly n-dimensional. The mapping ϕ defined in the statement of this proposition, is visibly linear. Clearly ϕ applied to the basis vectors $(0, \ldots, 1, \quad , 0)$ gives n independent elements of $\mathscr{D}_x$. Since the range as well as the domain of ϕ is an n-dimensional vector space, ϕ is an isomorphism, as desired.

Remarks   In this framework, a tangent vector field associates to each $x \in M^n$ a derivation L_x in $\mathscr{D}_x$ in a continuous way.

Obviously, in any coordinate chart, we may then express a tangent vector field (i.e., section of the tangent bundle) as

$$L: f(y) \to \sum_{i=1}^{n} \alpha_i(y)\frac{\partial f}{\partial x_i}(y), \qquad y \in O_\alpha.$$

Note also that in this context one may formulate the condition that L is analytic (or L is smooth) by demanding that all the $\alpha_i(y)$ be analytic (or smooth). A left invariant field is then seen to be analytic.

Note further that if $f: M^n \to N^k$ is analytic one may express df on a tangent vector at x, say $L \in \mathscr{D}_x$, by $(df)(L)(\alpha) = L(\alpha \circ f)$.

One advantage of this description of tangent vectors is that it allows us to describe, in easy language, the *bracket product*, which we now define, of two (analytic) vector fields on an (analytic) manifold.

Proposition 9.2 Let L and M be two analytic vector fields on the analytic manifold M^n. That is, for each $x \in M^n$, L and M are derivations from $\mathscr{A}(x)$ to $\mathbb{R}$ Then the function $\mathscr{A}(x) \to \mathbb{R}$ given by

$$[M, L](f) = L(M(f)) - M(L(f)),$$

is a derivation, called the bracket product of M and L. (Note that because M, for example, is a vector field, rather than merely a simple tangent vector at x, $M(f)$ is a function on the manifold, so that $L(M(f))$ makes sense.) $[M, L]$ is an analytic vector field on M^n.

Proof In a coordinate chart $\phi_\alpha : O_\alpha \to \mathbb{R}^n$, with $x \in O_\alpha$, we may write (as before)

$$L(f) = \sum_{i=1}^{n} \alpha_i \frac{\partial f}{\partial x_i}.$$

At a specific point $y \in O_\alpha$, $L(f)(y)$ is given by this formula where the α_i are functions (analytic) of y and the partial derivatives are to be evaluated at y. Similarly, we may write

$$M(f) = \sum_{i=1}^{n} \beta_j \frac{\partial f}{\partial x_j},$$

and thus we calculate

$$L(M(f)) = \sum_{i=1}^{n} \alpha_i \frac{\partial}{\partial x_i} \left(\sum_{j=1}^{n} \beta_j \frac{\partial f}{\partial x_j} \right) = \sum_{i=1}^{n} \alpha_i \left(\sum_{j=1}^{n} \frac{\partial \beta_j}{\partial x_i} \frac{\partial f}{\partial x_j} + \beta_j \frac{\partial^2 f}{\partial x_i \partial x_j} \right)$$

$$= \sum_{i,j=1}^{n} \alpha_i \frac{\partial \beta_j}{\partial x_i} \frac{\partial f}{\partial x_j} + \sum_{i,j=1}^{n} \alpha_i \beta_j \frac{\partial^2 f}{\partial x_i \partial x_j}.$$

Naturally, a similar calculation may be made to show that

$$M(L(f)) = \sum_{i,j=1}^{n} \beta_i \frac{\partial \alpha_j}{\partial x_i} \frac{\partial f}{\partial x_j} + \sum_{i,j=1}^{n} \alpha_j \beta_i \frac{\partial^2 f}{\partial x_j \partial x_i}.$$

If we subtract and cancel the second terms, we get

$$L(M(f)) - M(L(f)) = \sum_{i,j=1}^{n} \alpha_i \frac{\partial \beta_j}{\partial x_i} \frac{\partial f}{\partial x_j} - \sum_{i,j=1}^{n} \beta_i \frac{\partial \alpha_j}{\partial x_i} \frac{\partial f}{\partial x_j}$$

$$= \sum_{i,j=1}^{n} \left(\alpha_i \frac{\partial \beta_j}{\partial x_i} - \beta_i \frac{\partial \alpha_j}{\partial x_i} \right) \frac{\partial f}{\partial x_j}.$$

It follows at once that $[M, L](f)$ has the local form

$$\sum_{j=1}^{n} \gamma_j \frac{\partial f}{\partial x_j}$$

and hence is a derivation as claimed. It is visibly an analytic vector field on M^n.

Remark Suppose M^n is a Lie group. Then because the translation maps $\mathscr{L}_y$ are analytic diffeomorphisms (with inverses $\mathscr{L}_{y^{-1}}$) $d\mathscr{L}_y$ takes an entire analytic vector field (rather than simply a tangent vector) to another analytic vector field. One easily checks out the formula

$$d\mathscr{L}_y[M, L] = [d\mathscr{L}_y(M), d\mathscr{L}_y(L)].$$

Proposition 9.3 Let L, M, and N be analytic vector fields on the analytic manifold M^n. Then

(a) $[M, L]$ is linear, in each variable (e.g., $[M, L_1 + L_2] = [M, L_1] + [M, L_2]$ etc.),

(b) $[M, L] = -[L, M]$, and

(c) $[N, [L, M]] + [L, [M, N]] + [M, [N, L]] = 0$.

Proof (a) is immediate.

To check (b), note that it is sufficient to verify $[L, L] = 0$ for any L because one then applies this to $L + M$. But $[L, L] = 0$ follows immediately from the formula in the statement of Proposition 9.2 (check).

To prove (c), we note that

$$[N, [L, M]](f) = ([L, M])(N(f)) - N([L, M](f))$$
$$= MLN(f) - LMN(f) - NML(f) + NLM(f).$$

Naturally, by $MLN(f)$, etc., we mean the threefold repeated application of the derivations $M(L(N(f)))$. Then formula (c) follows at once by permutation and addition.

THE IDEA OF LIE ALGEBRAS

We are now in a firm position to look at the notion of a Lie algebra and to define the Lie algebra of a Lie group.

Definition 9.5 A *Lie algebra* is a vector space V over the real numbers on which an additional operation is defined. This operation associates with every X and Y an element $[X, Y] \in V$ satisfying the following laws:

(a) For $\alpha, \beta \in \mathbb{R}$ and $X, Y, Z \in V$, $[\alpha X + \beta Y, Z] = \alpha[X, Z] + \beta[Y, Z]$.

(b) $[X, Y] = -[Y, X]$ (this law is called anticommutativity).

(c) $[X, [Y, Z]] + [Y, [Z, X]] + [Z, [X, Y]] = 0$ (this is the Jacobi identity).

When V is finite dimensional, we then have a finite-dimensional Lie algebra over the real numbers.

Remarks (1) Clearly, (a) and (b) combine to yield

$$[Z, \alpha X + \beta Y] = \alpha [Z, X] + \beta [Z, Y].$$

Also, (b) implies $[X, X] = 0$.

(2) A similar definition is easily constructed, using other fields (of characteristic 0) such as the complex numbers.

(3) Any vector space may be trivially made into a Lie algebra, by defining $[X, Y] = 0$. Such a Lie algebra is often referred to as *abelian*, because the product $[\ ,\ ]$, as in Proposition 9.2, is a measure of deviation from commutativity (of the operations of partial derivative and multiplication by a function).

(4) The most familiar Lie algebra, which is not abelian, is that generated by the traditional vectors i, j, and k from vector analysis. Specifically, consider a 3-dimensional real vector space with a basis consisting of the symbols i, j, and k. We set

$$[i, j] = k, \qquad [j, k] = i, \qquad \text{and} \qquad [k, i] = j,$$

so that the bracket product is merely the usual cross product. In general, one may define the bracket product of two arbitrary elements by laws (a) and (b) above. The Jacobi identity is easy to check.

When we see (shortly) that a Lie algebra is associated with a Lie group, we shall be able to check that this Lie algebra comes from the Lie group of rotations of Euclidean space $\mathbb{R}^3$.

(5) A linear map of Lie algebras $\phi : L_1 \to L_2$ is called a *Lie algebra homomorphism* if we have

$$\phi[X, Y] = [\phi(X), \phi(Y)].$$

We are now ready to assemble, from the material that has now been developed, the basic result on the passage from Lie groups to Lie algebras.

Theorem 9.1 Let G be a Lie group of dimension n. Then the set of left-invariant analytic tangent vector fields on G is an n-dimensional vector space that is isomorphic to the tangent space to G at the identity element, g_e.

Given two left-invariant analytic tangent vector fields, say L and M, the vector field $[L, M]$ is also left invariant. Therefore, the bracket product makes these vector fields into an n-dimensional Lie algebra, denoted g. This Lie algebra g is referred to as the *Lie algebra* of the Lie group G.

Proof It follows immediately from Definition 9.3 that a left-invariant tangent vector field on a Lie group is completely determined by its given vector at e. The first assertion is then immediate.

Recall (from the remark after Proposition 9.2) that if L and M are two analytic vector fields

$$d\mathscr{L}_y[L, M] = [d\mathscr{L}_y(L), d\mathscr{L}_y(M)].$$

If we denote a vector field, say L, at a specific point, say x, by L_x, then when L and M are left invariant we have

$$d\mathscr{L}_y[L, M]_x = [d\mathscr{L}_y(L), d\mathscr{L}_y(M)]_{yx} = [L, M]_{yx}.$$

But this means that $[L, M]$ is left invariant, in the sense of Definition 9.3. It then follows directly from Proposition 9.3 that we have a Lie algebra g.

Remarks (1)   The converse of this theorem is also true, namely, every Lie algebra is the Lie algebra of some (by no means unique) Lie group. This is one consequence of the rather difficult theorem of Ado on the representations of Lie algebras (see [21]).

(2)   We shall see that if a Lie algebra is a subalgebra (to be defined) of a Lie algebra that arises from a Lie group, then it, too, is the Lie algebra of a Lie group.

(3)   Some elementary examples of the above theorem are easy to calculate. Consider, for example, the real numbers as a Lie group. Its Lie algebra g is 1 dimensional. But, we have already observed, from Definition 9.5b, that $[X, X] = 0$. Then g must necessarily be abelian, because in a 1-dimensional g every element is a multiple of any fixed nonzero X.

It is not difficult to see that the Lie algebra of a Cartesian product of two Lie groups is the direct product of their Lie algebras, consisting of pairs of elements with the obvious coordinatewise operations. This may then be applied inductively to show that if $G_1, \ldots, G_k$ have abelian Lie algebras, so does $G_1 \times \cdots \times G_k$. In particular, $\mathbb{R}^k = \mathbb{R} \times \cdots \times \mathbb{R}$ has an abelian Lie algebra.

There is precisely one abelian Lie algebra of a given dimension. With little more work, one may prove that if G is an abelian Lie group ($xy = yx$, for all x, $y \in G$), then the Lie algebra of G is abelian. Therefore, the choice of name is totally reasonable.

We shall calculate the much less trivial Lie algebra of the general linear group $\mathrm{Gl}(n; \mathbb{R})$ shortly. First we must consider subgroups and subalgebras.

Definition 9.6 (a)   Let G be a Lie group. A subgroup $H \subseteq G$ is called a *Lie subgroup* if the set of points of H is a submanifold of the analytic manifold G (see Definition 1.5).

(b)   Let g be a Lie algebra. $h \subseteq g$ is a *Lie subalgebra* of g, if

   (i) h is a vector subspace of g, and
   (ii)   if X, $Y \in h$ then $[X, Y] \in h$.

Remarks (1) Examples are easy to construct. Any factor in a finite cartesian product of Lie groups is a Lie subgroup. If G is not connected (such as $Gl(n; \mathbb{R})$), the connected component of the identity is a Lie subgroup.

(2) A Lie subgroup is itself visibly a Lie group, with the structure inherited from the bigger group. A Lie subgroup is also clearly closed. Conversely, it is known that the image of a 1–1 open analytic homomorphism from one Lie group to another Lie group is a Lie subgroup (see [20]). We shall prove later in this chapter the slightly different result that a closed subgroup of a Lie group is a Lie subgroup.

(3) If g is an abelian Lie algebra, then any vector subspace of g is a Lie subalgebra. Any 1-dimensional subspace of a Lie algebra is trivially a Lie subalgebra.

Many more sophisticated examples of these phenomena will come from the following theorem. The second half is sketched up to the Frobenius theorem on distributions on manifolds (see Theorem 10.2), which is proved in detail in the next chapter.

Theorem 9.2 (a) Let G be a Lie group and H a Lie subgroup. Then the Lie algebra of H, say h, is a Lie subalgebra of the Lie algebra of G, say g.

(b) Let G be a Lie group with Lie algebra g. Suppose h is a Lie subalgebra of g. Then there is a Lie group H whose Lie algebra is isomorphic to h and a 1–1 map of H to G.

Proof For (a), it is easy to check that a left-invariant analytic tangent vector field on H extends uniquely to one on G. Indeed, such a vector field is determined, as we have seen, by the vector that it assigns to e. Referring to Theorem 9.1 and the immediate fact that the bracket product of two tangent vector fields to H is a tangent field to H, we see that h is a subalgebra of g.

To prove (b), we note that we shall show (Theorem 10.2) that whenever we have k nonvanishing, everywhere linearly independent tangent vector fields $X_i, 1 \leq i \leq k$, on G such that every $[X_i, X_j]$ always lies in the subspace of the tangent space spanned by the $X_1, \ldots, X_k$, then in a neighborhood containing the point e (in particular) there is a maximal submanifold H of G whose tangent space at any point of that neighborhood is spanned by the vectors $X_1, \ldots, X_k$. We choose for the X_i the invariant fields that make up a basis of h.

To construct the entire manifold H, one glues together overlapping copies of the original neighborhood of e by successive left translations. H is easily checked to be a Lie group, because there is a natural 1–1 map from H into G from which H inherits the analytic map $\mu(x, y) = x \cdot y^{-1}$. The fact that the tangent space of H in a neighborhood of e is spanned by the vector fields

X_i implies at once that the Lie algebra of H is h, and the map of H into G yields a 1–1 map of Lie algebras.

Remarks (1) Unfortunately, the image of H in G Theorem 9.2b needs not be a submanifold of G; thus H is not necessarily a Lie subgroup of G. Refer to Remark (1) after Definition 9.7. There are subgroups of the torus $S^1 \times S^1$ that are 1–1 images of $\mathbb{R}$ but are dense proper subspaces ($\mathbb{R}$ wraps around $S^1 \times S^1$ infinitely many times).

If the image of H is closed, then (as we shall soon see) it will be a Lie subgroup of G.

(2) The reader may wish to consult the standard references [20, 54] concerning these points. It is, however, unfortunate that Chevalley [20] and others have bent the notion of a submanifold to try to make the theory look more esthetic at this point. The price for this is too high, for then a submanifold of a manifold is not necessarily so much a manifold but rather the image of a map. With this and a few other notable exceptions, [20] is generally a good reference. The recent editions of [92] are excellent.

We now turn to a concrete, nontrivial example of a Lie algebra, that of the general linear group.

Proposition 9.4 If $G = \text{Gl}(n; \mathbb{R})$, then the Lie algebra g of G is the Lie algebra made up of all $n \times n$ matrices over the real numbers, with the bracket product given by

$$[X, Y] = YX - XY,$$

that is, the difference between the products of the matrices X and Y, taken in opposite orders.

Proof Let M be the vector space of all $n \times n$ matrices. We define a linear map

$$\phi : g \to M$$

as follows: we let $x_{ij} : G \to \mathbb{R}$ be the map that associates to every matrix $A \in G$ the element in the ith row and jth column of A. Each x_{ij} is trivially analytic. Let $X \in g$ be a left-invariant analytic tangent vector field on G. We define

$$\phi(X) = ((X(x_{ij})_e);$$

that is $\phi(X)$ is the matrix whose entry in the ij place consists of the value of (the derivation) X on the function x_{ij} at the identity element e.

ϕ is visibly linear. We suppose that $\phi(X)$ is the zero matrix; i.e., $X(x_{ij})_e = 0$ for all i and j; then we note that the x_{ij} are a coordinate system (basis) for a coordinate neighborhood of e (with $x_{ii}(e) = 1$, while all other $x_{ij}(e) = 0$). Thus, any analytic function from G to $\mathbb{R}$ may be expressed in a neighborhood of e by a Taylor series whose linear terms are simply the constant multiples of the x_{ij}. Because any derivation annihilates both the constant as well as the quadratic or higher terms (compare the proof of Proposition 9.1), we conclude that if $X(x_{ij}) = 0$ for all i and j then $X(x) = 0$ for any analytic function x from a neighborhood of e to $\mathbb{R}$. In other words, if $X(x_{ij}) = 0$, for all i and j, then $X = 0$. Hence, ϕ is 1–1.

Now, we know that $\mathrm{Gl}(n;\mathbb{R})$ is an open subset of the set of all $n \times n$ matrices (described by $\det A \neq 0$). Hence, $\dim(\mathrm{Gl}(n;\mathbb{R})) = \dim g = n^2$. Because $\dim M = n^2$ and ϕ is linear, we conclude at once that ϕ is an isomorphism of vector spaces.

It remains to calculate the effect of ϕ on a bracket product $[X, Y]$. First we note that by the definition of left invariance, for any vector field $X \in g$, we have the following formula:

$$(X(x_{ij}))_a = (d\mathscr{L}_a)(X(x_{ij})_e),$$

where $a \in \mathrm{Gl}(n:\mathbb{R})$ is itself a (nonsingular) $n \times n$ matrix. Now, it is a trivial exercise in linear approximation to verify that $d\mathscr{L}_a$ consists of left matrix multiplication by a with respect to the coordinates at hand.

Therefore, we may write

$$X(x_{ij})_a = \left(\sum_{k=1}^{n} a_{ik}(X(x_{kj})_e) \right) = \left(\sum_{k=1}^{n} x_{ik}(a)(X(x_{kj})_e) \right).$$

Regarding the equation

$$X(x_{ij})_a = \left(\sum_{k=1}^{n} x_{ik}(a)(X(x_{kj})_e) \right)$$

as an equation in (analytic) functions of the variable a, we may apply the derivation $Y \in g$ and evaluate at e to calculate for each pair of indices i and j:

$$Y(X(x_{ij})_a)_e = \sum_{k=1}^{n} Y(x_{ik})_e(X(x_{kj}))_e.$$

Referring to the definition of ϕ, we see that this is simply the ij place of $\phi(Y)\phi(X)$.

Applying this to invariant vector fields X and Y in the other order and subtracting, we get

$$\phi([X, Y]) = (Y(X(x_{ij}))_e - X(Y(x_{ij}))_e) = \phi(Y)\phi(X) - \phi(X)\phi(Y),$$

which completes the proof of the proposition.

Remarks (1) The above proposition may be used in showing that any (finite-dimensional) Lie algebra is the Lie algebra of some Lie group. Ado's theorem (see [21, 92]), which is too difficult to be included here, shows that such a Lie algebra is isomorphic to a Lie subalgebra of g with the bracket product $[X, Y] = YX - XY$. This Lie algebra being the Lie algebra of a Lie group, namely $\text{Gl}(n; \mathbb{R})$, Theorem 9.2b may be used to get the desired Lie group.

(2) The correspondence between Lie groups and Lie algebras of Theorem 9.1 is not unique. For example, all 1-dimensional Lie groups (S^1 or $\mathbb{R}$) have the same Lie algebra. Similarly, the plane, torus, and cylinder, $S^1 \times \mathbb{R}$, all have the same (abelian) Lie algebra.

(3) It is not hard to see that two Lie groups that have isomorphic Lie algebras are themselves isomorphic in a sufficiently small neighborhood of their identity elements. This becomes more precise if one formulates a notion of a *local Lie group*, a sort of germ of a Lie group in a neighborhood of e. The correspondence between such local Lie groups and Lie algebras is indeed 1–1 and onto.

(4) One might be tempted to think that any given Lie group is isomorphic to a Lie subgroup of $\text{Gl}(n; \mathbb{R})$ for a suitable large n. This is not the case. Examples may be constructed starting from groups of 3×3 matrices. Of course there are Lie groups with the same Lie algebra that are mapped to some $\text{Gl}(n; \mathbb{R})$ by a 1–1 map (by Theorem 9.2b), but the Lie group may not be the original and the image need not necessarily be a Lie subgroup (e.g., [54, Chapter 18]).

(5) The famous theorem of Peter and Weyl (e.g., [20]) assures us that a compact Lie group is isomorphic to a subgroup of some $\text{Gl}(n; \mathbb{R})$ for some large enough n. Assuming this, we shall show (shortly), by integration theory, that a compact Lie group is isomorphic to some Lie subgroup of $\text{O}(m)$, the group of $m \times m$ dimensional orthogonal matrices. (Recall that A is orthogonal if $A^{-1} = A^t$.)

THE EXPONENTIAL MAP

We now take a closer look at closed subgroups of Lie groups; as a preliminary topic, which is of substantial importance in itself, we discuss the exponential map (generalizing the power series for e^x when x is a matrix). We begin with some of this preliminary material.

I. Let G be a Lie group, g its Lie algebra. Let $X \in g$ be nonzero, so that X is a nowhere vanishing, analytic, left-invariant tangent vector field. Then we claim that there is a *unique analytic homomorphism*

$$\rho : \mathbb{R} \to G$$

such that, if Y is the element given by d/dx, as an element in the Lie algebra of $\mathbb{R}$, then

$$d\rho(Y) = X.$$

We first tackle the problem in a neighborhood of the identity element of each group. We simple need a local map $\rho : U \to G$, U being a neighborhood of 0 in $\mathbb{R}$, with $d\rho(Y) = X$. But near $e \in G$, G is just Euclidean space, and the vector field X may be broken into its components. Then the condition $d\rho(Y) = X$ may be immediately translated into a set of first-order linear differential equations whose data are analytic, since X is an analytic vector field (check!). The local existence (and uniqueness) of ρ is thus a consequence of the basic theorem from ordinary differential equations.

To see that ρ, as yet only defined near 0, is a homomorphism, at least where it makes sense to ask if it is, we shall compare the two maps

$$\rho(x + y) \qquad \text{and} \qquad \rho(x)\rho(y)$$

near $0 \in \mathbb{R}$. Of course when $y = 0$ they are the same map. Taking differentials with respect to y, say when $y = 0$, and observing that the data in the system of differential equations for ρ are invariant under left translation, we conclude that the differentials of these two maps when $y \in U$ must be equal. Hence, by the uniqueness of solutions of such first-order systems of differential equations,

$$\rho(x + y) = \rho(x)\rho(y)$$

near 0.

To handle the passage from local to global, we are going to use a procedure that is actually valid for any local homomorphism from $\mathbb{R}$ to a topological group G. Suppose U is an open connected neighborhood of the identity 0 in $\mathbb{R}$, and $\rho : U \to G$ is a continuous map that satisfies

(i) $\rho(0) = e$ and
(ii) if x, y, and $x + y$ belong to U, $\rho(x)\rho(y) = \rho(x + y)$.

Then I claim that ρ extends to a unique continuous homomorphism

$$\rho : \mathbb{R} \to G;$$

that is, there is such a homomorphism agreeing with the original map. To see this, suppose $z \in \mathbb{R}$, $z \notin U$. We choose a positive integer n so that $(z/n) \in U$. Define

$$\rho(z) = \left(\rho\left(\frac{z}{n} \right) \right)^n.$$

We note that this is well-defined, because

$$\left(\rho\left(\frac{z}{n}\right)\right)^n = \rho\left(\frac{z}{mn}\right)^{nm} = \rho\left(\frac{z}{m}\right)^m.$$

(Recall that we are writing $\mathbb{R}$ additively and G multiplicatively, and that ρ is a homomorphism on U.) I claim that ρ so defined is a homomorphism, because for sufficiently large n,

$$\rho(x + y) = \rho\left(\frac{(x+y)}{n}\right)^n = \left(\rho\left(\frac{x}{n}\right)\rho\left(\frac{y}{n}\right)\right)^n$$

$$= \rho\left(\frac{x}{n}\right)^n \rho\left(\frac{y}{n}\right)^n = \rho(x)\rho(y).$$

It is clearly continuous; since U is connected, naturally when $x \in U, (x/n) \in U$, so there is no question that the map agrees with the original one on U.

II.   If g and h are the Lie algebras of the Lie groups G and H, respectively, and $\phi: g \to h$ is a homomorphism of Lie algebras, we observe that ϕ may not be the differential of any analytic homomorphism from G to H. (Indeed, the only such homomorphism may be constant at the identity; for example, consider $G = S^1$, $H = \mathbb{R}$, and ϕ any isomorphism of 1-dimensional Lie algebras.)

Nonetheless, we claim that a local version of this theorem is true. In a neighborhood of e, G may be regarded (as a vector space, not a Lie group) as a direct sum of copies of $\mathbb{R}$. Using Section I above, we shall construct a map from a neighborhood of the identity in G to H whose differential agrees with ϕ at e. We should, however, first bring our terminology into line with the (very standard) usage at this point. The construction of the homomorphism (local) will then be seen in Remark (2) following this definition.

Definition 9.7   Let G be a Lie group with Lie algebra g. Suppose $0 \neq X \in g$. There is then a unique linear $1-1$ map

$$i_X: \mathbb{R} \to g$$

defined by $i_X(1) = X$. (Actually, one might think of $\mathbb{R}$ as the Lie algebra of $\mathbb{R}$, which it is indeed, at this point.) Then, we use Section I to build an analytic homomorphism

$$\rho_X: \mathbb{R} \to G$$

satisfying $d\rho_X = i_X$.

We define the *one-parameter subgroup* generated by X to be the image of ρ_X.

We define the *exponential map*

$$\exp: g \to G$$

by $\exp(X) = \rho_X(1)$.

Remarks   (1)   The one-parameter subgroup generated by X, i.e., $\rho_X(\mathbb{R})$, is clearly a subgroup of X but not necessarily closed nor a Lie subgroup. In the torus, $S^1 \times S^1$, most one-parameter subgroups have images that are dense proper subsets. In fact, a one-parameter subgroup will have a closed image only when the slope of X with respect to two equal axes, one for each factor, at e, is rational. (Check!)

In a compact connected Lie group G, every point lies on a one-parameter subgroup (in this case, it is not difficult to see that those points lying on one-parameter subgroup are both an open and closed set in the group). But in general, even if G is connected, not every point lies on a one-parameter subgroup [see Remark (4) below].

(2)   Clearly $\exp(0) = e$. In case g is abelian, it is not hard to check

$$\exp(X + Y) = \exp(X)\exp(Y).$$

But no such simple formula could be expected to hold in an arbitrary Lie group when the right-hand side depends on the order in which the elements X and Y are taken but the left-hand side does not. In general, this is well-understood, though rather complex. It is the subject of the basic Baker–Campbell–Hausdorff formula (see [54]).

It is relatively easy to check that exp is analytic, and by the inverse function theorem, a diffeomorphism from an open neighborhood of 0 to an open neighborhood of e.

If $\phi: g \to h$, note that the differential of $\exp \circ \phi \circ \exp^{-1}$ will be ϕ.

(3)   In the case of groups of matrices, we can easily get an explicit formula. Let $G = \mathrm{Gl}(n; \mathbb{R})^+$, the subgroup of $\mathrm{Gl}(n; \mathbb{R})$ whose elements have positive determinants. G is the connected component of the identity in $\mathrm{Gl}(n; \mathbb{R})$. Let g be the Lie algebra of G (which is the same as that of $\mathrm{Gl}(n; \mathbb{R})$). Recall (Proposition 9.4) that g is the Lie algebra of all $n \times n$ matrices and the bracket product $[X, Y] = YX - XY$.

If $M \in g$, we define

$$e^M = \sum_{k=0}^{\infty} \frac{1}{k!} M^k.$$

It is easy to see that such a series converges in each place, due to the rate of growth of $k!$ Various properties of e^M are then quickly established:

(i)   If A is invertible, $e^{AMA^{-1}} = Ae^M A^{-1}$, because $AM^k A^{-1} = (AMA^{-1})^k$.

(ii)   If the eigenvalues (or characteristic values) of M are $\lambda_1, \dots, \lambda_n$, then the eigenvalues of e^M are $e^{\lambda_1}, \dots, e^{\lambda_n}$. In fact, that e^{λ_1} is an eigenvalue of e^M would follow easily if the upper left corner of M were λ_1 and the rest of the first column were zero. But using (i), one may assume that is the case.

(iii)   The determinant of e^M is e raised to the trace of M (the trace is the sum of the diagonal elements). Simply observe that the determinant is the product of the eigenvalues while the trace is the sum of the eigenvalues.

(iv) e^M is always nonsingular and belongs to G (those matrices with positive determinant).

(v)   If M and N commute, $e^{M+N} = e^M e^N$ (this is formally the same proof as in the case where M and N are just real numbers).

(vi)   The map $t \to e^{tM}$ is an analytic homomorphism (because, setting $\phi_M(t) = e^{tM}$, $\phi_M(t_1 + t_2) = e^{(t_1+t_2)M} = e^{t_1 M + t_2 M} = \phi_M(t_1)\phi_M(t_2)$, etc.

We may now make our explicit calculation. At the origin 0, we have

$$\phi'_M(0) = \lim_{h \to 0} \frac{\phi_M(h) - \phi_M(0)}{h} = \lim_{h \to 0} \frac{e^{hM} - I}{h} = \lim_{h \to 0} \frac{1}{h} \sum_{k=1}^{\infty} \frac{(hM)^k}{k!} = M.$$

Now, ϕ_M is an analytic homomorphism with $\phi'_M(0) = M$ (compare Definition 9.7) so $\phi_M = \rho_M$ and

$$\exp(M) = \phi_M(1) = e^M.$$

This shows that for the group G (as well as $Gl(n; \mathbb{R})$) the exponential map exp agrees with the Taylor series for e^x, evaluated on M. This is the historical justification for the terminology at hand.

(4)   We may now quickly check that $\exp: g \to G$ need not be onto, even when G is connected. Let $G = Gl(2; \mathbb{R})^+$, the 2×2 matrices with positive determinant; then g is all 2×2 matrices. Because the determinant is the product of the eigenvalues, and M has real entries, $\lambda_1 \lambda_2$ is real. Write θ_i for the argument of λ_i and $\|\lambda_i\|$ for its length or modulus. Then

$$\lambda_1 \lambda_2 = \|\lambda_1\| \|\lambda_2\| e^{i(\theta_1 + \theta_2)}.$$

For this to be positive and real, $\theta_1 + \theta_2$ is 0, and we conclude that we may write $\lambda_1 = \alpha \bar{\lambda}_2$, with α positive and real.

Note that $\mu_i = e^{\lambda_i}$ are the eigenvalues of e^M. Once again, if $\mu_1 \mu_2$ is to be positive and real, an easy calculation shows

$$\det(e^M) = e^{(\alpha+1)\operatorname{Re}(\lambda_2)}$$

with $\operatorname{Re}(\lambda_2)$ the real part of λ_2.

Consider the matrix

$$x = \begin{pmatrix} -2 & 0 \\ 0 & -\frac{1}{2} \end{pmatrix}$$

whose eigenvalues are precisely -2 and $-\frac{1}{2}$. Suppose $x = e^M$. Writing $\operatorname{Im}(\lambda_2)$ for the imaginary part of λ_2, we have

$$-2 = e^{\alpha \operatorname{Re}(\lambda_2)}(\cos(\alpha \operatorname{Im}(\lambda_2)) - (\sin \alpha \operatorname{Im}(\lambda_2))i)$$

and

$$-\tfrac{1}{2} = e^{\mathrm{Re}(\lambda_2)}(\cos(\mathrm{Im}(\lambda_2)) + (\sin(\mathrm{Im}(\lambda_2)))i).$$

But the two right-hand sides must be real and negative. This means that we must have $\mathrm{Im}(\lambda_2) = n\pi$ and $\alpha\, \mathrm{Im}(\lambda_2) = m\pi$, with both n and m odd and positive. (One could take $n = m = 1$.) Thus $\alpha = m/n$ and

$$2 = e^{(m/n)\,\mathrm{Re}(\lambda_2)}, \qquad \tfrac{1}{2} = e^{\mathrm{Re}(\lambda_2)}.$$

This implies

$$2^n = e^{m\,\mathrm{Re}(\lambda_2)}, \qquad 1/2^m = e^{m\,\mathrm{Re}(\lambda_2)},$$

an obvious contradiction. We may therefore conclude that

$$\begin{pmatrix} -2 & 0 \\ 0 & -\tfrac{1}{2} \end{pmatrix}$$

is not in the image of the exponential map.

This completes II. We may now tackle the closed subgroups of Lie groups.

CLOSED SUBGROUPS OF LIE GROUPS

Theorem 9.3 A closed subgroup of a Lie group is a Lie subgroup (in the sense of Definition 9.6a).

Proof From the remarks in II above, exp is an analytic map from the Lie algebra g, which is an n-dimensional vector space over $\mathbb{R}$, $n = \dim G$, to the analytic manifold G. We have also seen that for every tangent vector at the identity, $X \in g_e = (TG)_e$, there is a one-parameter subgroup $\rho_X(\mathbb{R})$ tangent to X at e. It follows at once that the differential of exp, that is, the map on tangent spaces defined by exp, is onto, and because the range and domain have the same dimension, it is an isomorphism. Using Corollary 1.1, we see (as we have already remarked above) that exp is a diffeomorphism of a neighborhood of 0 in g_e to an open neighborhood of $e \in G$, which we shall call U.

In this neighborhood of e, one may speak meaningfully of $\exp^{-1}$ (some authors go so far as to call this log).

Because the question of whether H is a Lie subgroup is simply a question of whether H is a submanifold of the analytic manifold G, and since any open neighborhood of any point $x \in H$ may be translated (by an analytic diffeomorphism) to an open neighborhood of e (using $\mathscr{L}_{x^{-1}}$), it suffices to show that H is a submanifold in a neighborhood of e. That is, we must find a coordinate chart at e, say V, compatible with the given analytic structure on G (see Definition 1.5), in which H is described by the vanishing of the last r coordinates for some $1 \leq r \leq n$. We shall actually show that in a suitably small neighborhood of e, $\exp^{-1}$ takes H to an (analytic) submanifold of g, which will then complete the proof.

There are now three basic steps.

(a)   If H does not contain an entire one-parameter subgroup, there is an open neighborhood of e that contains no point of H except e. To prove this, let x_n be a sequence of points in H, $x_n \neq e$, that converges to e. (If no such sequence exists, our claim is immediate.) We may assume that our neighborhood is homeomorphic to $\mathbb{R}^n$ and $\exp^{-1}$ is defined on it. We shall speak freely as if this neighborhood is $\mathbb{R}^n$ and e is the origin.

The vectors $x_n/\|x_n\|$ lie on the unit sphere $S^{n-1} \cong \mathbb{R}^n$. Since this is compact, we may select a subsequence x_{n_i} for which $x_{n_i}/\|x_{n_i}\|$ converges to a point of S^{n-1}, say v. Now, if we multiply each x_{n_i} by itself sufficiently many times, we arrive at a point whose distance from 0 is near and approaching 1. It follows at once that v is actually a limit of a sequence of points in H, and thus $v \in H$.

Now, it is easy to see that there is no loss in generality if we assume that the one-parameter subgroup through e and v (recall that exp is onto this neighborhood) is the straight line from e (or 0) to v. I claim that this one-parameter subgroup itself contains a sequence of points in H that converges to e.

For each x_{n_i} we observed that there is a power (in the sense of iterated multiplication) that is near v. But in the same fashion, if i is large, we can find a power of x_{n_i} near $\frac{1}{2}v$, the midpoint of the segment from 0 to v. Thus, $\frac{1}{2}v$ may also be shown to belong to the (closed) subgroup H. In precisely the same fashion, one may show that $\frac{1}{4}v$ belongs to H, etc. The sequence $v, \frac{1}{2}v, \frac{1}{4}v, \ldots$ is the desired sequence in the one-parameter subgroup that converges to e, or 0 in the terminology of the neighborhood.

Now, H is a subgroup that contains a sequence of points in a one-parameter subgroup through 0 and v that converges to 0. By taking an element in this sequence near 0 and multiplying it by itself a large number of times, we may quickly find an element of H as near as we wish to any point in the one-parameter subgroup from 0 to v. Because H is closed, we have that H contains the entire one-parameter subgroup from 0 to v. But then it follows at once (elementary group theory) that H contains the entire one-parameter subgroup. This establishes the claim in (a).

In this case (an open neighborhood of e contains no other point of H) H is a discrete 0-dimensional submanifold, and the proof is complete.

(b)   We now suppose that H contains an entire one-parameter subgroup of G. We set M to be the set of points $x \in U \cap H$, where U was our original neighborhood of e, that may be joined to e by a one-parameter subgroup that from e to x lies within $U \cap H$. I claim that $\exp^{-1}(M)$ is a vector subspace of g in a neighborhood of 0. It is clear that $\exp^{-1}(M)$ is (locally) closed under scalar multiplication. We must show that $\exp^{-1}(M)$ is closed under vector addition.

Suppose ρ_X and ρ_Y are two one-parameter subgroups that locally lie within $U \cap H$. Assume X and Y are linearly independent. I claim that $\rho_X(x)\rho_Y(x)$ is tangent to a one-parameter subgroup, and if we consider those values which lie in U, it lies within $U \cap H$. Furthermore, I claim that $\exp^{-1}(\rho_X(x)\rho_Y(x))$ is tangent to $X + Y$ at e. This may be checked by observing that $\exp^{-1}\rho_X + \exp^{-1}\rho_Y = X + Y$ and looking at difference quotients, or alternatively, expressing the one-parameter subgroups as power series and considering the linear terms. Thus, the one-parameter group $\rho_{X+Y}(x)$ and $\rho_X(x)\rho_Y(x)$ lie on a tangent set of points within U, though they may not be precisely the same functions from $\mathbb{R}$ to G. Nevertheless, since $\rho_X(x)\rho_Y(x)$ lies in $H \cap U$, so does $\rho_{X+Y}(x)$ locally. This proves that $\exp^{-1}(M)$ is a vector subspace of g.

Since exp is locally an analytic diffeomorphism, and a vector subspace is trivially an (analytic) submanifold, we conclude that near e, M is a submanifold of U. Of course, as we noticed, we need only show that H is a submanifold in some U.

(c) I claim, finally, that in a possibly smaller open neighborhood of e, H and M are the same set of points. When proved, this, coupled with the last remarks in (b) above, shows that H is a submanifold.

Of course, M is a subset of H. If there is an open set O, with $M \quad O \cap U$, and O contains no points of $H - M$, then we will be finished. (It is a good idea to sketch a picture of the relationships of these sets.)

Assuming the contrary (for the purposes of having a contradiction) we take a sequence of points in $U \cap H$ that do not lie in M but converge to a point in M. We shall use an argument similar to (a) above. To be precise, x_n converges to $m \in M$, with each $x_n \in (U \cap H) - M$. For each x_n, choose $m_n \in M$ so that the segment from m_n to x_n is perpendicular to M. Call the vector from m_n to x_n translated to e by the name v_n. This is sketched in Fig. 9.1.

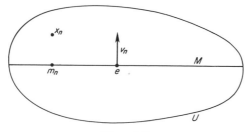

Figure 9.1

Although the translation to v_n may not necessarily yield a v_n perpendicular to M, for large enough n we may assume that the v_n make an angle θ with M, with $\pi/2 - \varepsilon \le \theta \le \pi/2 + \varepsilon$, ε small. Hence, the vectors

$$w_n = v_n/\|v_n\|$$

will have a subsequence that converges to a vector w *not* tangent to M.

Following the same argument as in (a), there is a one-parameter subgroup in H that runs through e and w in our coordinate neighborhood. This contradicts the fact that M is the set of all those points that may be joined to e by segments of one-parameter subgroups lying in $U \cap H$.

Corollary 9.1   A closed subgroup of a Lie group is a Lie group.

Proof   A Lie subgroup is clearly a Lie group.

Remarks   The first remark following Definition 9.6 shows that the image of a 1–1 analytic homomorphism need not be a Lie subgroup. Nonetheless, it is known that if $f: H \to G$ is a 1–1 continuous homomorphism from a locally compact topological group H to a Lie group G, then H is a Lie group (see [20, 54]). Because a closed subgroup of a Lie group is locally compact, this result immediately implies our Theorem 9.3. On the other hand, Theorem 9.3, combined with reasoning as in Theorem 9.2b, may be used to recover this theorem by a slightly different argument.

INVARIANT FORMS AND INTEGRATION

We now turn our attention to analysis on Lie groups. First, we shall look at the theory of integration. Referring to Definition 5.9, we see that in order to have a meaningful theory of integration of functions over a Lie group (or more generally, any orientable manifold), one must have a nowhere vanishing n-form defined on the Lie group (or manifold). But we wish to have more here, and we shall actually construct an integral that is invariant under the action of the group on itself (say, by left multiplication). We address the invariance in a trivial way after we complete the construction of the invariant (n-form).

Definition 9.8   Let $\omega \in D^i(G)$ be an i-form (differential) on the Lie group G. We say that ω is left invariant if, given any $y \in G$,

$$D(\mathcal{L}_y)(\omega) = \omega.$$

(Recall that $\mathscr{L}_y : G \to G$ is given by $\mathscr{L}(X) = yx$ and $D(f)$ is the (contravariant) map defined on differential forms by the analytic map f (as in Definition 5.6).)

Remarks (1) There are some easy examples. In $\mathbb{R}$ (or more generally $\mathbb{R}^n$), dx_1 (or more generally $dx_1 \wedge \cdots \wedge dx_n$) is left invariant. In S^1, using angular coordinates, one may express a differential 1-form as $d\theta$. On the multiplicative group of complex numbers (the plane with the origin removed), one may take $(1/r) dr \wedge d\theta$ (polar coordinates).

(2) One may give an alternative description of an invariant form in terms of its action on vector fields. (Details are not difficult.)

Proposition 9.5 Given a basis for the tangent space at e in a Lie group G, say $v_1, \ldots, v_n$, one defines a 1-form ω by setting

$$\omega(v_1) = 1 \quad \text{and} \quad \omega(v_j) = 0 \quad \text{for} \quad j \neq 1.$$

This may be extended to a left invariant 1-form on all of G by setting

$$\omega(v)_x = \omega(d(\mathscr{L}_{x^{-1}})(v))_e,$$

where the subscript is to refer to the point at which the form is to be evaluated, and v is any vector field.

Proof This is basically an exercise in recalling definitions, in order to verify left invariance. If v is a vector field and $x, y \in G$, we calculate

$$(D(\mathscr{L}_y)(\omega))(v)_{yx} = \omega(d(\mathscr{L}_y)(v))_{yx} = \omega(d(\mathscr{L}_{x^{-1}y^{-1}})d(\mathscr{L}_y)(v))_e$$

by the definition of ω. But this is clearly equal to

$$\omega(d(\mathscr{L}_{x^{-1}})(v))_e = \omega(v)_x.$$

(Recall that for the purposes of testing a map on forms, one needs the map induced on vector fields. There is no difficulty getting a well-defined map on vector fields here, because $\mathscr{L}_z$ is always a diffeomorphism.)

Remark In the classical terminology, as in [20], a 1-form is called a form of Pfaff, while a left-invariant 1-form on a Lie group is called a form of Maurer–Cartan.

We are now ready to get to integration.

Definition 9.9 Let G be a Lie group of dimension n; we let μ be the nowhere vanishing left-invariant n-form defined by

$$\mu = \omega_1 \wedge \cdots \wedge \omega_n,$$

where each ω_j is the left-invariant 1-form constructed as in Proposition 9.5 by the formula

$$\omega_j(v_i)_e = \delta_{ij} \quad \text{and} \quad \omega_j(v)_x = \omega_j(d(\mathcal{L}_{x^{-1}})(v))_e.$$

Since μ is left invariant and does not vanish at e ($\{v_1, \ldots, v_n\}$ is a basis) μ is nowhere vanishing. We shall use μ to develop our integration.

We recall that if G, a Lie group, is covered by coordinate charts U_i, with given homeomorphisms $\phi_i : U_i \to \mathbb{R}^n$, and ω is some n-form with compact support, we have defined (compare with Definition 5.9 and related material)

$$\int_G \omega = \sum_i \int_{\phi_i(U_i)} (\zeta_i g_i) \circ \phi_i^{-1} \, dx_1 \wedge \cdots \wedge dx_n,$$

with $\{\zeta_i\}$ a partition of unity associated with the $\{U_i\}$, presumed to be locally finite, and μ used to write

$$D(\phi_i)dx_1 \wedge \cdots \wedge dx_n = f_i \mu, \qquad f_i > 0,$$

and

$$\omega = g_i D(\phi_i)dx_1 \wedge \cdots \wedge dx_n.$$

Recall also [Remark (7) following Definition 5.9] that if f is a continuous function, with compact support in G, one may define its integral to be

$$\int_G f\mu.$$

Since μ is nowhere vanishing, any n-form ω may be expressed as an $f\mu$, if its support is compact.

Proposition 9.6   Given any fixed $y \in G$ and continuous $f : G \to \mathbb{R}$ with compact support, we shall write

$$f_y(x) = f(yx).$$

Then

$$\int_G f_y \mu = \int_G f\mu.$$

Proof $\int_G f_y \mu$ may be written in terms of a variable of integration x as

$$\int_G f_y(x)\mu(x).$$

In fact, the reader may prefer to write the customary $d\mu$, instead of μ, to achieve a formal similarity with basic calculus. Alternatively, one may think of such a formula as expressing the integral as a limit of sums, where $\mu(x)$ is the "volume of a small box around x."

But

$$\int_G f(yx)\mu(x) = \int_G f(yx)\mu(y^{-1}yx)$$

trivially. Since μ is left invariant, $\mu(y^{-1}yx)$ is the same as $\mu(yx)$, and the proposition results immediately from replacing the variable yx, y fixed, with the variable x.

This proposition asserts that our integration is unchanged by left translation. It is rather remarkable that when G is compact the integration is automatically unaffected by right translation.

Proposition 9.7 If G is a compact Lie group, $f: G \to \mathbb{R}$ is a continuous function, and we set $f^y(x) = f(xy)$, then

$$\int_G f^y\mu = \int_G f\mu.$$

Proof For $z \in G$, we define $a_z(x) = z^{-1}xz$. In other terms, a_z means conjugation by z^{-1}. a_z is clearly an analytic isomorphism from G to itself. On the other hand, $D(a_z)\mu$ is a left-invariant differential form (check). It is therefore determined by its value at the identity, so that we set

$$(D(a_z)\mu)(v_1, \ldots, v_n)_e = c(z)\mu(v_1, \ldots, v_n)_e$$

(treating the forms as multilinear alternating functions on vector fields, as in Proposition 5.1.). Since both sides of the equation are left invariant, the equation holds at all points x, rather than merely e. Alternatively, there is a constant $c(z)$ such that

$$D(a_z)\mu = c(z)\mu.$$

Observing that $D(a_z)$ takes forms at $a_z(x)$ to forms at x, we have

$$\int_G f(x)D(a_z)\mu(a_z(x)) = c(z)\int_G f(x)\mu(x).$$

Using Proposition 9.6 and the basic formulas for change of variable (material preceding Definition 5.9), we write the left side

$$\int_G f(x)D(a_z)\mu(a_z(x)) = \int_G f(za_z(x)z^{-1})D(a_z)\mu(a_z(x))$$

$$= \int_G f(a_z(x)z^{-1})D(a_z)\mu(a_z(x)) = \int_G f(xz^{-1})\mu(x).$$

Therefore,

$$\int_G f(xz^{-1})\mu(x) = c(z)\int_G f(x)\mu(x).$$

Since G is compact, we set $f \equiv 1$ and recover $c(z) = 1$. Whence, taking z^{-1} to be y, we get

$$\int_G f(xy)\mu(x) = \int_G f(x)\mu(x),$$

as desired.

Remarks (1) When a Lie group is compact, one frequently normalizes μ by dividing by

$$\int_G 1 \cdot \mu$$

as a matter of convenience.

(2) If G is *not* compact, our left-invariant integral (Definition 9.8) does not need to be right invariant. The situation on locally compact groups is rather more difficult.

REPRESENTATIONS OF LIE GROUPS

Naturally, the reader can supply various potential applications of integration on Lie groups, involving standard interpretations of averages or moments. We propose a different, more mathematical, sort of application. First, we discuss the basics of the theory of representations (where groups are viewed as groups of matrices). We then give a striking application of integration theory (Theorem 9.4) to the reduction of representations of compact Lie groups.

Definition 9.10 Let G be a Lie group. A *representation* of G means a continuous homomorphism

$$\rho: G \to \mathrm{Gl}(n; \mathbb{R}),$$

for some positive integer n, called the dimension of ρ.

A representations ρ is called *trivial* if ρ is constantly the identity matrix; it is *faithful* if ρ is $1-1$.

Remarks and Examples *1.* Many Lie groups are presented as groups of matrices, such as $\mathrm{Gl}(n; \mathbb{R})$, $\mathrm{Gl}(n; \mathbb{R})^+$, $\mathrm{O}(n)$, the orthogonal group, $\mathrm{Sl}(n; \mathbb{R})$, $n \times n$ matrices of determinant $+1$, etc. These groups come equipped with natural representations, the inclusion maps into $\mathrm{Gl}(n; \mathbb{R})$. They also have many other important representations.

2. The real numbers are not, in any particular way, a matrix group. There is a 1-dimensional representation given by

$$\rho(x) = e^x.$$

A two-dimensional representation is given by

$$x \to \begin{pmatrix} 1 & x \\ 0 & 1 \end{pmatrix}.$$

3. If G is any Lie subgroup of $Gl(n; \mathbb{R})$, the determinant furnishes a 1-dimensional representation of G.

4. Given two representations

$$\rho_1 : G \to Gl(n_1; \mathbb{R}) \qquad \text{and} \qquad \rho_2 : G \to Gl(n_2; \mathbb{R}).$$

one can build a direct sum of these two representations, of dimension $n_1 + n_2$, by sending each x to the $((n_1 + n_2) \times (n_1 + n_2))$-dimensional matrix

$$\left(\begin{array}{c|c} \rho_1(x) & 0 \\ \hline 0 & \rho_2(x) \end{array} \right).$$

Of course, 0 means the matrix, of the appropriate size, all of whose entries are zeros.

5. Let SU(2) be the group of those complex matrices of determinant 1 such that the conjugate of the transpose of M is equal to the inverse of M. (Conjugate means the matrix resulting from taking the conjugate in each place, transpose from the interchange of rows and columns.) SU(2) acts, as a group, on 2-dimensional complex space $\mathbb{C}^2$ (which is basically the same as $\mathbb{R}^4$) by left matrix multiplication. There is an inner product in $\mathbb{C}^2$ given by

$$(x, y) = x_1 \overline{y_1} + x_2 \overline{y_2},$$

where $x = (x_1, x_2)$, $x_i \in \mathbb{C}$, etc., and $\overline{x}_1$ is the conjugate of x_1, etc. It is easy to check that a matrix $M \in SU(2)$ preserves inner product, i.e., $(Mx, My) = (x, y)$. Thus, M preserves lengths and angles in $\mathbb{R}^4$, yielding a map

$$\rho : SU(2) \to SO(4),$$

SO(4) being the matrices in O(4) of determinant $+1$. ρ is clearly faithful. Now, represent $\mathbb{R}^3$ as 2×2 complex matrices

$$S = \begin{pmatrix} -z & x + iy \\ x - iy & z \end{pmatrix}$$

with $x, y, z \in \mathbb{R}$. If $M \in SU(2)$, an easy calculation shows that MSM^{-1} is a matrix of the same form as S, defining a representation

$$Sp : SU(2) \to SO(3),$$

which is not faithful. It is called the *spin representation*, and it takes pairs of distinct matrices in SU(2) to single matrices in SO(3).

Definition 9.11   Given two representations of the same group G having the same dimension, ρ_1 and ρ_2, we say that ρ_1 and ρ_2 are *similar* if there is a square nonsingular matrix of that dimension, A, such that

$$\rho_1(x) = A\rho_2(x)A^{-1}$$

for all $x \in G$.

Definition 9.12   A representation is called *orthogonal* if each $\rho(x)$ is an orthogonal matrix $(\rho(x)^{-1} = \rho(x)^t)$.

Our goal is then

Theorem 9.4   If G is a compact Lie group and

$$\rho: G \to \mathrm{Gl}(n; \mathbb{R})$$

is any representation, ρ is similar to an orthogonal representation.

Proof   A square matrix M is *symmetric* if it equals its own transpose; it is positive definite if, for any vector v,

$$(v, Mv) \geq 0$$

and $(v, Mv) = 0$ if and only if $v = 0$. Of course, (v, w) is the inner product $v_1 w_1 + \cdots + v_n w_n$. If v and w are column vectors, $(v, w) = v^t w$.

If A is any nonsingular matrix, it is virtually trivial to check that $A^t A$ is both symmetric and positive definite.

For each $x \in G$, we set

$$b(x) = \rho(x)^t \rho(x).$$

Write $b(x) = (b_{ij}(x))$. Let μ be the left-invariant form on G constructed above (Definition 9.9), normalized so that

$$\int_G 1\mu = 1.$$

Define

$$b_{ij} = \int_G b_{ij}(x)\mu$$

and form the matrix

$$B = (b_{ij}).$$

B is clearly symmetric, and because $b(x)$ is positive definite and the integrals of positive functions with respect to this μ are visibly positive, we see that B is positive definite.

Now, it is well known from linear algebra that a symmetric matrix is orthogonally similar to a diagonal one; i.e., if M is symmetric, there is an orthogonal A with the matrix

$$AMA^{-1}$$

being 0 in all places except possibly the diagonal. In other words there is a change of basis, preserving angles, for which we may write

$$B = \begin{pmatrix} \beta_{11} & 0 & \cdots & 0 \\ 0 & \beta_{22} & & 0 \\ \vdots & \vdots & & \vdots \\ 0 & 0 & & \beta_{nn} \end{pmatrix}.$$

Clearly the β_{ii} are all positive, so that in this basis (as well as any other) we may take the square root of A. In our original basis, denote such a square root by C; i.e., write

$$B = CC.$$

Observe that C is symmetric in either basis we have considered.

Now, for each $x \in G$ we have

$$\rho(x)^t B \rho(x) = \int_G \rho(x)^t \rho(y)^t \rho(y) \rho(x) \mu(y),$$

where we write $\mu(y)$ to indicate that we are integrating with respect to y, holding x temporarily constant.

We rewrite and use Proposition 9.7

$$\int_G \rho(yx)^t \rho(yx) \mu(y) = \int_G \rho(y)^t \rho(y) \mu(y) = B.$$

Thus

$$\rho(x)^t B \rho(x) = B.$$

Define $\sigma(x) = C\rho(x)C^{-1}$, and we may then calculate

$$\sigma(x)^t \sigma(x) = (C\rho(x)C^{-1})^t (C\rho(x)C^{-1}) = (C^{-1})^t \rho(x)^t C^t C \rho(x) C^{-1}$$
$$= (C^{-1})^t \rho(x)^t B \rho(x) C^{-1}$$

as $C^t = C$ and $CC = B$.

But this gives, by our earlier remark,

$$(C^{-1})^t B C^{-1} = (C^{-1})^t CCC^{-1} = (C^{-1})^t C^t CC^{-1} = (CC^{-1})^t (CC^{-1}) = I,$$

the identity matrix.

Summarizing, $\sigma(x)^t \sigma(x) = I$ or $\sigma(x)^{-1} = \sigma(x)^t$, so $\sigma(x)$ is always orthogonal. Because

$$\sigma(x) = C\rho(x)C^{-1},$$

our proof is complete.

To make a few further comments, we introduce the following definition:

Definition 9.13   Let $\rho: G \to \mathrm{Gl}(n; \mathbb{R})$ be a representation of a Lie group.

(a)   Every $\rho(x)$ acts as a linear transformation on $\mathbb{R}^n$, by left matrix multiplication. A linear subspace $V \subseteq \mathbb{R}^n$ is called *invariant* if whenever $x \in G$, $v \in V$, we have

$$\rho(x)(v) \in V.$$

(b) ρ is *irreducible* if it has no invariant subspace other than 0 and $\mathbb{R}^n$.
(c) ρ is *completely reducible* if it is similar to a (finite) direct sum of irreducible representations [in the sense of Remark (4) following Definition 9.10].

Examples   *1.* A nonzero 1-dimensional representation is automatically irreducible, for example, the determinant

$$\det: \mathrm{Gl}(n; \mathbb{R}) \to \mathrm{Gl}(1; \mathbb{R}).$$

2. As in Example 2 following Definition 9.9, consider the two-dimensional representation of $\mathbb{R}$,

$$x \to \begin{pmatrix} 1 & x \\ 0 & 1 \end{pmatrix}.$$

The subspace of $\mathbb{R}^2$ consisting of all column vectors

$$\begin{pmatrix} u \\ 0 \end{pmatrix}$$

is clearly invariant, and the equation

$$\begin{pmatrix} 1 & x \\ 0 & 1 \end{pmatrix}\begin{pmatrix} u \\ v \end{pmatrix} = \begin{pmatrix} u + xv \\ v \end{pmatrix}$$

shows that no other (proper) subspace is invariant. Hence, this representation could not be completely reducible, for then there would be two independent one-dimensional invariant subspaces.

We can now give a remarkable corollary of Theorem 9.4.

Corollary 9.2   If $\rho: G \to \mathrm{Gl}(n; \mathbb{R})$ is a representation of a compact Lie group, ρ is completely reducible.

Proof   By an easy induction, it will suffice to check that if $V \subseteq \mathbb{R}^n$ is an invariant subspace with

$$0 < \dim V < n$$

then there is an invariant subspace W, $\dim W = n - \dim V$, with $\mathbb{R}^n = V \oplus W$. ρ will then decompose as a sum [Remark (4) after Definition 9.10].

By Theorem 9.4, we may assume that in our basis $\rho(x)$ is orthogonal for all $x \in G$. If V is invariant, I claim that the orthogonal complement

$$V^\perp = \{w \in \mathbb{R}^n \,|\, \text{for all } v \in V, \quad (v, w) = 0\}$$

is also invariant. We simply note that if $w \in V^\perp$, then when $v \in V$,

$$(\rho(x)w, v) = (w, \rho(x)^t v) = (w, \rho(x)^{-1} v) = (w, \rho(x^{-1})v) = 0,$$

because V is invariant under $\rho(y)$, $y \in G$, so $\rho(x^{-1})v \in V$. Clearly $V \oplus V^\perp = \mathbb{R}^n$, finishing our proof.

Remarks One of the most important applications of representation theory is to quantum mechanics and the theory of atomic spectra ([112, 118]).

The theory of representations is much more elaborate than we have been able to indicate here. The interested reader should pursue this up to the theory of characters, the orthogonality relations, and the Peter–Weyl theorem. These may be found in various places, such as [20, 54]. This latter result shows that compact Lie groups always have faithful representations.

LIE GROUPS ACTING ON MANIFOLDS

Finally, we wish to touch upon the subject of a Lie group acting as a group of transformations on a manifold. We begin by reviewing some general facts, occasionally touched upon in Chapter 4, etc. Suppose that G is a topological group and $H \subseteq G$ a closed subgroup. We may form the coset space G/H, which is the quotient space of G under the equivalence relation "$x \sim y$, if and only if $x^{-1}y \in H$." If G is compact and Hausdorff and H is closed, then G/H is a compact Hausdorff space. To see that G/H is Hausdorff, let $\rho : G \to G/H$ be the projection map, and take $u, v \in G/H$. Then $\rho^{-1}(u)$ and $\rho^{-1}(v)$ are disjoint closed (and hence compact) subsets of G. One easily constructs disjoint open sets, around $\rho^{-1}(u)$ and $\rho^{-1}(v)$, that are invariant under multiplication by elements of H. This immediately implies G/H is Hausdorff. Since G is compact and ρ is onto, G/H is compact.

Now suppose G is a compact Lie group and H is a closed (and hence Lie) subgroup. I claim that G/H is much more than merely a compact Hausdorff space. Consider a small neighborhood U of $e \in G$ such that the exponential map is a diffeomorphism of a neighborhood of $O \in g$ onto U. The Lie algebra of H, say h, is then a vector subspace of g, so we may let V be the orthogonal complement of h in g, i.e., $g = h \oplus V$. V need not be a Lie subalgebra.

Define
$$\phi = \rho \circ \exp \circ i \colon V \to G/H,$$

where $i(a) \to (0, a)$ is the inclusion of V in g and $\rho \colon G \to G/H$ is the projection.

It is clear that near e any element of G differs from an element in the image of exp i by multiplication by an element in H. (Near e, $H = \exp h$.) But ρ sends all of H to a single point. It follows at once that ϕ is onto near e. It is also easy to see that $\exp V$ is orthogonal to H at e, from which it follows quickly that ϕ must be 1–1 near O. ϕ is then seen to be a diffeomorphism, either by use of the inverse function theorem or by observing that if W is open then

$$H \cdot W = \{h \cdot w \,|\, h \in H, \quad w \in W\}$$

is the union of open sets (and hence open), so that one quickly gets that $\phi(S)$ is open when S is open.

This all shows that $\rho(e) \in G/H$ has a neighborhood homeomorphic to an open set in Euclidean space of dimension equal to dim G − dim H. By the usual translation argument, this extends to any point of G/H.

In fact one may show more. G/H is actually an analytic manifold, and ρ is an analytic map. The group G acts on G/H by a map

$$\psi \colon G \times G/H \to G, \qquad \psi(x, \{y\}) = \{xy\}.$$

The map ψ may also be shown (without terrible difficulty) to be analytic.

There are many examples of manifolds of this form, frequently referred to as homogeneous spaces. For example, consider $O(n + 1)$ acting on $\mathbb{R}^{n+1}$ by right matrix multiplication; i.e., consider

$$xg$$

with $g \in O(n + 1)$, $x = (x_1, \dots, x_{n+1})$. Because g is orthogonal, $\|x \cdot g\| = \|x\|$. Consider $O(n)$ as the subgroup of $O(n + 1)$ made of matrices

$$\begin{pmatrix} 1 & 0 & \cdots & 0 \\ 0 & & & \\ \vdots & & \left(M \right) & \\ 0 & & & \end{pmatrix}$$

with $M \in O(n)$. Action by an element of $O(n)$ leaves the first coordinate fixed, so it sends $v_1 = (1, 0, \dots, 0)$ to itself. Define

$$\phi \colon O(n + 1) \to S^n$$

by $\phi(A) = v_1 A$. It easily seen that the subgroup of these orthogonal matrices that send v_1 to itself is $O(n)$, and there is a homeomorphism

$$O(n + 1)/O(n) = S^n.$$

More generally, we may define a *Stiefel manifold* $V_{m,k}$ to be $O(m)/O(k)$, with $0 < k < m$ and $O(k)$ regarded as the matrices in $O(m)$ with a $(m - k) \times (m - k)$-dimensional identity matrix in the upper left corner, a k-dimensional orthogonal matrix in the lower right corner· and all else 0. Our remarks above show, for example,

$$V_{n+1, n} = S^n.$$

For further material in this interesting area of homogeneous spaces, consult [47, 103].

PROBLEMS AND PROJECTS

1. If G is abelian and ρ_X, $\rho_Y: \mathbb{R} \to G$ are one-parameter subgroups, then we know of course that

$$(\rho_X \rho_Y)(t) = \rho_X(t)\rho_Y(t) = \rho_Y(t)\rho_X(t)$$

is a one-parameter subgroup. In this case, show that exp is a homomorphism from g, as an additive abelian group, to G.

2. Let G be a Lie group and ρ_X a one-parameter subgroup, $x \in G$. Then the elements $x\rho_X(t)$ are called a translate, by x, of the one-parameter subgroup. Show that if G is connected, $y \in G$, then there is a finite succession of closed intervals of translates of one-parameter subgroups, lying end-to-end, connecting e to y. (From Theorem 9.3, we know that exp is locally onto.)

3. Combine the methods of Problems 1 and 2 to show that if G is both abelian and connected, any point lies on a one-parameter subgroup. Hence, in this case exp is onto.

4. Show that if G is abelian and connected, G is a product of 1-dimensional subgroups, each of which is either S^1 or $\mathbb{R}$. (If X generates a 1-dimensional subalgebra and if ρ_X is not 1–1, then exp tX fills up a circle in G, say S^1, *and* $G = S^1 \times H$, for dim $H = $ dim $G - 1$.)

5. Use Proposition 9.4 and the map exp to calculate some Lie algebras of well-known subgroups of $Gl(n; \mathbb{R})$. For example, show that the Lie algebra of $O(n)$ is made up of skew symmetric $n \times n$ matrices with diagonal 0, with the product as in Proposition 9.4. Compare with Remark c, number 2, preceding Theorem 9.3.

6. Let G be a compact Lie group and μ the (left) invariant n-form. Prove that if f is continuous

$$\int_G f(x)\mu(x) = \int_G f(x^{-1})\mu(x).$$

(Let $i: G \to G$ be $i(x) = x^{-1}$. Observe that the map of i takes μ to some multiple of μ, and $i \circ i = 1$.)

7. Write Grassman manifolds (Definition 2.4) in the form G/H, where G is a Lie group and H a closed subgroup. Consult [103, 53] for material on this and on flag manifolds.

8. Look up the theory of quaternions [20, 103]. Exhibit the sphere S^3 as the multiplicative group of quaternions of length 1.

As a topological space, SO(3) is homeomorphic to $\mathbb{R}P(3)$, and is covered, in a twofold manner, by S^3; consult [103].

CHAPTER

10

Dynamical Systems

A one-dimensional differential equation, such as $x'' + x = 0$ (where $x'' = d^2x/dt^2$), will have solutions such as $x = A \sin t + B \cos t$. One may give a pair of first-order equations representing virtually the same phenomena, such as

$$dx/dt = u, \qquad du/dt = -x.$$

One pair of solutions would be

$$x = \sin t, \qquad u = \cos t.$$

This pair of functions may in turn be rethought as a function

$$\phi: \mathbb{R} \to S^1$$

given by $\phi(t) = (\sin t, \cos t)$, or in complex notation, $\phi(t) = e^{it}$. ϕ is then a homomorphism from $\mathbb{R}$ to S^1 with kernel being the numbers $2\pi n + (\pi/2)$, for $n \in Z$. Finally, this very same idea can be redescribed as an "action" of the group $\mathbb{R}$ on the space S^1, given by

$$\bar{\phi}: \mathbb{R} \times S^1 \to S^1$$

with $\bar{\phi}(x, t) = \phi(xt)$, this latter being clearly well defined.

Naturally, it is not my idea to look blindly at all possible ways of viewing a given phenomena, but rather to show the variety of viewpoints that are

289

available for treating a given idea. Basically, it is the latter point of view that we wish to emphasize in this chapter. That is, we intend to study a topological group acting as transformations on a topological space, just as $\mathbb{R}$ acted on S^1 in our example. Of course in general we shall be faced with considerable complexity, not generally revealed by the real numbers acting on a circle through rotation.

We shall begin with a discussion of the "abstract" theory of transformation groups. This will give us the opportunity to introduce many important classical concepts, such as orbit, recurrence, periodicity, invariance, and minimal set. Then we shall move towards "dynamics"; that is, we plan to look more closely at a transformation group consisting of the real numbers acting on a Euclidean space. Here, we shall be back to a system of first-order linear differential equations in a variable t whose solutions give the orbits of t acting on the Euclidean space. We take a closer look at the case of the plane, where there is a complete and beautiful theory, especially with respect to limit orbits. Our main goal will be the famous theorem of Poincaré–Bendixson. This will also be a logical place to look at the basic notions of stability.

An interesting question, indirectly connected with the actions of groups, asks when a submanifold is determined by a family of subspaces of the tangent space at each point. The matter is elegantly resolved by the famous theorem of Frobenius (Theorem 10.2). We have already indicated one application of this result in Chapter 9 (Theorem 9.2), and our presentation is totally independent of that earlier material.

We shall continue to assume the basic existence and uniqueness results from ordinary differential equations (good references would include [22, 52]).

To begin, we shall review some earlier concepts involving topological groups acting on spaces (see Chapter 4).

TRANSFORMATION GROUPS; INVARIANT
AND MINIMAL SETS

Definition 10.1   Let G be a topological group and X a topological space. Suppose that there is a continuous map

$$\mu: G \times X \to X$$

such that

(a)   if $g_1, g_2 \in G$, $x \in X$, then

$$\mu(g_1 g_2, x) = \mu(g_1, \mu(g_2, x)).$$

Here we write $g_1 g_2$ for the product of these elements in G.

(b)   For the identity element $e \in G$, and for any $x \in X$,

$$\mu(e, x) = x.$$

Then G is called a (left) *transformation group* acting on X. The map μ is occasionally referred to as the action. The entire structure is written (G, X, μ).

Remarks   (1)   If $G = \mathbb{R}$ and X is some Euclidean space, one may look at a first-order system of differential equations (vector notation)

$$dx/dt = A(x, t)x.$$

If $A(x, t)$ does not depend on t, or more restrictively is a constant, the solutions $x(t)$ define a flow by sending the pair (t, x_0) to the point $x(t)$, where $x(t)$ is the solution with $x(0) = x_0$. When A is constant this solution is $x(t) = e^{At}x_0$. This type of flow is often referred to as a *dynamical system*. (More details are in Proposition 10.3 below.)

(2)   If $X = G$ one has a trivial sort of flow by setting

$$\mu(g, x) = gx.$$

(3)   If G is a Lie group and $H \subseteq G$ a Lie subgroup, then we defined (at the end of Chapter 9)

$$\mu: G \times G/H \to G$$

by $\mu(g, \{k\}) = \{gk\}$. Here, the brackets refer to the equivalence class. It is trivial to verify that this example (of a Lie group and homogeneous space) is a transformation group.

(4)   The group $Gl(n; \mathbb{R})$ acts on $\mathbb{R}^n$, viewed as n-dimensional column vectors, by left matrix multiplication, yielding an important basic example.

(5)   Recall from Chapter 4 that a transformation group is *effective* if e is the only element of G that leaves everything fixed (i.e., e is the only element for which $\mu(e, y) = y$, for all y). It is *transitive* if, given $x, y \in X$, there is $g \in G$ such that $\mu(g, x) = y$. Clearly, the example in (4) immediately above is both effective and transitive, omitting the origin.

If $G_0 = \{g \in G \,|\, \mu(g, x) = x, \text{ for all } x\}$, then G_0 is normal (when $g \in G_0$, $\mu(h^{-1}gh, x) = \mu(h^{-1}, \mu(h, x)) = x$) and G_0 is closed (if $g_n \in G_0$, $g = \lim_n g_n$, then $\mu(g, x) = \lim \mu(g_n, x) = x$). The quotient group G/G_0 acts effectively on X by $\mu(\{g\}, x) = \mu(g, x)$. (Check!)

(6)   Observe that for any fixed $g \in G$ (where G is a transformation group acting on X), $f(x) = \mu(g, x)$ is a homeomorphism of X to itself. To see this, set $g(x) = \mu(g^{-1}, x)$ and note

$$g(f(x)) = g(\mu(g, x)) = \mu(g^{-1}, \mu(g, x)) = \mu(g^{-1}g, x) = x,$$

etc. The meaning of g as a map or an element should be clear from context.

Definition 10.2   Let (G, X, μ) be a transformation group.

(a)   For a given $x \in X$, the *orbit* of x is the set

$$\{\mu(g, x) \,|\, g \in G\}.$$

The *orbit closure* of x simply means the closure of the orbit of x.

(b)   Let $S \subseteq X$ be a subset. The *stability group* of S is the set of $g \in G$, such that, for all $x \in S$, $\mu(g, x) = x$.

If S is a single point x_0, write this as G_{x_0}, as above.

We shall frequently adopt the following shorthand convention. When there is no real danger of confusion, we write gx for $\mu(g, x)$. This will simplify (and frequently clarify) the exposition in what follows.

Remarks   (1)   The structure of orbits and their closures may be very complex. For example [following Remark (1) after Definition 9.6], let H be a one-parameter subgroup of the torus $T = S^1 \times S^1$ whose orbit is dense (H is not a Lie subgroup). H acts on T in the way any subgroup of a topological group acts on the group; i.e., if $h \in H$, $g \in T$, set $\mu(h, g) = hg$, the product of h and g in T. The orbit of e is H itself, while the orbit closure (of e) is all of T.

On the other hand, if H is a closed subgroup and we take the same action, the orbit of e is closed and homeomorphic to a circle. (I suggest some pictures of possible subgroups of T.)

(2)   We recall the material from Proposition 4.1. Let (G, X, μ) be a transitive transformation group. For any $x \in X$, we have a continuous map $\tilde{\rho}: G \to X$ defined by $\tilde{\rho}(g) = \mu(g, x)$, which (since G acts transitively) must be onto. This obviously passes to quotient, yielding a map

$$\rho: G/G_x \to X$$

with $G_x = \{g \in G \,|\, \mu(g, x) = x\}$, which we have called the stability group of x. When G/G_x is compact and X is Hausdorff, this is automatically a homeomorphism.

(3)   Finally, we note an important extreme case of a group acting on a space, namely that of a *free action*. It involves the requirement that for all $x \in X, g \in G, gx \neq x$. (We write gx for $\mu(g, x)$.) A free action is always effective, although not necessarily transitive. All stability groups are trivial. A classic example is the action by multiplication of a subgroup on a larger group, in which case the orbits are cosets. Another important example consists of the group with two elements Z_2 acting on a sphere $S^n = \{X \in \mathbb{R}^{n+1} \,|\, \|X\| = 1\}$. If $g \in Z_2, g \neq e$, set $gX = -X$.

Definition 10.3   Given a transformation group (G, X, μ), a subspace of X, say A, is called *invariant* if whenever $x \in A$, $g \in G$, then $gx \in A$.

If A is a single point, and invariant, it is called a *fixed point*.

Remarks   When A is invariant, the group G may be thought of as acting, as a transformation group, on A. This situation is often referred to as an *induced* action of the transformation group on the invariant subspace A.

Among invariant sets, we have the very important minimal sets, which we now define.

Definition 10.4   Let (G, X, μ) be a transformation group. A set $S \subseteq X$ is called a *minimal set*, or a minimal set of X, if it is nonempty, closed, and invariant, and if it contains no nonempty, closed, invariant, proper subset.

Observe that S is a minimal set precisely when the closure of the orbit of any point of S is S itself. One may thus generate some easy examples. For example, if $\mathbb{R}$ acts as a transformation group on the plane by $\mu(t, (x, y)) = (x, t + y)$, all orbits are minimal sets. If G is a topological group acting on itself by left translation, the only minimal set is G itself. More interesting examples are easily generated by the techniques which we develop in this chapter. For now, we handle the question of existence.

Proposition 10.1   Let (G, X, μ) be a transformation group, with X a compact Hausdorff space (nonempty). Then X has a minimal set.

Proof   Let $\mathscr{C}$ be the collection of compact, nonempty, invariant subsets. $\mathscr{C}$ is clearly nonempty, because at least X itself belongs to $\mathscr{C}$.

If S_1 and S_2 belong to $\mathscr{C}$, so does $S_1 \cap S_2$. This means that if $\{S_\alpha\}$ is a linearly ordered subset of $\mathscr{C}$ by inclusion, i.e., $S_2 < S_1$ precisely when $S_1 \subseteq S_2$, then the intersection of all the S_α, which is nonempty by compactness, must also be an element of $\mathscr{C}$.

Now, using this order, $\mathscr{C}$ is visibly a partially ordered set. A linearly ordered subset $\{S_\alpha\}$ has an upper bound (or supremum), namely $\bigcap_\alpha S_\alpha$. Zorn's lemma assures us that there is a maximal element with respect to this order. Such a maximal element $T \in \mathscr{C}$ has the property that there is no $U \in \mathscr{C}$, $T \lneq U$, i.e., $U \subsetneq T$. But then T is clearly a minimal set.

The theory of minimal sets is connected with the theory of almost periodic points under a transformation group. These are a natural generalization of the basic concepts of fixed point and periodic points. The intuitive background comes from functions of a real variable. A periodic function is one like $f(x) = \sin x$, or even

$$f(x) = A \sin \alpha \pi x + B \sin \beta \pi x,$$

with α and β rational. The behavior of the function repeats itself when x is increased by a certain amount (easily determined from the denominators of α and β). On the other hand, if α is rational and β is irrational, $f(x)$ is not quite periodic; this is then a basic example behind the theory of almost periodic functions (see [9, 10]). Before proceeding to our specific definitions, notice that these examples of functions may easily recast in the terminology of transformation groups. Let $B(-\infty, \infty)$ be the vector space of bounded real functions $f: \mathbb{R} \to \mathbb{R}$ with the metric

$$d(f, g) = \text{l.u.b.} |f(x) - g(x)|.$$

$\mathbb{R}$ acts as a transformation group on this metric space by $\mu(g, f) = f_g$, with

$$f_g(x) = (x + g).$$

In this context, a periodic point (almost periodic point) is a periodic function (almost periodic function). Here are the definitions.

Definition 10.5   Let (G, X, μ) be a transformation group

(a)   If $x \in X$, recall that x is a *fixed point* if $\mu(g, x) = x$, for all $g \in G$.

(b)   If $x \in X$, call x a *periodic point* when there is $g \in G$, $g \neq e$, with $\mu(g, x) = x$. The term is most frequently used when $G = \mathbb{R}$, in which case the smallest positive such g would be called the *period*.

The generalization is now easy.

Definition 10.6   Let (G, X, μ) be a transformation group. $x \in X$ is an *almost periodic point* if given an open neighborhood of x, say U, and writing

$$A = \{g \in G \mid gx \in U\},$$

then there is a compact $K \subseteq G$ such that

$$KA = G.$$

We note that gx is short for $\mu(g, x)$, while $KA = \{ka \mid k \in K, a \in A\}$.

Remarks   (1)   Suppose we are given a transformation group (G, X, μ), X being compact and $x \in X$. Then given an open neighborhood U of x, the set

$$\{g \mid gx \in U\}$$

is the same as the inverse image of U under the map f_x defined by

$$f_x(g) = \mu(g, x).$$

If we write $g_1 U = \{g_1 u \mid u \in U\}$, then whenever our action is transitive the set of all $\{g_i U\}$ cover X. By compactness, a finite number of $g_i U$ will cover X, say $g_1 U, \ldots, g_k U$. But then a finite number of the open sets

$$f_x^{-1}(g_1 U), \ldots, f_x^{-1}(g_k U)$$

cover G. Since

$$f_x^{-1}(g_1 U) = \{g \mid gx \in g_1 U\} = \{g \in G \mid g_1^{-1} gx \in U\}$$
$$= \{g_1 g \mid gx \in U\} = g_1(f_x^{-1}(U),$$

we see that there is a finite set of $g_1, \ldots, g_k$ such that

$$g_1 f_x^{-1}(U) \cup \cdots \cup g_k f_x^{-1}(U) = G.$$

Putting $K = \{g_1, \ldots, g_k\}$, we see that any $x \in X$ (G acting transitively) is almost periodic when X is compact.

If G is not transitive, one may study the same thing in the orbit closure of x (see Proposition 10.2 below).

(2)   Given $B = B(-\infty, \infty)$ defined before Definition 10.5, let $f \in B$ be periodic; i.e., there is $\omega \neq 0$ such that $f(x + \omega) = f(x)$ for all x; then

$$\{g \in \mathbb{R} \mid gf = f_g \in U \quad \text{with} \quad f_g(x) = f(g + x)\}$$

is clearly an open set containing all integral multiples of ω. If one chooses $k = [0, \omega]$, f is obviously almost periodic. With little more difficulty, one checks that functions such as

$$A \sin \alpha \pi x + B \sin \beta \pi x$$

are also almost periodic in the sense of Definition 10.6 above.

Proposition 10.2   Given a transformation group (G, X, μ), where X is locally compact, then a necessary and sufficient condition for x to be an almost periodic point is that the orbit closure of x, $\overline{Gx}$, be a compact minimal set.

Proof   For the necessity, let x be the almost periodic point. Let U be a neighborhood of x whose closure is compact. We may assume that there is a compact $K \subseteq G$ such that

$$K\{g \mid gx \in U\} = G.$$

But then the orbit of x is clearly contained in

$$\{kax \mid ax \in \overline{U}, \quad k \in K\} = \{ky \mid y \in \{ax\} \subseteq \overline{U}, \quad k \in K\}.$$

Since the continuous image under the map μ of the compact set $K \times \bar{U}$ is compact, the orbit of x belongs to a compact set. Therefore, its closure is compact. It is easy to see that if $y \in \{gx \,|\, g \in G\}$, then

$$\overline{Gx} = \overline{Gy}.$$

This shows us at once that the orbit closure in question is minimal.

For sufficiency, assume that for a given $x \in X$ the set

$$\overline{Gx}$$

is both compact and minimal. Choose an open neighborhood of x, say U, which we may even take with compact closure. It is clear that

$$\overline{Gx} \subseteq GU,$$

for if there were points in $\overline{Gx}$ that did not lie in GU they would make up a *closed*, invariant, proper subset of the minimal set $\overline{Gx}$.

Now, by compactness, we may find a finite set $F \subseteq G$ such that

$$\overline{Gx} \subseteq FU.$$

Of course, if $x \in X$ and $g \in G$, we have

$$gx = fu$$

for some $f \in F$, $u \in U$. But this means that

$$f^{-1}gx \in U \qquad \text{or} \qquad f^{-1}G \subseteq \{h \,|\, hx \in U\}.$$

Therefore,

$$G = F\{h \,|\, hx \in U\}.$$

Since F is compact, we have met the conditions for x to be an almost periodic point.

Corollary 10.1 If (G, X, μ) is a transformation group with X compact, then X has an almost periodic point.

Proof By Proposition 10.1, X has a minimal set, so X clearly has a minimal (invariant) orbit closure. Then Corollary 10.1 follows at once from Proposition 10.2.

Corollary 10.2 If X is locally compact and $x \in X$ is an almost periodic point for a transformation group (G, X, μ), then x is an almost periodic point for any topology on G that makes (G, X, μ) a transformation group.

Proof   The proof of Proposition 10.2 is quite independent of the topology on G. The orbit closure

$$\overline{Gx} = \overline{\{\mu(g, x) \mid g \in G\}}$$

does not depend on the topology on G, but just on the map μ.

LINEAR DIFFERENTIAL EQUATIONS IN EUCLIDEAN SPACE

We now propose to specialize our discussion and discuss some cases where the space is $\mathbb{R}^n$. We wish to emphasize that the origins of many transformation groups lie in the theory of systems of differential equations. Before considering specific examples, we note that if $X(t)$ is an n-dimensional column vector with entires $x_i(t)$, $i \le i \le n$, then if $\phi_t(X_0)$ is the solution to

$$dX(t)/dt = f(X(t))$$

with f a given C^1-function, with the property that $\phi_0(X_0) = X_0$, then where defined, we have $\phi_{s+t}(X) = \phi_s(\phi_t(X))$. For the proof, as in our construction of one-parameter subgroups in Chapter 9, we note that for any fixed t

$$(\phi_{s+t}(X))_{s=0} = (\phi_s(\phi_t(X)))_{s=0}$$

by trivial calculation. The desired equation follows from uniqueness of solutions of the system. In many cases, such as that where $f(X) = AX$ for some fixed $n \times n$ matrix A, the solutions are (as we shall now see) well defined for all values of t.

Proposition 10.3   Set

$$X(t) = \begin{pmatrix} x_1(t) \\ \vdots \\ x_n(t) \end{pmatrix}$$

and let A be an $n \times n$ matrix of real numbers. We note

$$e^A = \sum_{n=0}^{\infty} \frac{1}{n!} A^n$$

[compare Definition 9.6 and Remark (3) following].

Then for any fixed vector $V \in \mathbb{R}^n$ there is one and only solution to the system of linear differential equations

$$X'(t) = AX(t), \quad t \geqq 0,$$

that satisfies $X(0) = V$. This solution is

$$X(t) = e^{tA}V.$$

(We note that $X'(t)$ is the column vector with entries $x_i'(t)$.)

Proof   As in Remark (3) following Definition 9.6, we may immediately calculate

$$\frac{d}{dt}(e^{tA}V) = Ae^{tA}V.$$

Hence $e^{tA}V$ satisfies the given equation, and clearly when $t = 0$ we get V. To give an elementary proof of uniqueness, take another solution $\bar{X}(t)$ and calculate

$$\frac{d}{dt}(e^{-tA}\bar{X}(t)) = -Ae^{-tA}\bar{X}(t) + e^{-tA}A\bar{X}(t) = 0.$$

This implies that

$$e^{-tA}\bar{X}(t) = W,$$

for some fixed constant vector W. If $\bar{X}(0) = V$, set $t = 0$ and recover $W = V$, showing that

$$\bar{X}(t) = e^{tA}V,$$

as desired.

Remarks   (1)   In this example, we have

$$e^{(t_1 + t_2)A} = e^{t_1 A}e^{t_2 A},$$

so that the map

$$(t, V) \rightarrow e^{tA}V$$

trivially defines a transformation group with $X = \mathbb{R}^n$ and $G = \mathbb{R}$. Such an action is usually referred to as a *flow* on $\mathbb{R}^n$.

(2)   There are two extreme cases of this kind of system of equations, and thus two extreme cases of flows on $\mathbb{R}^n$, that are noteworthy. First, if every eigenvalue of A has negative real part, then $\mathbb{O} \in \mathbb{R}^n$, the origin, is called a *sink*, and the flow is called a *contraction*. Secondly, if every eigenvalue of A has positive real part, then $\mathbb{O}$ is called a *source* and the flow is referred to as an *expansion*.

To get some insight into these cases, suppose $\mathbb{O}$ is a sink. Then an easy calculation shows that there are positive constants C and a such that

$$\|e^{tA}X\| \le Ce^{-ta}\|X\|,$$

for all $t \geq 0$, $X \in \mathbb{R}^n$. The proof comes from choosing $-a$ be a negative number larger than the real parts of all the eigenvalues of A and then looking at the Jordan canonical form for A. The inequality actually characterizes a sink.

Reversing the inequality and changing the sign, one easily characterizes a source.

The geometric significance of these two extreme cases is either a family of orbits all pointing in towards $\mathbb{O}$, in the case of a sink, or pointing away from $\mathbb{O}$ in the case of a source.

We shall leave the details to the reader here.

We analyze in more detail the interesting in-between case, where some eigenvalues have negative and some have positive real part. In such a case we have a *hyperbolic flow*.

Proposition 10.4 Consider an $n \times n$ matrix, and the corresponding system of linear, first-order differential equations

$$X'(t) = AX(t)$$

in $\mathbb{R}^n$. Suppose that every eigenvalue of A has nonzero real part, some positive and others negative.

Then we have a vector space decomposition

$$\mathbb{R}^n = \mathbb{R}^s \oplus \mathbb{R}^t,$$

where $s + t = n$, so that s is the number of eigenvalues with negative real part. Furthermore, the flow arising from the system of equations, specifically,

$$(t, V) \to e^{At}V,$$

leaves both $\mathbb{R}^s$ and $\mathbb{R}^t$ invariant; that is, orbits beginning in one of these subspaces lie entirely in that subspace. The flow restricted to $\mathbb{R}^s$ has $\mathbb{O}$ as a sink, while the flow restricted to $\mathbb{R}^t$ has $\mathbb{O}$ as a source.

The decomposition of $\mathbb{R}^n$ is unique.

Proof We shall use some well-known results from linear algebra involving $\phi_A(x)$, the minimal polynomial of A, and the rational cannonical form. Any polynomial defined over the real numbers factors into irreducible polynomials of degree 1 or 2, so we may write

$$\phi_A(x) = (\phi_1(x))^{p_1} \cdots (\phi_k(x))^{p_k},$$

where the ϕ_i are relatively prime and of degree 2 or less, and the p_i are the multiplicities of the irreducible factors. Note that

$$\deg(\phi) = \sum p_i \deg(\phi_i).$$

$\mathbb{R}^n$ is then the direct sum of invariant subspaces V_i on each of which A acts with minimal polynomial $(\phi_i(x))^{p_i}$. We rearrange the basis such that the ϕ_i corresponding to roots with negative real parts come first, while those with positive real parts come second. This gives the desired decomposition.

A is in this basis the sum of two commuting matrices

$$\left(\begin{array}{c|c} A|V_1 & 0 \\ \hline 0 & 0 \end{array}\right) \quad \text{and} \quad \left(\begin{array}{c|c} 0 & 0 \\ \hline 0 & A|V_2 \end{array}\right).$$

Therefore, e^A breaks up into a product of two matrices each of which acts invariantly on the two subspaces. Set $\dim V_1 = s$, $\dim V_2 = t$.

The remarks following Proposition 10.3 show that $\mathbb{O}$ is a sink (or source) on the appropriate subspace.

Observe that the source and sink behavior are not compatible; i.e., a point on a given orbit cannot be both attracted towards and repelled away from $\mathbb{O}$. This quickly leads to the desired uniqueness of the decomposition.

Remarks (1)   A flow of this form is called *linear hyperbolic*, linear because A is a constant matrix, and hyperbolic because it contains two separate regions (subspaces) over which one has contraction or expansion. (Draw a picture in two dimensions.)

(2)   Sinks and sources, which we have just analyzed in the case of a linear, first-order system (A constant), are special cases of *equilibria*. Given a C^1-flow or transformation on $\mathbb{R}^n$, or an open subset of $\mathbb{R}^n$, say

$$\mu : \mathbb{R} \times \mathbb{R}^n \to \mathbb{R}^n,$$

then $X \in \mathbb{R}^n$ defines an *orbit* $\{\mu(t, X)\}$, which we usually abreviate $\mathbb{R}X$. One generates a vector field by assigning to a point X, the derivative of the flow, with respect to t, at that point. In other words, writing $\phi_t(X) = \mu(t, X)$, one assigns the vector

$$f(X) = d\,\phi_t(X)/dt|_{t=0}.$$

Of course, by construction our flow here is given by solutions to

$$dX(t)/dt = f(X).$$

Again, we use $X(t)$ to denote the column vector with entries $x_i(t)$, $1 \le i \le n$.

For a given vector X_0, if $f(X_0) = 0$, then the solution or orbit through X_0 must be constant (by the usual uniqueness for solutions to systems of differential equations). We may thus define an *equilibrium point* in terms of the vanishing of the associated vector field.

(3)   An equilibrium point (or just equilibrium) is of great importance when it is stable. There are various notions of stability, but the basic ideas (due to Liapunov) are as follows, if we are given a C^1-vector field defining a flow on $\mathbb{R}$ (or an open subset):

(a) An equilibrium X_0 is *stable* if whenever $X_0 \in U$, U open, there is an open V with

$$X_0 \in V \subseteq U$$

such that the orbit of any point in V, for $t > 0$, lies entirely within U.

(b) An equilibrium X_0 is *asymptotically stable* if it is stable and for some V as in (a) above every orbit originating in V approaches X_0 as $t \to \infty$.

For example, in $\mathbb{R}^2$, define $\mu(t, X)$ to be the rotation of X by an angle t about the origin $\mathbb{O}$. Then the origin is a stable equilibrium that is *not* asymptotically stable.

But a linear sink, as in Remark (2) preceding Proposition 10.4, where all orbits move in towards a fixed point, is clearly an example of an asymptotically stable equilibrium.

(4) We should mention another very important idea in the general area of stability, that of being *generic*. The concept of being generic often refers to a property, but equally well refers to a set of differential equations (or transformation groups) or to a set of matrices A (which are turned into differential equations as $dX/dt = AX$). In the interest of simplicity, we will explain what is meant by saying that a set of matrices is generic here.

A set of matrices (or linear transformations) is called *generic* if it contains an open, dense subset of the set of all such linear transformations, this being topologized in the case of $n \times n$ square matrices by considering it as an n^2-dimensional real vector space.

For example, the set of all such matrices with distinct eigenvalues is generic. It is obviously open, and one easily sees that the complementary set of matrices with at least two eigenvalues the same is a closed set of lower dimension. Hence, it is also dense. Alternatively, the rational canonical form is

$$
\begin{pmatrix}
\lambda_1 & & & & & & 0 \\
& \ddots & & & & & \\
& & \lambda_k & & & & \\
& & & D_1 & & & \\
& & & & \ddots & & \\
0 & & & & & D_{n-1} &
\end{pmatrix},
$$

where the λ_i are numbers and the D_i are 2×2 matrices

$$
\begin{pmatrix}
0 & 1 \\
c & d
\end{pmatrix}.
$$

Clearly, arbitrarily small changes in the c and d will assure that the eigenvalues are distinct. (For the rational cannonical form, consult [56, 60].)

Without much difficulty, one may show that the set of all matrices with nonzero real parts is generic.

Looking at Proposition 10.4 as well as the remarks that preceed it, the reader should sketch what it means, by some examples, for a set of systems of equations

$$dX/dt = AX$$

to be generic when A varies through a generic family of matrices (as in the two above examples).

We note that these concepts are *not* at all restricted to systems of first-order linear equations, but rather that we have tried to illustrate them in a straightforward case.

(5)   Lastly, we would like to point out that (as the reader probably knows) a single higher-order equation may be rewritten as a system of first-order equations, by a classical trick. Consider

$$\frac{d^n y}{dx^n} + a_{n-1} \frac{d^{n-1} y}{dx^{n-1}} + \cdots + a_0 y = 0.$$

We rewrite

$$y_1 = y;$$

$$\frac{dy_1}{dx} = y_2,$$

$$\frac{dy_2}{dx} = y_3,$$

$$\vdots$$

$$\frac{dy_n}{dx} = -a_{n-1} y_n - \cdots - a_0 y_1.$$

This is vaguely similar in formal structure to a block in the rational canonical form. It offers a way of carrying over much of what we have said to properties of higher order differential equations.

PLANAR FLOWS: THE POINCARÉ–BENDIXSON THEOREM

We now wish to specialize all this to dynamical systems in the plane. We consider a C^2-flow in the plane, that is, a transformation group

$$\mu: \mathbb{R} \times \mathbb{R}^2 \to \mathbb{R}^2$$

that is C^2, arising as solutions to a system of first-order linear differential equations $dX/dt = f(X)$, where f is a C^1- (vector) function. We frequently write $\phi_t(X) = \mu(t, X)$; for fixed X, the set $\{\phi_t(X)\}$ is the orbit of X.

We shall find the transformation group viewpoint more interesting in this context. We begin with some basic definitions, all of which are in fact valid when we replace $\mathbb{R}^2$ by any open subset of $\mathbb{R}^n$.

Definition 10.7   Given a dynamical system described by

$$\mu: \mathbb{R} \times \mathbb{R}^2 \to \mathbb{R}^2,$$

where we write $\mu(t, X) = \phi_t(X)$, a point X is a $+$ *limit point* of $Y \in \mathbb{R}^2$ if there is an increasing, unbounded sequence of real numbers t_n such that

$$\lim_{n \to \infty} \phi_{t_n}(Y) = X.$$

Similarly, X is a $-$ *limit point* of Y if the same condition holds for an unbounded decreasing sequence of real numbers.

The set of $+$ limit points of Y (resp. $-$ limit points of Y) is denoted $L_+(Y)$ (resp. $L_-(Y)$).

We set

$$L(Y) = L_+(Y) \cup L_-(Y);$$

this is the set of *limit points of Y*.

Remarks and Examples   *1.* $+$ limit points are sometimes called ω limit points; $-$ limit points are sometimes α limit points.

2. A sink is clearly a $+$ limit point; a source is clearly a $-$ limit point. [See Remark (2) after Proposition 10.3.]

3. Let X be a point in $\mathbb{R}^2$ regarded as the complex numbers and set

$$\phi_t(X) = e^{2\pi i t} X.$$

Then any point at distance $\rho > 0$ from the origin is both a $+$ limit point and a $-$ limit point of any other point at distance ρ from the origin.

For example, if $X = (0, 1)$ and $Y = (1, 0)$

$$X = \phi_{1/4}(Y) = \phi_{1/4 + 1}(Y) = \cdots.$$

4. There is no particular reason why an arbitrary flow should have limit points. For example, we can set

$$\phi_t(X) = X + (t, t),$$

so that the flow consists of translations parallel to the vector $(1, 1)$ and orbits are simply straight lines (unbounded).

The reader should sketch various possibilities in the plane.

There are many easy properties of limit sets, whose verifications are routine. For example,

(i)   If X and Y belong to the same orbit, i.e., they both belong to $\{\phi_t(Z)\}$ for some fixed Z, then

$$L_+(X) = L_+(Y) \qquad \text{and} \qquad L_-(X) = L_-(Y).$$

(ii)   Suppose K is a closed invariant set, in the sense that for each $a \in K$ all $\phi_t(a) \in K$. If $X \in K$, then $L_+(X)$ and $L_-(X)$ both belong to K, etc.

Proposition 10.5   Let ϕ_t be a C^2-flow in the plane $\mathbb{R}^2$ and let $X \in \mathbb{R}^2$ be *not* an equilibrium point (i.e., the tangent vector to the orbit through X is not zero at X).
Then there is a neighborhood (open) U of X and a diffeomorphism

$$\psi : U \rightarrow (-\alpha, \alpha) \times (-\beta, \beta),$$

where $(-\alpha, \alpha) = \{x \in \mathbb{R} \mid -\alpha < x < \alpha\}$, etc., $\alpha > 0$, $\beta > 0$, such that

(a) $\psi(X) = \mathbb{O} = (0, 0)$, and
(b)   if $Y \in U$, then ψ maps the orbit in U through Y to a line segment $\{(t, y_0)\}$, for some $y_0 \in (-\beta, \beta)$ in the range of ψ. In particular, $\phi_t(X) \cap U$ is mapped to $\{(t, 0)\}$, with $t \in (-\alpha, \alpha)$.

Proof   Let l be the line perpendicular to the orbit $\{\phi_t(X)\}$ through X at the point X (i.e., $t = 0$). Choose a positive direction along the orbit to be that of the tangent vector

$$\lim_{t \to 0}([\phi_t(X) - \phi_0(X)]/t) \neq 0.$$

Of course, near X every tangent vector is nonzero and l crosses every orbit transversally (though not necessarily orthogonally).

We may thus map an open neighborhood of X onto an open neighborhood of $(0, 0)$ in $\mathbb{R}^2$ by sending Y to that pair of points whose first coordinate is the directed distance from l to Y along the orbit through Y and whose second coordinate is the distance from $\mathbb{O}$ to the point on l, where the orbit through Y meets l. The idea is embodied in Fig. 10.1.

Call this map ψ. Observe that ψ is clearly differentiable (this follows very easily with our initial assumptions of differentiability). Note also that $d\psi$ is onto at X (and hence near X). We may then use the inverse function theorem (Theorem 1.1) to get the desired diffeomorphism in a possibly smaller neighborhood of X.

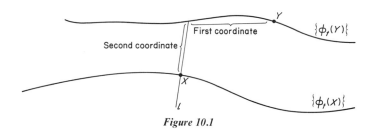

Figure 10.1

Remarks The neighborhood U endowed with the diffeomorphism ψ is often called a *flow box*. A line segment l, with X in its interior, that is transverse to all the orbits which cross it is usually called a *section* or a *local section*. We shall assume, in particular, that the tangent vector to an orbit at a point where it crosses a section is nonzero and does not lie in the section (as in the proof of Proposition 10.5).

The reader should have no difficulty in formulating the generalizations to flows on $\mathbb{R}^n$ here.

The next two propositions are basically lemmas to the Poincaré–Bendixson theorem (our Theorem 10.1).

Proposition 10.6 Let ϕ_t be a C^2-planar flow, and let Y_0, Y_1, and Y_2 be three points on the same orbit $\{\phi_t(X)\}$ for some given X. We assume that these three points are in order, in that they either represent increasing or decreasing values of t on the orbit. Suppose that l is a local section and that these three points lie on l.

Then Y_0, Y_1, and Y_2 are monotone on l in that Y_1 lies between Y_0 and Y_2.

Proof We consider the diagram (Fig. 10.2) for the orbit $\{\phi_t(X)\}$, with the arrows indicating the direction in which t is increasing.

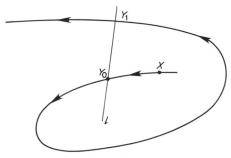

Figure 10.2

We claim that this picture represents the general case, for if the orbit crosses to the left at Y_0, it cannot cross to the right at Y_1, because the set of points where the orbit crosses a section in a given direction is visibly open, and the closed interval form Y_0 to Y_1 along l is a connected set (and so cannot be a disjoint union of two relatively open sets).

Now, let L denote the region bounded by the line l from Y_0 to Y_1, and the part of the orbit from Y_0 to Y_1. Let K be the complement or outside of L. It is trivial to see that K is invariant under all ϕ_t, $t > 0$. In particular, $Y_2 \in K$.

We have therefore seen that Y_2 cannot lie in the part of l that is below Y_0. It surely cannot lie on the part of l from Y_0 to Y_1, for all orbits pass out from the region L to the region K along this segment.

Since Y_2 is assumed to lie on l, we conclude that Y_2 lies above Y_1, completing the proof.

Note that we have implicitly used the Jordan curve theorem, which states that the homeomorphic copy of a circle divides the plane into two complementary regions. In fact, this sort of case, where the homeomorphic image of the circle is the union of two smooth curves, is rather simple to prove. For the general result, consult [85, 25].

Proposition 10.7 Let ϕ_t be a C^2-flow in the plane. Let X belong to either $L_+(Y)$ or $L_-(Y)$ for some Y. Then the orbit of X crosses any local section l in at most one point.

Proof Suppose $X \in L_+(Y)$, the other case being entirely similar. Consider l, and suppose—to be definite—that when an orbit crosses l it does so from left to right. We assume that the orbit $\{\phi_t(X)\}$ crosses l twice and use arrows to indicate the direction (increasing t) at a point, so that the ideas are faithfully represented by Fig. 10.3. We shall derive a contradiction from the existence of the two points Y_1 and Y_2.

Since $X \in L_+(Y)$, the orbit of Y (not shown in the picture) must come arbitrarily close to X as t runs through some increasing unbounded sequence. We choose disjoint flow boxes, say V_1 and V_2, around the respective

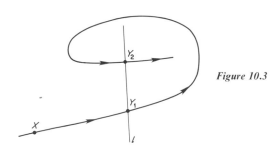

Figure 10.3

points Y_1 and Y_2, by Proposition 10.5. Because X is a limit point, and the orbits (or solutions) depend continuously on the initial data, we see that the orbit of X must enter both V_1 and V_2 infinitely many times (check!).

Thus, there is an increasing sequence t_i, unbounded, for which the points $\phi_t(X)$ alternatively lie in V_1 or V_2. But such a sequence directly contradicts the order property of points where an orbit meets a section, as shown in Proposition 10.6.

We are now ready to look at the famous theorem of Poincaré and Bendixson on the orbits of planar flows.

Theorem 10.1 Let ϕ_t define a flow or transformation group on the plane $\mathbb{R}^2$; that is, there is a C^1-vector field $f(X)$ and the ϕ_t are solutions to $dX/dt = f(X)$.

Then a nonempty compact limit set, i.e., either $L_+(X)$ or $L_-(X)$, that contains no equilibrium point is a closed orbit (that is, an orbit, not an equilibrium, for which there is a Y in the orbit and a $t \neq 0$ such that $\phi_t(Y) = Y$).

Proof Consider the case of $L_+(X)$, the other being entirely similar. $L_+(X)$ is compact. Suppose $Y \in L_+(X)$. We wish to show first that the orbit of Y is closed.

Of course $L_+(X)$ is also invariant under ϕ_t, $t > 0$. $L_+(Y)$ is clearly a non-empty subset of $L_+(X)$. For any $Z \in L_+(Y)$, we may invoke Proposition 10.5 to construct a flow box, i.e., a diffeomorphism

$$\psi : U \rightarrow (-\alpha, \alpha) \times (-\beta, \beta)$$

with a section l, through Z, that crosses the orbit of Z transversally at Z. However, Proposition 10.7 tells us that the orbit of Y crosses l at most at one point. Since the orbit of Y contains Z as a $+$ limit point, we conclude that the orbit of Y enters U infinitely many times, and therefore that there must be an increasing unbounded sequence t_n with

$$\phi_{t_n}(Y) = Z$$

(since any orbit which enters U must cross l, by Proposition 10.5).

Of course, we recall that there are no equilibrium points in the flow box, and when an orbit enters it must pass out before it can enter to cross l again (compare Proposition 10.5). Taking two adjacent values of t_n, we get

$$\phi_{t_{i+1}}(Y) = \phi_{t_i}(Y)$$

with $t_{i+1} - t_i > 0$. In other words, the orbit of Y is a closed orbit, since $\phi_{t_{i+1} - t_i}(Y) = Y$.

We contend, however, that this closed orbit must be all of $L_+(X)$. Call the closed orbit of Y, say Ω, and take $V \in \Omega$. Let U once again be a flow box, and section l at V so that Ω meets l at V alone. Take an increasing sequence $\bar{t}_i$, such that

$$\phi_{\bar{t}_i}(X) \in l, \qquad \phi_{\bar{t}_i}(X) \to V, \qquad \text{but} \qquad \phi_t(X) \notin l,$$

for all $t, \bar{t}_{i-1} < t < \bar{t}_i, i \geq 1$.

In general X is not a limit point, so $\phi_{\bar{t}_i}(X)$ may meet l at more than one point. But by Proposition 10.6, the $\phi_{\bar{t}_i}(X)$ converge to V monotonically in l.

The set of numbers $\bar{t}_{i+1} - \bar{t}_i$ is clearly bounded above, since there is some τ with $\phi_\tau(V) = V$ (Ω is a closed orbit), and for points nearby and converging to V, the minimum τ necessary to get back to the section l must be near τ.

On the other hand, from the fundamental theory of ordinary differential equations of our type $dX/dt = f(X)$ over a bounded interval $k_1 \leq t \leq k_2$,

$$\left| \phi_t(\phi_{\bar{t}_i}(X)) - \phi_t(V) \right|$$

converges uniformily to 0. In other terms, the limit set of X, i.e., $L_+(X)$, must lie arbitrarily close to the closed orbit Ω, which contains all $\phi_t(V)$. Hence, $L_+(X) = \Omega$, completing the proof.

Remarks (1) For the basic continuity of solutions to first-order systems of our type, consult [52, Chap. 8] or (for the more general theory) [22, Chap. 1].

(2) Theorem 10.1 works equally well in an open subset of the plane.

(3) A nonempty closed orbit contained in $L_+(X)$ (or $L_-(X)$) for some X is often called a *limit cycle*. One consequence of the Theorem 10.1 is that a nonempty compact set that contains no-limit cycle in its interior but whose boundary is a limit cycle must contain an equilibrium.

For further comments and discussion, one may consult [22, 52, 72].

FAMILIES OF SUBSPACES;
THE FROBENIUS THEOREM

As a final topic in this area, we wish to consider a higher-dimensional generalization of the process of integrating, or solving a system of first-order differential equations in $\mathbb{R}^n$, say

$$dX/dt = f(X),$$

where $f(X)$ is a C^1-vector function. A solution is a family of curves $\{\phi_t(X)\}$ filling $\mathbb{R}^n$ without intersections of different curves, such that at each point

X the vector

$$d\phi_t(X)/dt\big|_{t=0}$$

is precisely the vector $f(X)$.

This problem has an obvious higher-dimensional generalization. Suppose that $f(X)$ associates to each point X an i-dimensional linear subspace of $\mathbb{R}^n$ in a continuous (or differentiable) way. Locally, such a function may be described by i nonvanishing, mutually independent vector fields in $\mathbb{R}^n$. The continuity (or differentiability) of f may be formulated in terms of that of the vector fields. The generalized question is then the following:

Given such a function f and a point $X_0 \in \mathbb{R}^n$, is there an open i-dimensional submanifold $M^i \subseteq \mathbb{R}^n$ such that

(i) $X_0 \in M^i$, and
(ii) for each $X \in M^i$, $f(X)$ is (parallel to) the tangent space of the submanifold M^i at the point X?

Of course in the previous case, $i = 1$ and the submanifold is a curve.

One might be tempted to think that this problem always has an affirmative solution, at least if the collection of i-dimensional subspaces $f(X)$ depends on X in a sufficiently smooth way. Unfortunately, this is not the case. Consider a neighborhood of the identity in the Lie group SO(3). The Lie algebra of SO(3) [by Remark (4) after Definition 9.5, or more generally Problem 5 at the end of Chapter 9] is spanned by the usual basis vectors i, j, k, where the bracket product is the cross product. Let V_i and V_j denote the invariant tangent vector fields on SO(3), which are i and j in the Lie algebra. One might even choose any two independent vector fields here.

For each $x \in$ SO(3), let $f(x)$ be the subspace of $(T SO(3))_x$ spanned by V_i and V_j at x. Suppose there is a submanifold M, $e \in M \subseteq$ SO(3), dim $M = 2$, with V_i and V_j tangent to M at every point of M, say, in any open neighborhood of e. That is, $f(x)$ is tangent to M at x. Then by Proposition 9.2 $[V_i, V_j]$ must be a vector field in the tangent space of M at every point of this neighborhood. Since V_i and V_j span the tangent space to M (they are independent), we have

$$[V_i, V_j] = aV_i + bV_j,$$

or finally

$$V_k = aV_i + bV_j.$$

This is impossible, because V_i, V_j, and V_k span the tangent space to SO(3) at any point. Therefore, there is no such manifold M.

Now, the reader could object that our original problem was posed for i-dimensional subspaces of $\mathbb{R}^n$, not subspaces of the tangent space of a Lie group such as $SO(3)$. But since a neighborhood of e in $SO(3)$ is diffeomorphic to $\mathbb{R}^3$ (for example, using $\exp^{-1}$), we clearly have such an example in $\mathbb{R}^3$ as well. It turns out that the bracket product of vector fields (Proposition 9.2) plays a key role in determining whether the question at hand will have an affirmative solution. In the case of ordinary differential equations, $i = 1$, the difficulty does not arise, because the bracket product always vanishes in a 1-dimensional subspace of tangent vectors, such as multiples of a fixed vector field $f(x)$, for the problem $dX/dt = f(X)$.

We now state the theorem of Frobenius, which characterizes when such manifolds exist. We note that, by analogy with ordinary differential equations, many authors refer to such submanifolds as *integral manifolds*.

Theorem 10.2   Let us suppose that f associates with each point $x \in M^n$, with M^n a smooth manifold, an i-dimensional subspace of $(TM^n)_x$, for some fixed i, $1 \leq i < n$. Suppose that locally, $f(x)$ is spanned by i smooth vector fields $X_1, \ldots, X_i$.

Then every point x has a coordinate neighborhood $\phi: U \to \mathbb{R}^n$ such that the submanifold

$$\phi^{-1}(x_1, \ldots, x_i, c_{i+1}, \ldots, c_n),$$

with the c_j's constant, has at each point $\bar{x}$, for its tangent space, the subspace $f(\bar{x})$ if and only if $[X_p, X_q]$ lies in the subspace spanned by $X_1, \ldots, X_i$, for all $1 \leq p, q \leq i$.

Proof.   If we assume that the desired coordinate neighborhoods exist, then through any point there is a submanifold whose tangent space near that point if given by f. Referring back to Proposition 9.2, the bracket product of two tangent vector fields to a manifold (in this case the submanifold) is again a tangent vector field. Thus, each $[X_p, X_q]$ is tangent to the submanifold, i.e., lies in the subspace spanned by $X_1, \ldots, X_i$, which are a basis for the tangent space to the submanifold.

The other way around, assume the condition on $X_1, \ldots, X_i$. This will involve more work, and we will proceed by induction. If $i = 1$, then locally we are trying to find a neighborhood where the linear subsets with the last $n - 1$ coordinates constant, are tangent to the vector field X_1. Using x for a point in a coordinate chart, we use the basic theorems from ordinary differential equations to construct a flow $\phi_t(x)$ with

$$d\phi_t(x)/dt \big|_{t=0} = X_1(x),$$

as we have already done several times before. Then Proposition 10.5 on the existence of flow boxes (used in n dimensions) gives the desired coordinate chart at once.

Assume now the truth of the theorem for any $i - 1$ independent vector fields satisfying the condition on bracket products. The trick now will consist of introducing new coordinates and vector fields so that the vector fields span the space $f(x)$ for each x in question, the first vector field remains the same, and the last $i - 1$ vector fields satisfy the condition on bracket products. To this end, we assume (without loss of generality) that X_1 is a unit vector tangent to the first coordinate axis x_1 in our coordinate chart. The first coordinate of a point in our chart is a smooth function, so that we may apply a vector field to it, getting another smooth function. We may then set

$$Y_1 = X_1, \qquad Y_p = X_p - (X_p(x_1))X_1, \quad 2 \leq p \leq i.$$

It is clear that these i vector fields $\{Y_p\}$ span the same subspace of the tangent space as the $\{X_p\}$, that is, the subspace $f(x)$. They are then obviously linearly independent. Note also that if $p > 1$, $Y_p(x_1) = 0$.

Of course, the vector fields $Y_2, \ldots, Y_i$ lie in the tangent space to the submanifold given by $x_1 = 0$ in our chart. I claim, furthermore, that each $[Y_p, Y_q]$, for any $2 \leq p \leq q \leq i$, lies in the subspace spanned by $Y_2, \ldots, Y_i$. This follows at once because $Y_2, \ldots, Y_i$ lie in the tangent space to the submanifold given by $x_1 = 0$, and the bracket product of tangent vectors to a (sub)manifold is a tangent vector to that (sub)manifold. Now we use our induction on the submanifold $x_1 = 0$ to conclude that there is another (possibly smaller) coordinate chart of this new manifold in which the submanifolds of dimension $i - 1$, with last $n - i$ coordinates held constant, have tangent spaces spanned by the $Y_2, \ldots, Y_i$. (Check that this follows from the induction assumption.)

Now, introduce a new (and last) coordinate chart, on the original manifold, by taking x_1 to be the first coordinate, but with our new $n - 1$ coordinates, obtained from induction on $x_1 = 0$, as the last $n - 1$ coordinates. We now contend that the theorem can be completed using these new coordinates, which for clarity we shall name $y_1, \ldots, y_n$.

Observe first that we have not changed the first coordinate at all, so that clearly

$$Y_1 = \partial/\partial y_1.$$

In other words, $Y_1(y_p) = 0$, for $p > 1$. But also, when $p > 1$,

$$\frac{\partial}{\partial y_1}(Y_q(y_p)) = Y_1 Y_q(y_p) = Y_1(Y_q(y_p)) - Y_q(Y_1(y_p)) = [Y_q, Y_1](y_p).$$

We may write (by our hypothesis)

$$[Y_q, Y_1] = \sum_{r=1}^{i} \alpha_r Y_r,$$

so that

$$\frac{\partial}{\partial y_1}(Y_q(y_p)) = \sum_{r=1}^{i} \alpha_r Y_r(y_p).$$

Now, for $i < p \leq n$, these constitute $i - 1$ differential equations ($q = 2, \ldots, i$) for $Y_q(y_p)$ satisfying the initial condition $Y_q(y_p) = 0$ when $y_1 = 0$ (because, by our earlier use of induction, the $Y_2, \ldots, Y_i$ are tangent vectors to the subspace obtained by setting the last coordinates equal to constants). Therefore, by the standard material on uniqueness from the basic theory of differential equations, $Y_q(y_p)$ vanishes identically in our chart.

We conclude that the submanifolds obtained by setting the last $n - i$ coordinates equal to constants, in our new coordinate chart, have tangent spaces that are spanned by the i vector fields $Y_1, \ldots, Y_i$. Since we have already observed that the $Y_1, \ldots, Y_i$ and the $X_1, \ldots, X_i$ span the same subspace of TM^n, namely $f(x)$, the proof of our theorem is complete.

The theorem of Frobenius has various other formulations, which the reader may wish to examine in the literature. See, e.g., [20, 65, 104].

PROBLEMS AND PROJECTS

1. One point of view in the theory of transformation groups emphasizes the study of Lie groups acting on manifolds. This specific case is closely tied to the idea of differentiability, and it yields many interesting examples.

For the general theory, consult [80]; for the closer ties with differential geometry, see [7, 47].

2. Another point of view in transformation groups concerns the more abstract theory. See [34, 40]. The interested reader should explore the various concepts (such as *distal*) which underlie the modern theory.

3. There is an elegant, totally abstract theory of *symbolic dynamics*, due to Morse and Hedlund (see [83, 40]). The space is usually made of sequences, drawn from some finite set. The transformations are often shifts. Many interesting properties have been developed here.

4. Work out some specific examples of flows, in the spirit of Proposition 10.4. For example, examine the flow given by solutions to

$$dx_1/dt = -2x_1 - x_2, \qquad dx_2/dt = x_1, \qquad dx_3/dt = 5x_1 + x_3.$$

Sketch the orbits.

5. In a dynamical system in the plane, if x is almost periodic (by Proposition 10.2), the orbit closure of x a compact, minimal subset of the plane. Use Theorem 10.1 to give a complete analysis of such possible orbit closures.

6. Consult [22, 72] for Denjoy's theory of dynamical systems on the torus, $S^1 \times S^1$. Schwartz [98] generalizes this to arbitrary, two-dimensional manifolds.

7. There are many interesting consequences to Theorem 10.1. Consult [22, 52].

8. Construct examples to show that a compact, nonempty limit set of a flow in $\mathbb{R}^3$ may be complex and varied. Is such a set of a given dimension(s)? Is such a set a manifold?

9. Many properties of flows touch on measure theory and probability. Consult [40] or [45] to learn about mixing, ergodicity, etc.

10 Motions along geodesics in Riemann manifolds offer interesting types of flows, closely related to geometry. Consult [40, 46] and the references therein for the basic theory of geodesic flows on manifolds of negative curvature. This has led to major work by Anosov [3].

11

A Description of Singularities

and Catastrophes

If we think informally of a singularity of a smooth map $f: M^n \to N^k$ as a point where the differential or Jacobian has less than maximal possible rank, then earlier in this text we achieved a certain analysis of singularities. For example, if $M^n = \mathbb{R}^n$ and $N^k = \mathbb{R}$, a singularity means the vanishing of the gradient, or equivalently, the linear terms of the Taylor's polynomial of some degree greater than 1. If one has a singularity, one may look at the second-order terms in the Taylor's polynomial. One may ask whether the Hessian is nonsingular (see Definition 8.2), in which case the Morse lemma (Proposition 8.1) tells us that in a suitable coordinate neighborhood of the point one may write f according to the formula

$$f(x) - f(a) = -x_1^2 - \cdots - x_i^2 + x_{i+1}^2 + \cdots + x_n^2.$$

In this case i is precisely the number of negative eigenvalues of the Hessian matrix. One may of course begin to ask what happens if the Hessian matrix is singular. Here life becomes more complicated. In general, there are certain notions in the area of stability that are relevant, and with suitable control on the number of variables involved, there is a fascinating classification (Thom's classification of elementary catastrophes, our Theorem 11.3).

Our goal in this chapter is simply to provide a descriptive introduction to the area. We give very little in the way of proof, but we do consider most basic definitions and theorems in the area. All important concepts are introduced, although we cannot explore all the various ramifications of these concepts. We work out at some length various examples. Specific references to more detailed expositions are given at the appropriate places. Note, also, that we refrain from commenting on the applicability of this theory. While the potential applications of catastrophe theory make interesting reading, they are generally subject to scrutiny at the present time; one cannot yet make any final statements on whether these potential applications will be accepted by the relevant experts in various fields.

Specifically, we shall look at the following concepts: stability, genericity, finite determination, codimension, and unfolding. Without striving for maximum generality we state all the basic theorems concerning these concepts as they apply to catastrophe theory. To fix the terminology, we shall adopt the following as a definition of a singularity.

SINGULARITIES OF SMOOTH MAPS; STABILITY

Definition 11.1   Let M^n and N^k be smooth manifolds, and let $f : M^n \to N^k$ be a smooth map. A point $x \in M^n$ is called a *singularity* of f if the rank of the differential, or Jacobian, of f at the point x is less than the maximal possible value that it might have. In other terms, if $\phi_\alpha : U \to \mathbb{R}^n$ and $\psi_\beta : V \to \mathbb{R}^k$ are coordinate charts, with $x \in U$, $f(x) \in V$, then

$$\operatorname{rank}(J_{\psi_\beta \circ f \circ \phi_\alpha^{-1}}(\phi_\alpha(x))) < \min(n, k).$$

We now look at the idea of stability at a point.

Definition 11.2   A smooth map $f : M^n \to N^k$ is *stable at* x (or locally stable near x) if there is a coordinate neighborhood $\phi_\alpha : U \to \mathbb{R}^n$, $x \in U$, a coordinate neighborhood in N^k, $\psi_\beta : V \to \mathbb{R}^k$, with $f(U) \subseteq V$, and a small neighborhood of f in the space of maps $C^\infty(M^n, N^k)$ (see Definition 2.10 and Definition 3.3a) such that if g is any other map in this neighborhood of f, there are smooth diffeomorphisms

$$h_1 : U \to U \qquad \text{and} \qquad h_2 : V \to V$$

such that the diagram

$$
\begin{array}{ccc}
U & \xrightarrow{\;\;f\;\;} & V \\
{\scriptstyle h_1}\downarrow & & \downarrow{\scriptstyle h_2} \\
U & \xrightarrow[\;\;g\;\;]{} & V
\end{array}
$$

is commutative (that is to say, $h_2 \circ f = g \circ h_1$).

Remarks and Examples   *1.* Naturally, a similar definition may be given for C^k-maps.

2. For a global definition of a stability, one needs to consider diffeomorphisms h_1 and h_2 defined on all of M^n and N^k, respectively.

3. If in the above definition $x \in M^n$ is a singularity of f, then $h_1(x)$ is a singularity of g. One speaks of these singularities, x of f and $h_1(x)$ of g, as being *equivalent*.

4. Naturally, all of this can be restated, with a bit more elegance, in the language of germs of mappings (see Definition 1.11).

5. One also encounters a notion of *topological stability*, as distinct from the *differentiable stability* of Definition 11.2. This simply means that the maps h_1 and h_2 are only required to be homeomorphisms, rather than diffeomorphisms. The notions are actually different, particularly in higher dimensions. Other variants in the literature treat the same idea where h_1 and h_2 are translations, etc.

6. Let us now look at the simplest possible examples.

(a)   Let $f : \mathbb{R} \to \mathbb{R}$, $f(x) = x$. f is clearly stable at any point of its domain.

(b)   Suppose now that $f(x) = x^2$. I claim that f is stable at the singularity 0. Clearly if g is near f in the function space topology on $C^\infty(\mathbb{R}, \mathbb{R})$, g will have a minimum near 0, but because $f''(x) \equiv 2 > 0$, $g''(x) > 0$ with f near g, and then g will be obviously related to f as required in Definition 11.2. The idea in this second case is easily expressed by the graphs of Fig. 11.1.

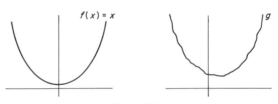

$f(x) = x$

g

Figure 11.1

7. On the other hand, $f : \mathbb{R} \to \mathbb{R}$, $f(x) = x^3$ has a singularity at 0, but it is *not* stable at 0. Consider the map

$$g_\varepsilon(x) = x^3 + \varepsilon x.$$

If $\varepsilon > 0$, $g_\varepsilon'(x) = 3x^2 + \varepsilon > 0$, so that g_ε is strictly increasing (and hence 1–1). If $\varepsilon < 0$, g_ε' vanishes at two points near 0, a local maximum and a local minimum. g_ε is clearly not 1–1 in this case.

Now if g_ε were equivalent to f, that is (as in Definition 11.2), there were diffeomorphisms h_1 and h_2, with $g_\varepsilon = h_2 \circ f \circ h_1^{-1}$, then we see that g_ε would have to have a single singularity that is not a local maximum or a local minimum at $h_1(0)$. But when $\varepsilon < 0$, g_ε does not have this property. Hence, f is not locally stable at 0.

We shall see later that this singularity can be embedded in a bigger singularity which is stable.

Actually, we have performed the analysis to show that a large number of singularities are indeed stable, in the sense of Definition 11.2.

Proposition 11.1 Let $f: M^n \to \mathbb{R}$ be a smooth Morse function (Definition 8.4). Then every singularity of f is stable in the sense of Definition 11.2.

Proof   Let $a \in M^n$ be a singularity, i.e., a critical point. The Morse lemma (Proposition 8.1) says that there is a coordinate neighborhood U of a in which f has the form

$$f(x) = f(a) - x_1^2 - \cdots - x_i^2 + x_{i+1}^2 + \cdots + x_n^2.$$

(Here we identify U directly with Euclidean space, in which the coordinates are named $x_1, \ldots, x_n$.) We must show that near a, for functions g sufficiently close to f, one has the same type of singularity.

Consider the map

$$\rho: U \to \mathbb{R}^n$$

defined by $\rho(x) = (\partial f/\partial x_1, \ldots, \partial f/\partial x_n)$ (the gradient). Clearly $\rho(a) = 0$. The Jacobian of ρ is precisely the Hessian of f (Definition 8.2), which is non-singular, since the critical points of a Morse function are nondegenerate. Then by the inverse function theorem (Theorem 1.1), ρ is a diffeomorphism on a possibly smaller neighborhood of a onto some open neighborhood of $0 \in \mathbb{R}^n$.

It follows immediately that any smooth map $g: M^n \to \mathbb{R}$ sufficiently near f in $C^\infty(M^n; \mathbb{R})$ will have the same property, that ρ maps a neighborhood of a diffeomorphically onto a neighborhood of $0 \in \mathbb{R}^n$. In particular, g will have a single, isolated singularity near a. Because the signs of the eigenvalues of nearby matrices remain the same, the Hessian of g at this critical point will have the same index (number of negative eigenvalues) as the Hessian of f did at a. In summary, g will have precisely one critical point, near a, and this critical point will be nondegenerate, with the same index as that of f at a.

To construct the diffeomorphisms required by Definition 11.2, we write

$$f(x) = f(a) - x_1^2 - \cdots - x_i^2 + x_{i+1}^2 + \cdots + x_n^2$$

and

$$g(u) = g(b) - u_1^2 - \cdots - u_i^2 + u_{i+1}^2 + \cdots + u_n^2,$$

where b is the single critical point of g near a and the u_j's are a coordinate neighborhood of b of the type given by the Morse lemma.

It is then easy to construct the desired diffeomorphisms sending the x_j's to the u_j's and adjusting h_2 to account for the difference between $f(a)$ and $g(b)$. (Check!)

Remark Results of Mather (see e.g., [13]) imply global stability in this case also.

The condition in Definition 11.2 may be thought of in terms of defining an equivalence on functions and their singularities. This idea has a slightly different expression in the notion of right equivalence. To simplify the notation, we shall state this in terms of germs. (Recall from Definition 1.11 that a germ of a smooth function at a point a means that we have an open neighborhood of that point, say U, and a smooth function $f : U \to \mathbb{R}$. (U, f) and (V, g) represent the same germ when there is an open set W, $a \in W \subseteq U \cap V$, with $f | W = g | W$.)

Definition 11.3 Let f and g be germs of mappings $\mathbb{R}^n \to \mathbb{R}^m$. Suppose $a = \mathbb{O}$ and $f(\mathbb{O}) = g(\mathbb{O}) = \mathbb{O}$. We say that f and g are *right equivalent*, if there is a germ $\phi : \mathbb{R}^n \to \mathbb{R}^n$, with $\phi(\mathbb{O}) = \mathbb{O}$ and $\det J_\phi(\mathbb{O}) \neq 0$, that is, ϕ is the germ of a local diffeomorphism, such that

$$f = g \circ \phi.$$

This is visibly an equivalence relation.

Although the classification of all germs having singularities at $\mathbb{O}$ is simply impossible, we shall see that under some reasonable assumptions, which we shall soon make precise, one can offer a classification up to right equivalence. Before discussing these assumptions, we should like to point out that there is a concept, which we have already introduced in Chapter 10, that in certain expositions of the theory of singularities is given a key role.

Definition 11.4 A set $F \subseteq C^\infty(M^n, N^k)$ is called *generic* if it contains an open dense subset of $C^\infty(M^n, N^k)$.

Remark Under suitable assumptions, there is a generic set of smooth functions whose singularities are of the type which we shall describe in Theorem 11.3. For more emphasis on this point of view, see [120]. For our purpose, of classifying germs with singularities, it will be more convenient to involve a slightly different language.

FINITE DETERMINATION AND CODIMENSION

We must now look at two very important and rather new concepts that relate to singularities. The first concept concerns the question of when a germ is determined by a finite piece of its Taylor series. For example, we have already seen that a Morse function is determined by the terms up to order 2 in its Taylor series, in a suitable neighborhood of a singularity. The second concept, codimension, is best understood in terms of elementary ring theory. We shall see that codimension involves the minimal number of additional dimensions that are necessary to create a universal stable germ from the given germ. Between these two definitions we shall need to review a little basic ring theory.

Definition 11.5 Let $U \subseteq \mathbb{R}^n$ be open, $\mathbb{O} \in U$, with $f: U \to \mathbb{R}$ representing a smooth germ, with $f(\mathbb{O}) = 0$. We then say that f is k *determined* if whenever g is another such germ, and f and g have the same Taylor's polynomials through degree k, then f and g are right equivalent.

If f is k-determined for some positive integer k, then we say that f is *finitely determined*.

Remarks (1) Even in the world of polynomials the notion of k-determined is subtle.

For example, consider $f: \mathbb{R} \to \mathbb{R}$, $f(x) = x^2$. Then 0 is a nondegenerate critical point with index 0. If g is another smooth function whose Taylor polynomial through order 2 is also x^2, then g also has 0 as a nondegenerate critical point of index 0. g may be written as u^2 in a suitable coordinate neighborhood of 0, and the right equivalence of these germs is easily verified (compare with Proposition 11.1 for $a = b = 0$ and $f(a) = g(b) = 0$).

But suppose f is a map $\mathbb{R}^2 \to \mathbb{R}$ given by $f(x, y) = x^2$. Of course $(0,0)$ is then a degenerate critical point, and if the function were k-determined,

$$g(x, y) = x^2 + y^{2k}, \qquad k > 1,$$

would have the same Taylor polynomial through degree 2 as $f(x, y)$. Then, were f k-determined, f and g would be right equivalent. But in fact f and g cannot be right equivalent, for if $\phi: \mathbb{R}^2 \to \mathbb{R}^2$ represents a germ of a diffeomorphism, $\phi(\mathbb{O}) = \mathbb{O}$ and

$$f(\phi(x, y)) = g(x, y),$$

then the left-hand side would be constant on the 1-dimensional subset of $\mathbb{R}^2$, which is the inverse image under ϕ of the y-axis, since f does not depend on y. Surely g is not constant on any 1-dimensional submanifold of a neighborhood of the origin.

(2)   As another extreme, consider the standard example of a germ $\mathbb{R} \to \mathbb{R}$, given by

$$f(x) = \begin{cases} e^{-1/x^2}, & \text{if} \quad x \neq 0 \\ 0, & \text{if} \quad x = 0. \end{cases}$$

It is easy to check (say, from l'Hôpital's rule) that for each $k \geq 0$, $f^{(k)}(0) = 0$. If f were finitely determined then f would be right equivalent to a germ that vanishes in a neighborhood of 0, i.e., the zero germ. But of course f is nowhere 0, except at 0, so that any germ that is right equivalent to f would have this property, too. Thus, f cannot be finitely determined.

(3)   In fact we shall soon come to many other examples, when we discuss elementary catastrophes. Mather's results (see [13, 73]) constitute a rather full analysis of the situation, and we shall shortly obtain necessary and sufficient conditions for a germ to be finitely determined. We observe that for a germ to be finitely determined says more of its Taylor polynomial than about the entire germ.

Before we can proceed to codimension, we need to review and fix some elementary terminology from ring theory.

Definition 11.6   Let $g(n)$ be the set of germs of smooth maps $f: \mathbb{R}^n \to \mathbb{R}$ at $\mathbb{O}$, with $m(n)$ being those germs f for which $f(\mathbb{O}) = 0$.

$g(n)$ is a commutative ring where germs are added and multiplied like ordinary functions (on the intersection of their domains). $m(n)$ is an ideal, because if $f \in g(n)$ and $h \in m(n)$, $fh \in m(n)$.

Remarks   (1) $m(n)$ is in fact the unique maximal ideal in the ring $g(n)$. For if $l \in m(n)$ does not vanish in a neighborhood of $\mathbb{O}$, i.e., l is not the 0 germ, $1/l$ exists as an element of $g(n)$. If $l \in m(n)$, an ideal, then

$$1 = \frac{1}{l} \cdot l, \qquad l \in m(n),$$

where 1 means the germ that is constant at 1 in some open neighborhood of $\mathbb{O}$. But if 1 belongs to an ideal, it follows trivially that the ideal must be the entire ring. It then is easy to see that $m(n)$ had to be a maximal ideal and that there can also be no other maximal ideal.

(2) $m(n)$ is generated by the germs $x_1, \ldots, x_n$, that is, the coordinate functions. To see this, we write

$$f(x) = \int_0^1 \frac{d}{dx} f(tx) \, dt = \int_0^1 \sum_{i=1}^n f_i(tx) x_i \, dt$$

$$= \sum_{i=1}^n \left(\int_0^1 f_i(tx) \, dt \right) x_i,$$

where f_i refers to the ith partial derivative. As each

$$\int_0^1 f_i(tx)\,dt$$

is an element of $g(n)$, this shows that the ideal is generated by the x_i's in the sense that every element is a linear combination of the x_i's with coefficients in the ring $g(n)$.

If $\mathscr{I}$ is an ideal in a ring and is generated by the elements $y_1, \ldots, y_n$, we write

$$\mathscr{I} = \langle y_1, \ldots, y_n \rangle.$$

And when there is no danger of possible confusion, we shall write, for simplicity,

$$\mathscr{I} = \langle y_i \rangle.$$

Definition 11.7 Let $f \in m(n)^2$; i.e., f is a finite sum of products of two elements, each from $m(n)$. (Clearly, f has a singularity at $\mathbb{O}$.) We define the *codimension* of f by

$$\mathrm{codim}\, f = \dim_{\mathbb{R}}\left(m(n)\Big/\left\langle \frac{\partial f}{\partial x_i}\right\rangle\right).$$

Here, $\langle \partial f/\partial x_i \rangle$ is short for $\langle \partial f/\partial x_1, \ldots, \partial f/\partial x_n \rangle$, where it is clear that each $\partial f/\partial x_i \in m(n)$. The dimension refers to the dimension as a real vector space of the quotient of the ring $m(n)$ by the ideal $\langle \partial f/\partial x_i \rangle$.

Remarks (1) It is not hard to see that $f \in m(n)^2$ precisely when $f \in m(n)$ has a singularity at $\mathbb{O}$. It is also elementary that the ideal $\langle \partial f/\partial x_i \rangle$ is independent of the choice of coordinate axes.

(2) One may quickly calculate some elementary examples. For example, if $f(x) = x^k$, $k > 1$, then

$$df/dx = kx^{k-1}.$$

In this case, the quotient ring $m(n)/\langle \partial f/\partial x_i \rangle$ will have a basis consisting of $x, x^2, \ldots, x^{k-2}$. Clearly codim $f = k - 2$.

On the other hand, if $f \in m(2)^2$ has the form

$$f(x, y) = x^3 + y^3,$$

then $\langle \partial f/\partial x_i \rangle$ is generated by x^2 and y^2. The quotient ring is generated by all monomials in x and y not containing x^2 or y^2. That is to say, the quotient ring is generated by x, y, and xy. Therefore, in this case

$$\mathrm{codim}\, f = 3.$$

But if $f \in m(2)^2$ is given by $f(x, y) = x^2y^2$, then $\langle \partial f/\partial x_i \rangle$ is generated by xy^2 and x^2y. The elements

$$y, \quad y^2, \quad y^3, \quad \dots$$

in the quotient ring are nonzero and linearly independent. Therefore, in this case codim $f = \infty$.

(3)   It is well known (see [120, 73]) that at a singularity either both codim f is finite and f is finitely determined or codim f is infinite and f is not finitely determined.

We are now in a position to state (but not prove) the basic results of Mather and Tougeron (see [13, 120] for details) on finitely determined germs.

Theorem 11.1   (a)   A germ $f \in m(n)$ is finitely determined if and only if

$$m(n)^k \subseteq \langle \partial f/\partial x_i \rangle$$

for some k. Here, $m(n)^k$ means finite sums of k-fold products of elements of $m(n)$.

(b)   A sufficient condition for f to be k determined is that

$$m(n)^{k+1} \subseteq m(n)^2 \langle \partial f/\partial x_i \rangle.$$

Here, the right-hand side means finite sums of products of elements, one from $m(n)^2$ and the other from $\langle \partial f/\partial x_i \rangle$.

UNFOLDINGS OF SINGULARITIES

For the interpretation of codimension, we shall need the basic facts about unfoldings. Intuitively, this means that we embed a singularity in a map with a higher-dimensional domain, so that the bigger map offers some advantages. The details now follow.

Definition 11.8   Suppose $f \in m(n)^2$, so f has a singularity at $\mathbb{O}$. An *unfolding* of f is a germ $\tilde{f} : \mathbb{R}^{n+r} \to \mathbb{R}$, with $\tilde{f}(\mathbb{O}) = 0$, such that if $x \in \mathbb{R}^n$,

$$\tilde{f}(x, \mathbb{O}) = f(x).$$

Here, of course, we mean $\mathbb{O} = (0, \dots, 0)$, with r entires. The unfolding $\tilde{f}$ is said to have r parameters. We shall sometimes write the unfolding by $(r, \tilde{f})$.

Remarks   (1)   One always has the *constant unfolding* $\tilde{f}$, defined by

$$\tilde{f}(x, y) = f(x).$$

(2)   If $b_1, b_2: \mathbb{R}^n \to \mathbb{R}$, both sending the origin to zero, and both smooth, then the formula

$$\tilde{f}(x, y_1, y_2) = f(x) + b_1(x)y_1 + b_2(x)y_2$$

defines an unfolding of f with two parameters.

(3)   If $(r_1, \tilde{f}_1)$ and $(r_2, \tilde{f}_2)$ are two unfoldings of f, with r_1 and r_2 parameters, respectively, then we may define their sum by setting

$$(\tilde{f}_1 + \tilde{f}_2)(x, y, z) = \tilde{f}_1(x, y) + \tilde{f}_2(x, z) - f(x),$$

where $y \in \mathbb{R}^{r_1}$ and $z \in \mathbb{R}^{r_2}$.

(4)   We shall see some sophisticated examples of unfoldings, when we discuss Thom's seven elementary catastrophes.

In general, the collection of all possible unfoldings of a singularity is too big to be of any interest. We wish to single out some special unfoldings, for which we shall need to look at the notion of a map of unfoldings of a given singularity.

Definition 11.9   Let $(r_1, \tilde{f}_1)$ and $(r_2, \tilde{f}_2)$ be unfoldings of a given germ $f \in m(n)^2$. Then a *map of unfoldings* $(r_1, \tilde{f}_1) \to (r_2, \tilde{f}_2)$ consists of

(a)   a germ $\phi: \mathbb{R}^{n+r_1} \to \mathbb{R}^{n+r_2}$ with $\phi(x, \mathbb{O}) = (x, \mathbb{O})$ whenever $x \in \mathbb{R}^n$.

(b)   A germ $\Phi: \mathbb{R}^{r_1} \to \mathbb{R}^{r_2}$ with $\pi_{r_2} \circ \phi = \Phi \circ \pi_{r_1}$. (Here, $\pi_j: \mathbb{R}^{n+j} \to \mathbb{R}^j$ is the projection.)

(c)   A germ $\alpha \in m(r_1)$, i.e., $\alpha: \mathbb{R}^{r_1} \to \mathbb{R}$, $\alpha(\mathbb{O}) = 0$, such that

$$\tilde{f}_1 = \tilde{f}_2 \circ \phi + \alpha \circ \pi_{r_1}.$$

(All germs are assumed smooth and defined in a neighborhood of the origin.)

One may naturally write down examples. But the following definition of an induced unfolding will show how to construct both new unfoldings and maps.

Definition 11.10   Let $(r_2, \tilde{f}_2)$ be an unfolding of $f \in m(n)^2$. Suppose we are given germs $\phi: \mathbb{R}^{n+r_1} \to \mathbb{R}^{n+r_2}$ and $\Phi: \mathbb{R}^{r_1} \to \mathbb{R}^{r_2}$ that satisfy (a) and (b) in Definition 11.9 above. For any $\alpha \in m(r_1)$ formula (c), i.e.,

$$\tilde{f}_1 = \tilde{f}_2 \circ \phi + \alpha \circ \pi_{r_1},$$

will then define a specific unfolding of f. The unfolding $\tilde{f}_1$ is then said to be *induced* from $\tilde{f}_2$.

The important cases are now the following:

Definition 11.11   (a)   Let $f \in m(n)^2$; i.e., f is a germ with $f(\mathbb{O}) = 0$ and f has a singularity at $\mathbb{O}$. Let $(r, \tilde{f})$ be an unfolding. We say that $(r, \tilde{f})$ is *versal* if any other unfolding of the germ f is induced from $(r, \tilde{f})$ in the sense of Definition 11.10.

(b)   A versal unfolding of a germ $f \in m(n)^2$ is called *universal* if r, the number of parameters, is minimal. (Naturally, if versal unfoldings exists, there will be a universal unfolding.)

The subject of universal unfoldings, which offers a beautiful and concise route to catastrophe theory, has been studied in depth by Mather and others (see [13, 120] for specific references). We collect the key results.

Theorem 11.2   (a)   A germ $f \in m(n)^2$ has a versal (and thus a universal) unfolding if and only if it is finitely determined. For $f \in m(n)^2$ this is equivalent to finite codimension.

(b)   If $(r, \tilde{f})$ is a universal unfolding of $f \in m(n)^2$, then $r = \text{codim } f$. All universal unfoldings are isomorphic.

(c)   The universal unfolding of a germ is stable (even if the germ is not).

ELEMENTARY CATASTROPHES

The basic theorem of Thom now classifies singular germs of codimension ≤ 4. These are the elementary catastrophes. We state this result in the language of germs, that is, as a local theorem, and then in the following remarks we discuss the relations with global theory, stability, and genericity.

Theorem 11.3   (Thom's Elementary Catastrophes)   Let $f \in m(n)^2$ be a smooth germ $(f(\mathbb{O}) = 0$, with $\mathbb{O}$ a singularity). Suppose that $c = \text{codim } f$ and $1 \leq c \leq 4$. Then f is 6-determined (so that we may look at terms up through degree 6).

Up to changes of sign and the addition of a nondegenerate quadratic form, f is right equivalent to one of the germs in the following table of 7.

Germ	Codimension	Universal unfolding	Popular name
x^3	1	$x^3 + ux$	fold
x^4	2	$x^4 + ux^2 + vx$	cusp
x^5	3	$x^5 + ux^3 + vx^2 + wx$	swallow-tail
$x^3 + y^3$	3	$x^3 + y^3 + wxy - ux - vy$	hyperbolic umbilic
$x^3 - xy^2$	3	$x^3 - xy^2 + w(x^2 + y^2) - ux - vy$	elliptic umbilic
x^6	4	$x^6 + tx^4 + ux^3 + vx^2 + wx$	butterfly
$x^2y + y^4$	4	$x^2y + y^4 + wx^2 + ty^2 - ux - vy$	parabolic umbilic

Remarks   (1)   There are several places where the details of this classification theorem are now carried out. Brocker's notes [13] are excellent, and
details, including comments on higher codimension, are available in Zeeman's expository article [120]. See also [111].

(2)   While germs such as x^3 are not locally stable, their universal unfoldings are (see [120]). The global point of view, including genericity for maps
$\mathbb{R}^n \to \mathbb{R}$, is also discussed in [120].

(3)   There are now some interesting extensions; Wasserman (see [110])
has developed a finer classification, the so-called (r, s)-unfoldings. This is done
with an eye towards relativity theory, where one must distinguish between
space and time parameters in deformations and unfoldings. In general,
Wasserman's writings on the subject are excellent.

The reader will probably want to get an idea of what these elementary
catastrophes look like. There is a nice literature here, but often tied to
(alleged) applications. Two mathematical sources in which one may see good
pictures are [13, 39].

We shall confine ourselves here to an examination of the fold

$$z = x^3 + yx,$$

because in this low-dimensional case one may draw a picture of the entire
unfolding (Fig. 11.2).

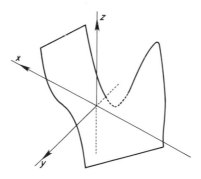

Figure 11.2

For fixed negative y, one sees a cubic with a local maximum and a local
minimum (sometimes called a local regime). As y increases to 0 these dis-
appear, and when $y > 0$, there is no local minimum. The rough idea of a
catastrophe is that it models the discontinuities in, or the disappearance of,
certain phenomena, such as the local minima in this case.

Finally, in closing, I would like to offer some hints to a reader who would like to pursue the details and specific proofs in this theory. To get the entire picture, one needs the following:

(i)   Some ring theory, including the Nakayama lemma, to get to the basic results of Mather *et al.* on finite determination and codimension. My personnal choice for a reference is [13].

(ii)   The Malgrange preparation theorem. This generalizes (to the smooth case) a famous theorem of Weierstrass from several complex variables (see [74, 87, 13]).

(iii)   Some basic algebraic geometry, well treated in most of the references.

As a final word of warning, catastrophe theory has brought with it a wealth of "applications." These run the distance from rather mathematical works (e.g., [41, 42]) all the way to Zeeman's rather tentative, though interesting, attempt [121] to explain riots in prisons. The theory may be exciting, but the risks of oversimplification in applications are enormous. No rational person would fault me if I suggested caution in this area.

Bibliography

1. Abraham, R., Transversality in manifolds of mappings, *Bull. Amer. Math. Soc.* **69**, 470–474 (1963).
2. Abraham, R., and Robbin, J., "Transversal Mappings and Flows." Benjamin, New York, 1967.
3. Anosov, D. V., Geodesic flows on compact Riemannian manifolds of negative curvature, *Trudy Mat. Inst. Steklov.* **90** (1967).
4. Atiyah, M., "K-Theory," Lecture Notes by D. W. Anderson. Benjamin, New York, 1967
5. Atiyah, M., Topology of elliptic operators *in* "Global Analysis" (S. S. Chern and S. Smale, eds.), Part III (Proc. Symp. Pure Mathematics, Vol. 16). Amer. Math. Soc., Providence, Rhode Island, 1970.
6. Atiyah, M., and Singer, I., The index of elliptic operators III, *Ann. of Math.* **87**, 546–604 (1968).
7. Auslander, L., Green, L., and Hahn, F., "Flows on Homogeneous Spaces," Annals of Math. Studies No. 53. Princeton Univ. Press, Princeton, New Jersey, 1963.
8. Battelle Rencontres (C. M. DeWitt and J. A. Wheeler, eds.), Univ. Press, Princeton, 1967. Lectures in Mathematics and Physics, Benjamin, New York, 1968.
9. Besicovitch, A. S., "Almost Periodic Functions." Dover, New York, 1954.
10. Bohr, H., Zur theorie der fastperiodischen funktionen, *Acta Math.* **45**, 29–127 (1924).
11. Bott, R., The stable homotopy of the classical groups, *Ann. of Math.* **70**, 313–337 (1959).
12. Bourbaki, N., Several Volumes on General Topology, Topological Vector Spaces, etc. Hermann, Paris.
13. Brocker, T., "Differentiable Germs and Catastrophes," London Mathematical Society Lecture Notes 17. Cambridge Univ. Press; London and New York, 1975.
14. Brockett, R. W., Nonlinear systems and differential geometry, *Proc. Inst. Electr. Engrs.* **64**, (1), 61–72 (1976).
15. Browder, W., "Surgery on Simply-Connected Manifolds." Springer-Verlag, Berlin and New York, 1972.

327

16. Cairns, S., The triangulation of the manifolds of class 1, *Bull. Amer. Math. Soc.* **411**, 549–552 (1935).
17. Cartan, E., "Leçons sur la Géométrie des Espaces de Riemann." Gauthier-Villars, Paris, 1951.
18. Cartan, H., Seminaire, 3, E. N. S., 2nd ed. Paris, 1955.
19. Chern, S. S., and Smale, S. (eds.), "Global Analysis" (Proc. Symp. Pure Mathematics, Vols. 15 and 16). Amer. Math. Soc., Providence, Rhode Island, 1970
20. Chevalley, C., "Theory of Lie Groups I." Princeton Univ. Press, Princeton, New Jersey, 1946.
21. Chevalley, C., "Theorie des Groupes de Lie." Hermann, Paris, 1968.
22. Coddington, E., and Levinson, N., "Theory of Ordinary Differential Equations." McGraw-Hill, New York, 1955.
23. deRham, G., "Variétés Différentiables." Hermann, Paris, 1955.
24. Dieudonné, J., "Foundations of Modern Analysis." Academic Press, New York, 1969.
25. Dold, A., "Lectures on Algebraic Topology." Springer-Verlag, Berlin and New York, 1972.
26. Dold, A., Partitions of unity in the theory of fibrations, *Ann. of Math.* **78**, 223–255 (1963).
27. Eells, J., "Singularities of Smooth Maps." Gordon & Breach, New York, 1967.
28. Eells, J., On the geometry of function spaces, Symp. *Internacional Topologia Algebraica*, Mexico, 1958.
29. Eells, J., and Elworthy, K. D., On the differential topology of Hilbert manifolds *in* "Global Analysis" (S. S. Chern and S. Smale, eds.), Part II (Proc. Symp. Pure Mathematics, Vol. 15). Amer. Math. Soc., Providence, Rhode Island, 1970.
30. Eells, J., and Kuiper, N., An invariant for certain smooth manifolds, *Annali Math. Italy* **60**, 93–110 (1962).
31. Ehresmann, C., Sur la topologie de certains espaces homogénes, *Ann. of Math.* **35**, 396–443 (1934).
32. Eilenberg, S., Singular homology theory, *Ann. of Math.* **45**, 407–447 (1944).
33. Eilenberg, S. and Steenrod, N., "Foundations of Algebraic Topology." Princeton Univ. Press, Princeton, New Jersey, 1952.
34. Ellis, R., "Lectures on Topological Dynamics." Benjamin, New York, 1969.
35. Elworthy, K. D., and Tromba, A. J. Differentiable structures and Fredholm maps on Banach manifolds *in* "Global Analysis" (S. S. Chern and S. Smale, eds.), Part II (Proc. Symp. Pure Mathematics, Vol. 15). Amer. Math. Soc., Providence, Rhode Island, 1970.
36. Gaal, S. A., "Point Set Topology." Academic Press, New York, 1964.
37. Gillman, L., and Jerison, M., "Rings of Continuous Functions." Van Nostrand Reinhold, Princeton, New Jersey, 1960.
38. Godbillon, C., "Géometrie Différentielle et Mécanique Analytique." Hermann, Paris, 1969.
39. Godwin, A. N., Three dimensional pictures for Thom's parabolic umbilics, *Inst. Hautes Etudes Sci. Publ. Math.* **40**, 117–138 (1971).
40. Gottschalk, W., and Hedlund, G., "Topological Dynamics." Amer. Math. Soc., Colloquium Publications 36, Providence, Rhode Island, 1955.
41. Guckenheimer, J., Bifurcation and catastrophe *in* "Dynamical Systems" (M. Peixoto, ed.). Academic Press, New York, 1973.
42. Guckenheimer, J., Catastrophes and partial differential equations, *Ann. Inst. Fourier* (*Grenoble*) **23**, 31–59 (1973).
43. Haeffliger, A., Plongements différentiables dans le domaine stable, *Comment. Math. Helv.* **37**, 155–176 (1963).
44. Halmos, P., "Measure Theory." Van Nostrand Reinhold, Princeton, New Jersey, 1950.

45. Halmos, P., "Lectures on Ergodic Theory." Mathematical Society of Japan, Tokyo 1956.
46. Hedlund, G., The dynamics of geodesic flows, *Bull. Amer. Math. Soc.* **45**, 241–260 (1939).
47. Helgason, S., "Differential Geometry and Symmetric Spaces." Academic Press, New York, 1962.
48. Hilbert, D., and Cohn-Vossen, S., "Geometry and the Imagination." Chelsea, Bronx, New York, 1956.
49. Hilton, P., and Wylie, S., "Homology Theory." Cambridge Univ. Press, London and New York, 1960.
50. Hirsch, M. W., Immersions of manifolds, *Trans. Amer. Math. Soc.* **93**, 242–276 (1959).
51. Hirsch, M. W., "Differential Topology." Springer-Verlag, Berlin and New York, 1976.
52. Hirsch, M. W., and Smale, S., "Differential Equations, Dynamical Systems, and Linear Algebra." Academic Press, New York, 1974.
53. Hirzebruch, F., "Topological Methods in Algebraic Geometry." Springer-Verlag, Berlin and New York, 1966.
54. Hochschild G., "The Structure of Lie Groups." Holden-Day, San Francisco, California 1965.
55. Hodge, W. V. D., "The Theory and Applications of Harmonic Integrals." Cambridge Univ. Press, London and New York, 1952.
56. Hoffman, K., and Kunze, R., "Linear Algebra." Prentice-Hall, Englewood Cliffs, New Jersey, 1961.
57. Hormander, L., "Linear Partial Differential Operators." Springer-Verlag, Berlin and New York, 1963.
58. Hormander, L., The Frobenius–Nirenberg theorem, *Ark. Mat.* **5**, 425–432 (1964).
59. Hu, S.-T., "Homotopy Theory." Academic Press, New York, 1959.
60. Jacobsen, N., "Lectures in Abstract Algebra II." Van Nostrand-Reinhold, Princeton, New Jersey, 1953.
61. Jänich, J., Vektorraumbundel und der Raum der Fredholm–Operatoren, *Math. Ann.* **161**, 129–142 (1965).
62. Kelley, J., "General Topology." Van Nostrand–Reinhold, Princeton, New Jersey, 1955.
63. Kervaire, M., A manifold which does not admit any differentiable structure, *Comment. Math. Helv.* **34**, 304–312 (1960).
64. Kirby, R., and Siebenman, L., "Foundational essays on topological manifolds," Annals of Math. Studies No. 88, Princeton Univ. Press, Princeton, New Jersey, 1977.
65. Kobayashi, S., and Nomizu, K., "Foundations of Differential Geometry," Vols. I and II. Wiley (Interscience), New York, 1963.
66. Koschorke, U., Infinite dimensional *K*-theory and characteristic classes of Fredholm bundle maps *in* "Global Analysis" (S. S. Chern and S. Smale, eds.), Part II (Proc. Symp. Pure Mathematics, Vol. 15). Amer. Math. Soc., Providence, Rhode Island, 1970.
67. Kuiper, N., The homotopy-type of the unitary group of Hilbert space, *Topology* **3**, 19–30 (1965).
68. Kuiper, N., and Burghelea, D., Hilbert manifolds, *Ann. of Math.* **90**, 379–417 (1969).
69. Lang, S., "Introduction to Differentiable Manifolds." Wiley (Interscience), New York, 1962.
70. Laudenbach, F., "Topologie de la Dimension Trois, Homotopie, et Isotopie", No. 12, L'Astérisque, Soc. Math. de France, 1974.
71. Levi-Civita, T., "Lezioni di calcolo differenziale assoluto", Stock, Rome, Italy, 1925.
72. Lefschetz, S., "Differential Equations: Geometric Theory." Wiley (Interscience), New York, 1957.

73. Lu, Y-C., "Singularity Theory and an Introduction to Catastrophe Theory." Springer-Verlag, Berlin and New York, 1976.
74. Malgrange, B., The Preparation Theorem for Differentiable Functions, Differential Analysis, Bombay Colloquium, Oxford, 1964.
75. Massey, W. S., "Algebraic Topology: An Introduction." Harcourt, New York, 1967.
76. Milnor, J., On manifolds homeomorphic to the seven sphere, *Ann. of Math.* **64**, 399–405 (1956).
77. Milnor, J., "Morse Theory." Lecture notes by M. Spivak and R. Wells, Annals of Math. Studies No. 51, Princeton Univ. Press, Princeton, New Jersey, 1963.
78. Milnor, J., and Stasheff, J., "Characteristic Classes," Annals of Math. Studies No. 56, Princeton Univ. Press, Princeton, New Jersey, 1974.
79. Mitchell, B., "Theory of Categories." Academic Press, New York, 1965.
80. Montgomery, D., and Zippin, L., "Topological Transformation Groups." Wiley (Interscience), New York, 1955.
81. Morse, A., The behavior of a function on its critical set, *Ann. of Math.* **40**, 62–70 (1939).
82. Morse, M., "The Calculus of Variations in the Large," Amer. Math. Soc. Colloquium Publications 18, Providence, Rhode Island, 1934.
83. Morse, M., and Hedlund, G., Symbolic dynamics, *Amer. J. Math.* **60**, 815–866 (1938).
84. Narasimhan, R., "Analysis on Real and Complex Manifolds." Advanced Studies in Pure Mathematics, Masson, Paris and North-Holland Publ., Amsterdam, 1968.
85. Newman, M. H. A., "Elements of the Topology of Plane Sets of Points." Cambridge Univ. Press, London and New York, 1954.
86. Nirenberg, L., Pseudo-differential operators *in* "Global Analysis" (S. S. Chern and S. Smale, eds.), Part III (Proc. Symp. Pure Mathematics, Vol. 16). Amer. Math. Soc., Providence, Rhode Island, 1970.
87. Nirenberg, L. A Proof of the Malgrange Preparation Theorem, "Proceedings of Liverpool Singularities-Symposium I" (C. T. C. Wall, ed.) (Lecture Notes in Math., Vol. 192). Springer-Verlag, Berlin and New York, 1971.
88. Palais, R. (ed.), "Seminar on the Atiyah–Singer Index Theorem," Annals of Math. Studies No. 57. Princeton Univ. Press, Princeton, New Jersey, 1965.
89. Peetre, J., Rectifications à l'article "Une characterization abstraite des operateurs différentiels", Math. Scand. **8**, 116–120 (1960).
90. Pitcher, E., Inequalities of critical point theory, *Bull. Amer. Math. Soc.* **64**, 1–30 (1958).
91. Pontrjagin, L., Smooth manifolds and their applications in homotopy theory, *Trudy Mat. Inst. Steklov* **45** (1955) [*Amer. Math. Soc. Transl.* Ser. 2 **11** (1959)].
92. Pontrjagin, L., "Topological Groups," 2nd Russian ed. Moscow, 1954.
93. Porteous, I., "Topological Geometry." Van Nostrand-Reinhold, Princeton, New Jersey, 1969.
94. Rudin, W., "Real and Complex Analysis." McGraw-Hill, New York, 1974.
95. Saks, S., "Theory of the Integral." Warsaw-Lvov, 1937.
96. Sard, A., The measure of the critical values of differentiable maps, *Bull. Amer. Math. Soc.* **48**, 883–890 (1942).
97. Schubert, H., "Kategorien." Springer-Verlag, Berlin and New York, 1970 [Engl. Transl. by J. Gray, 1972].
98. Schwartz, A., A generalization of the Poincaré–Bendixson theorem to closed two-dimensional manifolds, *Amer. J. Math.* **85**, 453–458 (1963).
99. Serre, J. P., Homologie singulière des espaces fibrés, *Ann. of Math.* **54**, 425–505 (1951).
100. Smale, S., The classification of immersions of spheres in Euclidean space, *Ann. of Math.* **69**, 327–344 (1959).
101. Smale, S., An infinite-dimensional version of Sard's theorem, *Amer. J. Math.* **87**, 861–866 (1965).

102. Spanier, E., "Algebraic Topology." McGraw-Hill, New York, 1966.
103. Steenrod, N., "The Topology of Fibre Bundles." Princeton Univ. Press, Princeton New Jersey, 1951.
104. Sternberg, S., "Lectures on Differential Geometry." Prentice-Hall, Englewood Cliffs, New Jersey, 1964.
105. Stong, R., "Notes on Cobordism Theory," Math. Notes, Princeton Univ. Press, Princeton, New Jersey, 1968.
106. Swan, R., "The Theory of Sheaves," Chicago Lecture Notes, Univ. of Chicago Press, Chicago, Illinois, 1964.
107. Thom, R., Quelques propriétés globales des variétés différentiables, *Comment. Math. Helv.* **28**, 17–86 (1954).
108. Thom, R., "Stabilité Structurelle et Morphogenèse." Benjamin, New York, 1972.
109. Wall, C. T. C., "Surgery on Compact Manifolds." Academic Press, New York, 1970.
110. Wasserman, G., (r, s)-stable unfoldings and catastrophe theory *in* "Structural Stability, the Theory of Catastrophes and Applications in the Sciences" (P. Hilton, ed.) (Lecture Notes in Math., Vol. 525), Springer-Verlag, Berlin and New York, 1976.
111. Wasserman, G., "Stability of Unfoldings" (Lecture Notes in Math., Vol. 393). Springer-Verlag, Berlin and New York, 1974.
112. Weyl, H., "The Theory of Groups and Quantum Mechanics." Dover, New York, 1949.
113. Whitehead, J. H. C., Combinatorial homotopy I, *Bull. Amer. Math. Soc.* **55**, 213–245 (1949).
114. Whitney, H., The singularities of a smooth n-manifold in $(2n-1)$-space, *Ann. of Math.* **45**, 247–293 (1944).
115. Whitney, H., The self-intersections of a smooth n-manifold in $2n$-space, *Ann. of Math.* **45**, 220–246 (1944).
116. Whitney, H., Differentiable manifolds, *Ann. of Math.* **37**, 645–686 (1936).
117. Whitney, H., "Geometric Integration Theory." Princeton Univ. Press, Princeton, New Jersey, 1957.
118. Wigner, E., "Group Theory." Academic Press, New York, 1959.
119. Yosida, K., "Functional Analysis." Springer-Verlag, Berlin and New York, 1966.
120. Zeeman, E. C., The Classification of Elementary Catastrophes of Codimension ≤ 5 *in* "Structural Stability, the Theory of Catastrophes, and Applications in the Sciences" (P. Hilton, ed.) (Lecture Notes in Math., Vol. 525), Springer-Verlag, Berlin and New York, 1976.
121. Zeeman, E. C., Prison Disturbances *in* "Structural Stability, the Theory of Catastrophes, and Applications in the Sciences" (P. Hilton, ed.) (Lecture Notes in Math., Vol. 525), Springer-Verlag, Berlin and New York, 1976.

Index